The Digital Transformation of Labor

Through a series of studies, the overarching aim of this book is to investigate if and how the digitalization/digital transformation process causes (or may cause) the autonomy of various labor functions, and its impact in creating (or stymieing) various job opportunities on the labor market. This book also seeks to illuminate what actors/groups are mostly benefited by the digitalization/digital transformation and which actors/groups that are put at risk by it.

This book takes its point of departure from a 2016 OECD report that contends that the impact digitalization has on the future of labor is ambiguous, as on the one hand it is suggested that technological change is labor-saving, but on the other hand, it is suggested that digital technologies have not created new jobs on a scale that it replaces old jobs. Another 2018 OECD report indicated that digitalization and automation as such does not pose a real risk of destroying any significant number of jobs for the foreseeable future, although tasks would by and large change significantly. This would affect welfare, as most of its revenue stems from taxation, and particularly so from the taxation on labor (directly or indirectly). For this reason, this book will set out to explore how the future technological and societal advancements impact labor conditions.

The book seeks to provide an innovative, enriching and controversial take on how various aspects of the labor market can be (and are) affected by the ongoing digitalization trend in a way that is not covered by extant literature. As such, this book intends to cater to a wider readership, from a general audience and students, to specialized professionals and academics wanting to gain a deeper understanding of the possible future developments of the labor market in light of an accelerating digitalization/digital transformation of society at large.

Anthony Larsson (Ph.D.) is a researcher at the Stockholm School of Economics Institute for Research (SIR), Sweden. He holds a Ph.D. from Karolinska Institutet, Sweden. He also holds an MBA and M.Sc. degrees in political science, social anthropology, and business administration and economics respectively, as well as an associate degree in psychology.

Robin Teigland (Ph.D.) is Professor of Management of Digitalization in the Entrepreneurship and Strategy Division at the Department of Technology Management and Economics at Chalmers University of Technology, Gothenburg, Sweden. She is also Professor of Business Administration with a specialization in strategic information systems management at the Stockholm School of Economics in Sweden and a member of the Royal Swedish Academy of Engineering Sciences.

Routledge Studies in Labour Economics

Youth and the Crisis
Unemployment, Education and Health in Europe
Edited by Gianluigi Coppola and Niall O'Higgins

Workers and the Global Informal Economy
Interdisciplinary Perspectives
Edited by Supriya Routh and Vando Borghi

The Political Economy of Employment Relations
Alternative Theory and Practice
Aslihan Aykac

The Economics of Trade Unions
A Study of a Research Field and Its Findings
Hristos Doucouliagos, Richard B. Freeman and Patrice Laroche

Young People and the Labour Market
A Comparative Perspective
Floro Caroleo, Olga Demidova, Enrico Marelli and Marcello Signorelli

Women's Economic Empowerment in Turkey
Edited by Onur Burak Çelik and Meltem İnce Yenilmez

A Comparative Perspective of Women's Economic Empowerment
Edited by Meltem İnce Yenilmez and Onur Burak Çelik

Mediterranean Migration and the Labour Markets
Policies for Growth and Social Development in the Mediterranean Area
Edited by Salvatore Capasso and Eugenia Ferragina

The Digital Transformation of Labor
Automation, the Gig Economy and Welfare
Edited by Anthony Larsson and Robin Teigland

The Digital Transformation of Labor

Automation, the Gig Economy and Welfare

Edited by Anthony Larsson and Robin Teigland

Routledge
Taylor & Francis Group
LONDON AND NEW YORK

First published 2020
by Routledge
2 Park Square, Milton Park, Abingdon, Oxon OX14 4RN

and by Routledge
52 Vanderbilt Avenue, New York, NY 10017

Routledge is an imprint of the Taylor & Francis Group, an informa business

First issued in paperback 2021

British Library Cataloguing-in-Publication Data
A catalogue record for this book is available from the British Library

Library of Congress Cataloging-in-Publication Data
A catalog record for this book has been requested

ISBN: 978-0-367-33070-5 (hbk)
ISBN: 978-1-03-208493-0 (pbk)
ISBN: 978-0-429-31786-6 (ebk)

Typeset in Times New Roman
by Apex CoVantage, LLC

Contents

Figures

Tables

Contributors

Editors

Editor-in-Chief

Anthony Larsson (Ph.D.) is a researcher at the Stockholm School of Economics Institute for Research (SIR), Sweden. He holds a Ph.D. from Karolinska Institutet, Sweden. Larsson also holds an MBA and M.Sc. degrees in political science, social anthropology, and business administration and economics respectively, as well as an associate's degree in psychology. His recent work includes the books, *The Rise and Development of FinTech: Accounts of Disruption from Sweden and Beyond* (2018), *Too Big to Fail? A Case Study of the Rise and Fall of a Medical Research Infrastructure* (2018) and *Digital Transformation and Public Services: Societal Impacts in Sweden and Beyond* (2020), as well a number of research papers. Larsson's research interests include innovation management, medical science, digitalization, customer loyalty, bank marketing, management, organizational behavior, organizational branding, stakeholder analysis, entrepreneurship and qualitative research methods. A "Routledge Featured Author", you may follow his research at www.routledge.com/authors/i18347-anthony-larsson or via his Amazon.com author page at www.amazon.com/author/anthonylarsson.

Editor

Robin Teigland (Ph.D.) is Professor of Management of Digitalization in the Entrepreneurship and Strategy Division at the Department of Technology Management and Economics at Chalmers University of Technology, Gothenburg, Sweden. She is also Professor of Business Administration with a specialization in strategic information systems management at the Stockholm School of Economics (SSE) in Sweden. She is also a member of the Royal Swedish Academy of Engineering Sciences. Teigland has more than 20 years of research experience within social networks, strategy, innovation, entrepreneurship and start-up ecosystems. In particular, she is interested in how the convergence of disruptive technologies influences value creation in society as well as disrupts long-standing institutional structures. Teigland is also an impact entrepreneur, having recently started a blue circular economy project and an ocean tech hub in Peniche, Portugal, that focuses on enabling regional development in coastal communities through digitalization. In 2017, 2018 and 2019, she was listed by the Swedish business magazine, *Veckans Affärer*, as one of Sweden's most influential women, primarily in technology. In her free time, she likes to "play" with her five kids as well as surf, and help run her surf school, Go4surf, at her favorite beach in Peniche.

Contributors (in order of appearance)

Kjell A. Nordström (Ph.D.) is a strategic management expert, economist, writer and international public speaker based in Stockholm, Sweden. He holds a Ph.D. in business from the Institute of International Business (IIB) at the Stockholm School of Economics, Sweden. He also has an undergraduate degree in engineering. His research focuses primarily on global strategies and the internationalization process of firms. Among his most popular work are the internationally renowned books *Funky Business – Talent Makes Capital Dance*, *Karaoke Capitalism – Management for Mankind* and *Urban Express – 15 Rules for Navigating a World Shaped by Cities and Women*. Nordström has more than 25 years' experience in working with multinational companies. He has served as an advisor and consultant to several governments around the world, along with a sizable number of the largest multinational firms in different countries. He suggests that our time is unique, arguing that we are in the beginning of the fastest business model transformation since the beginning of industrialization. The dynamic style of Nordström has made him a highly appreciated and sought-after speaker around the world. He has given keynote presentations in more than a hundred countries. The Global Gurus' *World Top 30 Management Professionals* ranking placed him among the top 20 in the world and among the top five in Europe. In his free time, Nordström likes to spend time with his family and managing Sweden's largest and most famous Shakespeare Theater, located on the Swedish island of Gotland. The latter of which is probably the one of his many business projects that lies closest to his heart.

Alexander Bard is a philosopher and futurologist, based in Stockholm, Sweden. Educated at the Stockholm School of Economics, he is the author of five books: *The Netocrats* (2000), *The Global Empire* (2002), *The Body Machines* (2009), *Syntheism – Creating God in The Internet Age* (2014) and *Digital Libido – Sex, Power and Violence in The Network Society* (2018) together with his co-writer Jan Söderqvist (also featured in this volume). Bard is a radical process philosopher, merging Hegel and Nietzsche with Whitehead and Deleuze, using human beings as the constant and technological disruption as the variable while working toward the deepest possible understanding of human history, contemporary society and how to approach the intensely technology-driven future that humanity is facing. He has also enjoyed a highly successful career spanning over 30 years as a producer, songwriter and artist in the international music industry and is a celebrity in his native Sweden as a "tough love", reality-checking judge on the popular TV-shows "Swedish Idol" and "Sweden's Got Talent".

Jan Söderqvist is a writer, futurologist and lecturer based in Stockholm, Sweden, and Lucca, Italy. Together with Alexander Bard (also featured in this volume), he has published five works of social philosophy during 20 years of collaboration: *The Netocrats* (2000), *The Global Empire* (2002), *The Body Machines* (2009), *Syntheism – Creating God in the Internet Age* (2014) and, most recently, *Digital Libido – Sex, Power and Violence in The Network Society* (2018). Educated at Stockholm University, Sweden, his main focus lies on interpreting different perspectives on how the digital revolution will transform any and all areas of society beyond recognition, while also changing the lives of everyone living in it. To keep himself busy between and during book projects, Söderqvist writes movie reviews and columns for the Stockholm daily newspaper *Svenska Dagbladet*. He also serves as an editor at *Axess magasin*, one of the leading Swedish liberal arts and social science publications.

Jochem van der Zande works as a strategy consultant at Danske Bank Group Strategy. He holds a M.Sc. in CEMS International Management from the Rotterdam School of Management. He has also studied at the Stockholm School of Economics and BI Norwegian Business School. Van der Zande's research interests are strategy, digitalization and M&A. Previous research projects include contributed chapter on banks and digitalization in the book, *The Rise and Development of FinTech: Accounts of Disruption from Sweden and Beyond* (2018). His master's thesis covered the role of external advisors during the acquisition process. Outside of his professional life, he is a passionate badminton player and enjoys travelling and cooking.

Karoline Teigland is a M.Sc. student in Mechanical Engineering with Industrial Design at the Faculty of Engineering LTH, Lund University, Sweden. She has worked on various digital technology projects including interactive screens with FlatFrog AB in Lund and virtual reality with the Stockholm School of Economics in Stockholm. In addition, she plays on Norway's National Women's Basketball team as well as for EOS Club in Sweden's top national women's league, while enjoying an active student life at Lund University.

Shahryar Siri is a research analyst and project manager at the Stockholm School of Economics Executive Education, Sweden. In his work, Siri collaborates with start-ups, multinationals, academia, and government organizations, to identify technology and marketing developments and opportunities. Siri has an M.Sc. in technology entrepreneurship from University College London, UK.

Alex J. Wood (Ph.D.) is a lecturer in the Sociology of Work at the University of Birmingham, UK. He is a sociologist of work and employment, focusing on the changing nature of employment relations and labor market transformation. Wood is currently researching new forms of worker voice and collective action in the remote gig economy as part of the iLabour project. Wood received his Ph.D. in 2016 from the Department of Sociology at the University of Cambridge. His Ph.D. focused on the changing nature of flexible and insecure forms of work such as zero hour contracts. A book, *Despotism on Demand: How Power Operates in the Flexible Workplace*, based on his thesis is due to be published in 2020. His research has appeared in leading academic journals, such as: *Work, Employment and Society*; *Human Relations*; *New Technology, Work and Employment*; *Industrial Relations Journal*; and *Employee Relations*.

Mark Graham (Ph.D.) is professor of internet geography at the Oxford Internet Institute, a Faculty Fellow at the Alan Turing Institute, a Senior Research Fellow at Green Templeton College, a research affiliate in the University of Oxford's School of Geography and the Environment and a research associate at the Centre for Information Technology and National Development in Africa at the University of Cape Town. Graham leads a range of research projects spanning topics between digital labor, the gig economy, internet geographies and ICTs and development. A full list of his publications can be found at www.markgraham.space.

Mohammad Amir Anwar (Ph.D.) is a post-doctoral researcher at the Oxford Internet Institute, University of Oxford, UK, and holds a research associate position at the School of Tourism and Hospitality, University of Johannesburg, South Africa. His research focuses on the political economy of globalization and its emerging geographies, with expertise on Africa and India. Anwar is currently working on the multi-year *Geonet* project, which examines the emergence of the information economy and related gig economy labor in Africa.

Antoine Maire (Ph.D.) is a research associate at the Center for International Studies (CERI) of Sciences Po, France. He graduated from Sorbonne University and *Ecole Normale Supérieure* and he holds a Ph.D. in political science from Sciences Po. His works have focused on public policies. Maire has done extensive research on Mongolia's policies toward economic development. He published, in 2013, *La Mongolie en quête d'indépendance* (Mongolia's quest for independence) in *L'Harmattan* and, in 2016, *Les Mongols, Insoumis* (Mongols) in *Ateliers Henry Dougier*. His works have now shifted to study public policies toward science. Maire has previously worked as an analyst for Orano, a French mining company, and for *Ministère des Armées* (the French Ministry for the Armed Forces). He is presently working for Conseil & Recherche, an independent research agency.

Victor Erik Bernhardtz is a policy analyst specializing in digital labor markets at *Unionen*, the world's largest white-collar private-sector union. In his work, Bernhardtz seeks to understand how digitalization disrupts labor, the organization of labor and the implications for the relationship between labor and society. In this, a key component involves strategic innovation for labor unions in the twenty-first century. Bernhardtz has a background in media, advocacy and contemporary theatre. He holds a B.Sc. in political science from Stockholm University and lives in Stockholm, Sweden.

Fernanda Torre embodies the interdisciplinary intersection between design, business and innovation, while exploring synergies between practice and research. Torre is the CEO of FT Innovation, a consultancy on innovation management and design strategy, and is a Visiting Teacher affiliated to the House of Innovation at the Stockholm School of Economics (SSE), Sweden. Torre lectures in the SSE MBA in Innovation and Entrepreneurship and is a faculty member in the Stockholm School of Entrepreneurship, where she teaches trendspotting and future forecasting. She is also a Design Industry Leader at the digital creative business school *Hyper Island*, in Stockholm, Sweden. With a background in graphic design (*Red Dot Design* award, 2014 with Gabor Palotai) and experience design (*Beam me up, Scotty!* award in Helsinki design fair, 2013), Torre has made her way into innovation management (*Stora Enso Innovation Award for Intelligent Packaging*, 2016). In terms of research, Torre is part of the Vinnova-funded project 4Boards.ai, aiming to support corporate boards in governing AI toward innovation and sustainability. Furthermore, Torre is a founding member of the Speculative Futures Chapter in Stockholm, a board member of *Innovationsledarna* (Association for Innovation Management Professionals in Sweden) and a Swedish expert for the ISO standard on innovation management. Torre has an MBA from SSE, awarded with the President's List for outstanding academic achievement, an MFA in experience design from Konstfack, Sweden, and a bachelor's in arts in graphic design from Porto University, Portugal.

Liselotte Engstam is an experienced international leader, chairman and non-executive board director with more than 20 years of general manager and executive leadership positions in major European, American and Indian companies in technology, engineering and professional services sectors. Including her current role as chairman and non-executive board director at three listed and two private companies and as board member at INSEAD International Directors Network, Engstam has eight years of independent board mandates at digitally transforming and sustainable-focused international companies. Currently, she is part of an expert group advising the European Commission on corporate governance and digitalization. Engstam is an affiliated researcher contributing to four academic research projects. Together with KTH Royal Institute of Technology (KTH) in Stockholm, Sweden, she is researching Boards' contribution to corporate renewal and innovation, with findings shared in the article "Corporate renewal and innovation also disrupt boards" in the Management of Technology and Innovation publication, and also researching Boards contribution to strategy and innovation for sustainable value creation. Together with Chalmers University of Technology, Sweden, she researches how boards can successfully govern and leverage AI in their sustainability and innovation efforts. Together with INSEAD she is researching the evolving

Chairman practices. Engstam is the executive chair at Digoshen, advising leaders on business realignment for a digital era and contributing to four academic research projects on board innovation, AI and chairman practices. She is a keynote speaker and educator at various chairman and board programs. Engstam is a partner in venture capital and coaching start-ups and scale-ups within AI, blockchain and SaaS. Engstam holds an M.Sc. in engineering from Chalmers University of Technology, Gothenburg, Sweden. She has also attended executive programs at Harvard and the International Institute for Management Development (IMD), board programs at local Institute of Directors (IoD) and at INSEAD, attaining a certificate as international board director (IDP-C). In her free time, she enjoys time with her family, watercolor painting and digital art creation, golfing and exploring roots in genealogy and the links to history.

Mårten Blix (Ph.D.) is a research scholar at the Research Institute for Industrial Economics in Sweden. He is frequently invited to give seminars and has recently been interviewed by the *New York Times*, the *Economist* and the *Süddeutsche Zeitung*. His work is focused on understanding the effects of digitalization and structural change. His recent work includes the book *Digitalization, Immigration and the Welfare State* (2017). Blix has a Ph.D. in economics from Stockholm University and an MSc in econometrics and mathematical economics from the London School of Economics (LSE). He has held several senior management positions, including director for the division of budget totals in the Swedish Ministry of Finance, chief economist at the Ministry of Industry and advisor at the Swedish Central Bank; he has also worked at the European Central Bank and been secretary to the Swedish prime minister in the Commission on the Future Challenges of Sweden.

Bent Greve (Ph.D.) is a professor of social science with an emphasis on welfare state analysis at the University of Roskilde, Denmark. His research interest focuses on the welfare state, and social and labor market policy, often from a comparative perspective. Greve has published extensively on social and labor market policy, social security, tax expenditures, public-sector expenditures and financing of the welfare state. He is the editor of the academic journal *Social Policy and Administration*. Recent books include *Long-Term Care for the Elderly in Europe. Development and Prospects* (Ed., 2017); *Handbook of Social Policy Evaluation* (Ed., 2017); *Technology and the Future of Work. The Impact on Labour Markets and Welfare States* (2017); *The Routledge Handbook of the Welfare State* (2nd ed., 2019); and *Welfare, Populism and Welfare Chauvinism* (2019).

Dominika Sabolová is a doctor of dental medicine (*medicinae dentalis doctor*). She graduated from Pavol Jozef Safarik University, Košice, Slovakia, in 2012. After having spent a little over two years working as a dentist in her native Slovakia, she relocated to Sweden and has been working as a general dentist at the Swedish Public Dental Service (Swe: *Folktandvården*) in Linköping, Sweden, since 2015. Sabolová's main research interests concern digital dentistry, endodontics, periodontics, prosthodontics and oral surgery. In her spare time, she enjoys travelling, hiking, dancing and horseback riding.

Pernilla Lilja is a senior business consultant at C.A.G MER Consulting. She holds an M.Sc. in business administration and economics from Stockholm Business School at Stockholm University, with additional studies at CERAM Business School, Sophia Antipolis, France. Prior to her current position, Lilja worked for several years at one of the leading Swedish banks. She has extensive experience from the banking and finance industry, where she has held various positions and titles over the years such as, business developer, scrum master, requirement analyst and business process developer etc. Some of Lilja's greatest professional interests include change management, business improvement and modeling business processes. She is passionate about investigating how the implementation of regulations can affect business and what impacts change has to different companies' working conditions as well as the ever-changing dynamics of organizational change. In her spare time, Lilja is always looking to challenge herself and push her boundaries. Being an avid runner, she has participated on annual basis in the Stockholm Marathon for the past several years. However, she also likes to relax with a good round of golf or by going for some cross-country skiing. In 2018, she completed her first Ironman Triathlon.

Edward "Ted" Castronova (Ph.D.) is a professor of media, specializing in games, technology, and society at the Media School, Indiana University – Bloomington. He has served in the past as director of the B.Sc. degree program in game design, and chair of the Department of Media Arts and Production. Books include *Wildcat Currency: The Virtual Transformation of the Economy* (2014), *Virtual Economies: Analysis and Design* (with Vili Lehdonvirta, 2014), *Exodus to the Virtual World* (2007), and *Synthetic Worlds: The Business and Culture of Online Games* (2005). Castronova's papers at the Social Science Research Network place him in the top one percent of downloads of over 300,000 academic authors. Castronova's game design work involves board games with a historical focus. Born "Edward Bird" in 1962, he took his wife's name on marrying in 2000. He has two sons, two god-children, and a beagle named Tilly. Castronova thinks God is a game designer: Get to Heaven for the win.

Yamit Viitaoja is an entrepreneur specializing in network formation and business talent development. She is currently working as a sales and marketing specialist at the Scandinavian Wine Academy, based in Stockholm, Sweden. She holds an MBA/M.Sc. in business administration from Stockholm University, Sweden, and an associate degree in sales and management from Nackademin, Sweden. She has extensive experience from the IT software and service industry and in working with customer relations, and holding various managerial titles such as a "Financial Services Customer Experience Manager" at Tieto, Sweden. Viitaoja also co-founded the female-exclusive career network

International Women Group (IWG). Some of Viitaoja's greatest research interests include business development, entrepreneurship, CRM, digitalization, marketing, financial services and project management. Her personal interests include traveling, wine degustation, yoga and power walking.

Nicole Andersson is a management consultant currently based in Sydney, Australia. She has previously worked as a management consultant at Bain & Company and OPX Partners in Stockholm, Sweden. She holds a B.Sc. in industrial engineering from the KTH Royal Institute of Technology, with studies at Zhejiang University, China, and the Hong Kong University of Science and Technology, Hong Kong. She has a background in aviation and holds a commercial pilot's license (CPL) from OSM Aviation, Sweden. Andersson has experience from working with a variety of industries, such as financial services, industrial goods and services, consumer products and logistics. She has worked with a range of companies, from start-ups to private equity owned companies to publicly listed companies on OMX Stockholm Large Cap. She is specialized in a variety of topics, ranging from merger and acquisitions (M&A) and strategy to various types of transformative processes.

Peter Markowski (Ph.D.) is a management consultant and researcher, currently residing in Stockholm, Sweden. He obtained his Ph.D. in operations management from Stockholm University, Sweden. He also holds an M.Sc. in industrial engineering and management from KTH Royal Institute of Technology in Stockholm. Markowski's research interests include organizational capabilities and digitalization. He currently advises organizations on issues such as digital transformation and has led several projects within organization design aimed at integrating knowledge work and digital technology.

Malin Nilsson is a Ph.D. student in applied mathematics at the University of Gothenburg, Sweden. She has previously worked as a consultant for several years, specializing in algorithm development, data science and business intelligence. With a penchant for gaining insights from the data she collects, Nilsson is passionate about mathematics in general and in algorithms and partial differential equations in particular. Her favorite programming languages are Python, MATLAB and C++. A curious and adventurous soul, Nilsson enjoys spending her spare time mountain climbing, riding her motorbike and exploring all corners of the world, in addition to running her own website (www.marlinnilsson.com), where she illustrates and explains various pedagogical mathematical proofs.

Ivy Mayor is a Ph.D. candidate and a graduate teaching assistant at the Department of Cell and Molecular Biology (CMB), Karolinska Institutet, Stockholm, Sweden. Her research explores the intersection of neuroscience and immunology in application for regenerative medicine. Mayor also has an extensive interdisciplinary background in the fields of mechatronic engineering, biophysics, physiology and neuroscience, after completing her tertiary education at the University of Lübeck, Germany, as well as at the University of Western

Australia in her native Australia. Her current areas of interest include STEM and history communication, AI, robotics, technology, innovation, space exploration and regenerative biology.

Linn Lindfred is an entrepreneur and consultant based in Stockholm, Sweden. She recently founded a sole proprietorship aimed at working with, and educating in, circular economy. Though this was not Lindfred's first entrepreneurial endeavor; in 2017, she co-founded a start-up initiative called "Waste2Value". This initiative, which won Lindfred two separate entrepreneurship awards, sought to establish a digital marketplace for trading with industrial waste. Linn is currently a board substitute for Cradlenet – Sweden's platform for knowledge and networking around circular economy. She also works as a consultant at Jayway by Devoteam, where she is responsible for the initiative of designing digital solutions for circular economy and sustainable innovation. She holds an M.Sc. in management and economics of innovation from Chalmers University of Technology with studies at Stanford University, University of California (Berkeley) and National Chiao Tung University, Taiwan.

John Øvretveit (B.Sc. [Hons], M.Phil., Ph.D., C.Psychol., AFBPsS, C.Sci.) is director of research for the Medical Management Center (MMC) at Karolinska Institutet, Sweden, and serves as professor of health improvement, implementation and evaluation at that university. Previously, he was professor of health policy and management at Bergen University Medical School, Norway, for ten years and was also an appointed chair of health organization and management for 15 years at the Nordic School of Public Health, Gothenburg, Sweden. Translations of some of his 400 scientific papers and books have been made into eight languages. Øvretveit was awarded the 2014 *Avedis Donabedian International Quality Award* for his work on quality economics. He is a founding member and board director of the Global Implementation Initiative and the Quality Improvement Research Network, and was a board director for 12 years of the Joint Commission International. Øvretveit is currently a visiting professor with: Australian Institute or Health Innovation Macquarie University, Kaiser Permanente RD&E, Stanford Medical School and CERC. His work is based on the belief that organization and management can bring out the best and worst in people, and that the right organization design is critical for effective health care. A theme underlying his work is how practical research can contribute both to better care for patients and to a "healthy work organization". Øvretveit has pioneered action evaluation and partnership research methods for giving rapid feedback for service providers and policy-makers to improve their services, and for assessing the role of context in implementation.

Foreword

Galaxy incognito

We are on an express train headed for the hubbub of future life. To be "disrupted" is the new normal in almost any human activity. Public, private and even non-profit organizations are today used to the fact that anything can – and indeed will – happen.

This book directly and indirectly addresses one of the most fundamental questions of our time – digitalization and its impact on employment and working conditions. Work is an old, well-known concept, but digitalization is not. We are still grappling with what it is both in philosophical and practical terms. And above all we struggle to understand what impact digitization will have on some of the most fundamental human activities. Work and the labor market certainly being one of them.

When Marshall McLuhan, the Canadian-born professor and communication theorist, described modern communication as expanded consciousness in his 1962 book, *The Gutenberg Galaxy*, he was almost prophetic. Our societies are defined by communication or the absence of it. Everything is communication. Today, we know that even we ourselves are information and data. Ask any friendly, local genetic researcher and they will tell you that we are physical manifestations of code. But if everything is information and anything can be digitalized, it should come as no surprise that almost any human institution is at risk of disruption. The labor market is not in any way an exception, nor is it protected. And it is one of the cornerstones of any society.

Digitalization alters and redefines. Whether it is "Labor", "Work", "Who owns what", "What is valuable or not", etc. It could today be argued at least theoretically that Karl Marx finally turned out to be right. We, the people, now own the critical resources of production. New fortunes are being created by brains, knowledge and information, not by muscle power or machinery. Maybe that is how it always been. Now, though, we can see that it is our intellect that is the most important production resource. The other production resources are indeed necessary but certainly not sufficient. The problem is this is not "the workers' heaven" Marx prescribed and foresaw. It is indeed something completely different. Digitization has transformed our planet into a flying bazaar. Everything can be bought and sold. Zeros and ones too.

The most critical production resource of our time – our intellect – is subject to a massive digitalization and hence the labor market is subject to a complete and fundamental redefinition. Some kind of fundamental deregulation beyond politics and driven by technology. Few, if any, of our traditional perspectives, concepts or institutions can and will encompass the labor market of tomorrow. The digitalized labor market lowers transaction costs and offers opportunities to reorganize anything we do. What is transport or taxi in an "uberified" world? Who are the winners and the losers when national boundaries increasingly provide limited protection for jobs? What is public and what is private when self-organized – for profit or not – systems conquer some of the public domains?

And how do we ensure basic rights and minimum pay when "gigification" of traditional jobs short circuits labor market institutions and sets more and more people on the one hand free and on the other hand makes their life more uncertain?

It is this new galaxy incognito – unknown galaxy – that this text brings us into.

Enjoy and buckle up. Bumpy ride ahead.

Dr. Kjell A. Nordström
Valencia, Spain, May, 2019

Letter from the editors/ acknowledgements

This edited volume constitutes the final phase of the research project, "The Innovative Internet", funded by the Internet Foundation in Sweden (IIS – Swe: *Internetstiftelsen i Sverige*). Released as part of the book trilogy by the same name, this volume concludes the research project preceded by *The Rise and Development of FinTech: Accounts of Disruption from Sweden and Beyond* (2018) (ISBN hardback: 978-0-8153-7850-1; ISBN eBook: 978-1-351-18362-8), and *Digital Transformation of Public Services: Societal Impacts in Sweden and Beyond* (2020) (ISBN hardback: 978-0-367-33343-0; ISBN eBook: 978-0-429-31929-7), both published by Routledge under Open Access.

The premise of the first book was to discuss the conditions in which financial technology (FinTech) could prosper and the ramifications that followed along with it. Having explored the impact of the digital transformation of the financial sector, the next books in the trilogy were set to explore the digital transformation of the public sector and the labor market respectively. However, the initial plan did not involve a trilogy at all. In fact, from the outset, this phase of the project had begun as a small-scale plan for a shorter report on the future of labor and the working conditions of tomorrow. Following the completion of the FinTech book, we soon discovered a need to cover the future of welfare and the public sector as well, as this is an area that is seeing rapid and profound effects of the digital transformation. Consequently, the project shifted focus, at least initially. However, the research project soon began expanding to also include the original plan of the digitalization of the labor market and a decision was made to write and release the books in separate installments.

We would like to take this opportunity to extend our deepest gratitude to IIS, and in particular to Danny Aerts and Jannike Tillå, who have supported us throughout the project while also allowing us full freedom to take ownership of our research and its direction. This volume would not have been possible without the experiences, insights and expertise from our dedicated team of co-authors, consisting of academics and their sponsors, practitioners, and industry experts. To all the contributors we have had the privilege to lead and guide, throughout this process, we wish to express our sincerest thanks for being the foundation of this book! We would also like to thank Executive Director Johan Söderholm at the Stockholm School of Economics Institute for Research (SIR) as well as Senior

Editor Kristina Abbotts and Editorial Assistant Christiana ("Chrissy") Mandizha, along with the rest of the assisting staff at Routledge/Taylor & Francis for their invaluable help and support throughout every step in the making of this volume. We also wish to acknowledge the diligent work carried out by the copyeditors, typesetters, and indexers of this book, while extending a special thanks to Kate Fornadel at Apex CoVantage, the project manager in charge of preparing our book for the printer. We would also like to extend our sincerest gratitude to IIS; Accenture; Cartina; Department of Education, Stockholm University; Sven and Dagmar Saléns' Foundation (Swe: *Sven och Dagmar Saléns stiftelse*), all those who have generously contributed toward securing Open Access for this book.

Last, but not least, we would also like to extend our special gratitude to all family and friends for their unwavering and invaluable moral support throughout this process.

We hope that you, the reader, will have as much satisfaction in reading this book as we had in bringing it together. As an Open Access volume, we invite you to spread and redistribute this book freely to any and all interested parties you may encounter.

Stockholm, September 27, 2019.
Anthony Larsson, Ph.D.
Editor-in-Chief
Robin Teigland, Professor
Editor

1 A journey of a thousand miles

An introduction to the digitalization of labor

Anthony Larsson

1. Introduction

It is held that the ancient Chinese philosopher Lao Tzu (604–531 BC) once said "a journey of a thousand miles begins with a single step" (Keyes, 2006, p. 107). In the same way, mankind has always looked for practical solutions to various problems throughout history, one step at a time, eventually advancing far beyond his original intention.

The extensive technological developments throughout the twentieth century have set the tone for how the future labor market of the twenty-first century has developed and will continue to develop. For instance, the Third Agricultural Revolution would increase agricultural production worldwide, and especially so in the developing world. During this period, many new technologies and techniques would develop, such as chemical fertilizers and agro-chemicals, along with controlled water-supply/irrigation and new methods of cultivation, including mechanization (Farmer, 1986; Esteva, 1996). This was followed by the scientific-technical revolution (circa 1940–1970) (Šmihula, 2010). This was subsequently followed by the digital/information and telecommunications revolution, also known as the "Third Industrial Revolution" (circa 1975–2010) (Esteva, 1996; Kheinman, 1978; Melnikov and Semenyuk, 2014; Vickery, 1999). These eras brought not only technological advancements that sought to improve people's everyday lives, but they would also fundamentally change the economics of society and the way in which the labor market operated. For instance, bureaucracy began expanding and industries began developing information-generating activities, specifically the so-called "Research and Development" (R&D) functions (Veneris, 1990). Moreover, information has become a factor of production much in the same way as with the case of capital, labor, property, economics etc., while also becoming a "commodity", i.e., a product/service that customers are willing to purchase on the open market. As information acquires a "use value" and "exchange value", it also nets itself a price (Repo, 1986; Vargo, Maglio and Akaka, 2008; Eggert et al., 2018).

Progressing beyond the "Third Industrial Revolution", is the "Information Age", or the "Fourth Industrial Revolution", also known as "Industry 4.0" (circa 2010s–present). Significant of this era are the emerging technology breakthroughs and developments in fields such as robotics, artificial intelligence (AI), nanotechnology, autonomous vehicles, biotechnology, Internet of Things (IoT), smartphones, Blockchain and 3D printing etc. (Walwei, 2016). The “Information Age” is intrinsically different from the technological eras, as the previous eras were mainly characterized by advancements in various types of technologies. As for the “Information Age”, its main advances lie not so much in the emerging of new technology per se, but rather in new means of communication and connectivity (Schwab, 2016; Schwab and Davis, 2018). Specifically, these new forms of communication technologies enable billions of more people worldwide to connect via the web, drastically improving the efficiency of business and organizations, while promoting better asset management by improved information access (Wisskirchen, 2017).

The definition of AI has many different variations. For instance, the Government of Canada (2019, para.28) defines AI as:

> Information technology that performs tasks that would ordinarily require biological brainpower to accomplish, such as making sense of spoken language, learning behaviours, or solving problems.

The European Commission (2018, para.6), on the other hand, (having gone through some minor revisions in the past few years) has a somewhat lengthier definition:

> Artificial intelligence (AI) refers to systems that display intelligent behaviour by analysing their environment and taking actions – with some degree of autonomy – to achieve specific goals.
>
> AI-based systems can be purely software-based, acting in the virtual world (e.g. voice assistants, image analysis software, search engines, speech and face recognition systems) or AI can be embedded in hardware devices (e.g. advanced robots, autonomous cars, drones or Internet of Things applications).

Naturally, other variations may occur in different countries around the world. Interestingly, while the Canadian definition stresses the complexity of the system in assisting humans with chores, the European definition places more emphasis on the intelligent and autonomous design and behavior of the system. In this way, the European definition of AI accounts for the possibility of an evolution of the behavior of AI, in a way that follows technological advancement. That is to say, AI is not a “fixed construct”, and to that end, this definition better encompasses the relevance of AI in the scope of this book.

It is, in this context, appropriate to also mention the terms “digitization”, “digitalization” and “digital transformation”, as they have become frequently used “buzzwords” in many different businesses. However, the terminologies are sometimes erroneously used interchangeably. The first term, “digitization, entails the conversion of analogue material (such as images, video and/or text etc.) into a digital format (Larsson and Viitaoja, 2017; Feldman, 1997; Brynjolfsson and McAfee, 2014). The second term, “digitalization”, refers to a process wherein the use of digital/computer technology (also mobile applications) is adopted, alternatively, increased by an actor (Wachal, 1971; Castells, 2010). More often than not, the digital technology implemented with the intent of establishing a communication infrastructure that connects various activities of the actor's various processes (Van Dijk, 2012; Larsson and Viitaoja, 2017). "Digital transformation", on the other hand, is a considerably broader term that signifies customer-driven strategic business transformation requiring far-reaching and cross-cutting organizational change in addition to the implementation of digital technologies (Bloomberg, 2018; Cochoy et al., 2017). Due to its scope, digital transformation is in reality not a matter of implementing *one* project, but rather a *whole series* of different projects, effectively necessitating the organization to deal better with change overall. In this way, digital transformation in and by itself essentially makes organizational change a core competency inasmuch that the venture seeks to become customer-driven end-to-end (Bloomberg, 2018).

For this reason, "digitalization" and "digital transformation" are the two most useful/significant terms when explaining the changes and impact that digital technology has had on society at large. That is to say intelligent algorithms make our day-to-day tasks easier, and it is in many cases near impossible to imagine how we could manage without them. The use of AI and robotics continues to gain momentum at a rapid pace, which prompts the question as to what the future of labor will look like once fully evolved. Extant literature suggests that digitalization has opposing effects on labor markets and that as such, it is still not clear what effects a digitalized society will ultimately have on the labor market (Bührer and Hagist, 2017). Will mass unemployment, poverty and social distortions be a given consequence of this development or may there be a different outcome?

This book will seek to explore these issues and many more through a series of different studies by scientists and industry professionals from Europe and the United States, with deep insight into their respective areas. It is true that the chapters in this volume are to a large extent inherently based on a speculative and/or predictive premise, given the fact that much of the digital transformation is still happening and is nowhere near completed and/or optimized. However, while the authors have sought to interpret near- and far-future developments, they have availed themselves to uphold scientific rigor by following proper academic protocol. This means using citations and basing their point of departure in extant issues/problems and undertaking due analytical procedure and research rather than conveying conjecture or personal opinions. As such, the chapters offer an array of methodological and thematic studies, with some studies presenting original, empirical material while others are more theoretically rooted, with some additional chapters basing their foundation on various forms of literature reviews or departing from the authors' personal, "best practice" experiences. To this end, the studies covered throughout the different chapters have based their assumptions in referenced facts, but while doing so, the studies may at times also transcend the conventional academic comfort zone by offering some foresight in how their subject area could transpire based on the current and expected developments due to digitalization and digital transformation.

The overall premise of this book takes its point of departure from a 2016 OECD report that targets the rapid structural transformations that have followed the digitalization process throughout the OECD countries (Berger and Frey, 2016). Specifically, this report lends support to the aforementioned academic notion that the impact digitalization has on the future of labor is ambiguous. That is to say that there is accumulating anecdotal evidence suggesting that the potential scope of automation has expanded beyond mere routine work, which would make technological change potentially increasingly labor-saving. On the other hand, there is evidence suggesting that digital technologies have not created new jobs on a scale that it replaces old ones.

Adding to this, an additional 2018 OECD report indicated that digitalization and automation as such does not pose a real risk of destroying any significant number of jobs for the foreseeable future (Nedelkoska and Quintini, 2018). Nevertheless, the report did contend that tasks by and large would change significantly, which in turn affects welfare, as most of its revenue stems from taxation, and particularly so from the taxation on labor (be it directly or indirectly). Taking its point of departure from the findings uncovered in these reports, the structure of this book seeks to explore some overarching themes in which digitalization and digital transformation can be expected to impact the labor conditions to some degree or another. The themes investigated are as follows:

1.1. Practical utilization of new technologies

These chapters discuss how the development of new technology can be applied in practice to enable people to work in ways they have not previously been able to.

1.2. The role of the digital welfare state

These chapters discuss how the transformation of labor markets affects the welfare state and the tax revenue system.

1.3. Digital disruption of status quo

These chapters discuss how digitalization and the digital transformation may be used by different groups or actors in ways to advance their positions on the labor market, or alternatively, how these developments may disrupt the status quo prompting these groups or actors to change their mode of operation in the future.

2. Chapter overviews

2.1. Practical utilization of new technologies

2 **Alexander Bard, Jan Söderqvist and Anthony Larsson – *Behind the history of labor: technology as the driving force***

This chapter explores the factor(s) that drive the organization of labor and how technology is used as a driving force, even in those instances where it may cause society to surrender its extant norms and routines. The chapter provides a narrative/philosophical discussion behind the historical development behind labor, while discussing the importance of communication and the exchange of information driving this development. The study postulates two research questions. RQ 1: *Is there a new paradigm shift taking place in the future organization of labor?* RQ 2: *In the event that a new paradigm shift can be expected, is it compatible with contemporary social norms or can such a paradigm shift of labor organization be expected to also transform the essence of society itself?*

3 Jochem van der Zande, Karoline Teigland, Shahryar Siri and Robin Teigland – *The substitution of labor: from technological feasibility to other factors influencing the potential of job automation*

This chapter provides a comprehensive yet perspicuous introduction to the area by illustrating an overview of how digitalization and automation, along with the three underlying technologies of artificial intelligence, machine learning (a subcategory of AI), and robotics may be used in the future to perform wide varieties of routine and non-routine work tasks. The chapter seeks to understand to what extent these technologies and digital developments have the potential to replace human capabilities in the workplace. The chapter proceeds by discussing the factors that influence the pace and scope of job automation.

4 Alex J. Wood, Mark Graham and Mohammad Amir Anwar – *Minimum wages for online labor platforms? Regulating the global gig economy*

This chapter investigates how the rise of the "gig economy" has served to enable internet users to find new job opportunities that have previously been unavailable to them. The authors describe the emergence of the phenomenon called "online labor platforms", which effectively constitutes a global remote gig economy that provides clients to access world-wide labor power. The authors provide a detailed account of how these platforms work, while providing some original empirical research by interviewing 250 remote gig economy workers across ten countries and four continents along with platform CEOs and government and trade union officials. In addition, a survey encompassing 679 Asian and African workers has been conducted in addition to an analysis of transaction data and observation studies.

5 Antoine Maire – *The digital disruption of science: governments and scientists toward an "Open Science"*

This chapter explores how digitalization affects the life cycle of research from a "bottom-up" approach. The study applies a qualitative approach using two case studies in order to investigate the European and the French policy toward the digitalization of science. The author argues that while digitalization is crucial to improve the competitiveness of a research community, it has led government institutions to take direction. The author discusses how this new direction is used in order to frame innovative policies to encompass and foster the evolution toward digitalization further.

6 Victor Erik Bernhardtz – *Black boxes of cognitive computers and the impact on labor markets*

This chapter considers two major changes to labor brought on by digitalization. The first is investigating how tools used in production can be connected to the internet and/or local networks in order to create opportunities for data gathering. The second is investigating the potential of improved and customized cognitive computing systems in analyzing gathered data in a more optimal manner than humans can. This chapter is thus a literature study of, and a theoretical discussion on, the impact of digitalization on the labor market. The chapter seeks to address various concepts and issues while investigating if digitalization can be integrated into existing labor markets while at the same time avoiding deterioration of labor quality.

7 Fernanda Torre, Robin Teigland and Liselotte Engstam – *AI leadership and the future of corporate governance: changing demands for board competence*

This chapter investigates how corporate boards, a subset of labor, are developing their capabilities to better govern and leverage AI in their innovation and sustainability efforts, while also having a defining impact on organizations' future labor. Corporate boards make complex strategic decisions in uncertain environments, such as mergers and acquisitions, new product launches, and digital transformation, and as such are not expected to be replaced by automation within the foreseeable future. The authors present preliminary results from a research project that includes a systematic literature review and expert interviews, while also touching on how AI could change the future of board work. Two areas for capability development at the board level are identified: (1) AI operational capabilities, including the guidance of gathering, harvesting and analyzing big data; innovating using AI; and implementing a digital business ecosystem, and (2) AI governance capabilities, including the stewarding of managing data, ethics and black box decision making; staying ahead of AI security threats, and leading the digital business ecosystem. The authors proceed to present their tool for board capability development: the "Boards 4 AI Leadership Matrix". This tool is intended to facilitate the development of the two aforementioned competence areas. The chapter also touches on how AI may change the future of board work, such as new board processes and augmenting board tasks.

2.2. The role of the digital welfare state

8 Mårten Blix – *Polarization, tax revenue and the welfare state: digital disruption or still standing strong?*

This chapter seeks to investigate the resilience of the welfare state in the face of rapid technological change, using Sweden as a case study. The author contends that while labor markets have become more polarized in other countries, the effects in Sweden have thus far been limited. The author analyzes the reason for why Sweden's labor market has been less polarized, while discussing how the country's social safety net and institutions have worked in the country's favor, but that digitalization at the same time poses a risk due to the taxation system and the shortage of skilled labor in various key segments.

9 Bent Greve – *Welfare states and digitalization*

The chapter looks closer at how the digital transformation of the labor markets can entail a risk of fewer people managing to sustain stable jobs, while the income of citizens continues to influence the welfare states and their development. The author illustrates the differences in impact of changes on the labor market and discusses the welfare states' ability to finance their sustenance in the future.

10 Anthony Larsson and Dominika Sabolová – *"Gig patients": health and dental care in the gig economy*

The "gig economy" provides for a new style of employment where workers (referred to as "giggers") sustain themselves by performing a number of "gigs" on a freelance basis (often for several different employers). This has given rise to a new phenomenon referred to as "gig patients", which entails "giggers" who cannot afford to take time off to seek medical or dental assistance until it is absolutely necessary, by which time the treatments needed are often extensive and expensive. These patients will put themselves in debt with no feasible way of paying their medical bills in full, which in turn affects the welfare system. Thus, this chapter seeks to investigate the wider ramifications that "gig patients" could have to the welfare society and the future of labor.

2.3. Digital disruption of status quo

11 Anthony Larsson and Pernilla Lilja – *GDPR: what are the risks and who benefits?*

In this chapter, the authors investigate the ramifications of the General Data Protection Regulation (GDPR) on the future labor market. The overarching research question this chapter seeks to answer is: *In what way may GDPR influence the labor market of tomorrow, and what businesses are at risk?* That is to say, can GDPR help stimulate certain types of business and will it have a stymieing effect on other types of business ventures? This chapter serves as a theoretical/speculative study that endeavors to look at some of the available literature and best practices in order to anticipate the future role of GDPR in a labor market that is becoming increasingly digitalized.

12 Edward Castronova – *Players for hire: games and the future of low-skill work*

In this chapter, the author uses different trends from automation and video game revenue models to make various predictions about the future of low-skill work based on a five-year, a ten-year, and a twenty-year time-frame. Specifically, the author argues that within five years, there will be game companies who pay players in some way to play their games. Within ten years, paying for players will become a standard revenue model in the game industry. Within twenty years, game playing will be a significant source of income of the low-skill workforce. The author contends that "wagc-playing" will be the primary means by which the extreme gains of the wealthy will trickle down to the poor, while analyzing the ramification this development has in the broader spectrum in regards to the future labor market.

13 Mark Graham and Mohammad Amir Anwar – *The global gig economy: toward a planetary labor market*

This chapter discusses the emergence of a "planetary labor market" for digital work. Building on a five-year study of digital gig work in some of the world's economic margins, the authors illustrate that a planetary labor market does not dismiss the notion of geography, but rather exists to take advantage of it. That is to say, digital technologies have been deployed in order to bring into being a labor market that can operate at a planetary scale, and has particular affordances and limitations that rarely bolster both the structural and associational power of workers. In this study, the authors seek to understand how gig work platforms are generating online labor markets that help clients/employers reconfigure the geography of their production networks for almost zero cost while risks are shifted towards workers, who can sell their labor power globally, but still are tethered to the locales in which they go to bed every night.

14 Anthony Larsson and Yamit Viitaoja – *Identifying the digital gender divide: how digitalization may affect the future working conditions for women*

This chapter draws upon existing research and studies, as well as on the authors' "best practice" insights in investigating the popularly-termed "digital gender divide" and how digitalization and the digital transformation impact on the future prospects for women on the labor market. Providing an analytical commentary on the present situation, this chapter discusses the relative lack of women in the Western countries undertaking science, technology, engineering and mathematics (STEM) subjects. This in turn leads to a lower turnout for women working with information and communication technology (ICT) related jobs. The authors explore the possible reasons behind this development while providing insights to how women can secure a more prominent role overall in a future digitalized labor market.

15 Anthony Larsson, Nicole Andersson, Peter Markowski, Malin Nilsson and Ivy Mayor – *Consulting in the digital era? The role of tomorrow's management consultants*

This chapter draws upon extant literature as well as the authors' own "best practice" experiences in exploring some of the most pressing issues of the digitalization process of management consulting of today. The chapter pursues the following two research questions: RQ 1: *How may digitalization influence the consultant's role of tomorrow?* RQ 2: *How may the profile of the "typical consultant" change in the future?* In seeking the answers to these questions, the authors anticipate how the role and profile of management consultants may come to develop in the near future as digitalization and the digital transformation ensues.

16 Anthony Larsson and Linn Lindfred – *Digitalization, circular economy and the future of labor: how circular economy and digital transformation can affect labor*

This chapter seeks to investigate how labor would be affected by a transition to a circular economy facilitated through digitalization. This study introduces a premise under which the introduction of a circular economy would become realized. The chapter discusses how this transformation would affect businesses, labor, industries, and society at large, while illustrating how digitalization is a tool to facilitate such a transformation. The authors do not seek to prognosticate answers or provide a "one-size-fits-all" solution, as a potential future circular scenario involves major uncertainties. Moreover, there is lack of real case studies on which to base predictions. Rather, this chapter offers a conceptual study that seeks to draw upon available literature and research findings in order to answer how the labor conditions are affected when digitalization is used to achieve circular businesses and societies in different ways.

References

Berger, T. and Frey, C.B., 2016. *Structural Transformation in the OECD: Digitalisation, Deindustrialisation and the Future of Work.* OECD Social, Employment and Migration Working Papers, No. 193. Paris, France.

Bloomberg, J., 2018. *Digitization, Digitalization, and Digital Transformation: Confuse Them at Your Peril.* [online] Forbes. Available at: <www.forbes.com/sites/jasonbloomberg/2018/04/29/digitization-digitalization-and-digital-transformation-confuse-them-at-your-peril> [Accessed 27 Sep. 2019].

Brynjolfsson, E. and McAfee, A., 2014. *The Second Machine Age: Work, Progress, and Prosperity in a Time of Brilliant Technologies.* New York, NY: W.W. Norton & Company.

Bührer, C. and Hagist, C., 2017. The effect of digitalization on the labor market. In: H. Ellermann, P. Kreutter and W. Messne, eds., *The Palgrave Handbook of Managing Continuous Business Transformation.* London, UK: Palgrave Macmillan, pp. 115–137.

Castells, M., 2010. *The Rise of the Network Society: The Information Age: Economy, Society, and Culture.* 2nd ed. Chichester, UK: Wiley-Blackwell.

Cochoy, F., Hagberg, J., McIntyre, M.P. and Sörum, N., 2017. Digitalizing consumption: Introduction. In: F. Cochoy, J. Hagberg, N. Sörum and M.P. McIntyre, eds., *Digitalizing Consumption: How Devices Shape Consumer Culture.* London, UK: Routledge.

Eggert, A., Ulaga, W., Frow, P. and Payne, A., 2018. Conceptualizing and communicating value in business markets: From value in exchange to value in use. *Industrial Marketing Management*, 69, pp. 80–90.

Esteva, G., 1996. Hosting the otherness of the other: The case of the Green Revolution. In: F. Apffel-Marglin and S.A. Marglin, eds., *Decolonizing Knowledge: From Development to Dialogue.* Oxford, UK: Clarendon Press, pp. 249–278.

European Commission, 2018. *Communication Artificial Intelligence for Europe.* [online] Digital Single Market. Available at: <https://ec.europa.eu/digital-single-market/en/news/communication-artificial-intelligence-europe> [Accessed 27 Sep. 2019].

Farmer, B.H., 1986. Perspectives on the 'green revolution' in South Asia. *Modern Asian Studies*, 20(1), pp. 175–199.

Feldman, T., 1997. *An Introduction to Digital Media.* New York, NY: Routledge.

Government of Canada, 2019. *Directive on Automated Decision-Making.* [online] Available at: <www.tbs-sct.gc.ca/pol/doc-eng.aspx?id=32592> [Accessed 27 Sep. 2019].

Keyes, R., 2006. *The Quote Verifier: Who Said What, Where, and When.* New York, NY: St. Martin's Press.

Kheinman, S.A., 1978. The scientific-technical revolution: Essence, directions, and stages. *Problems in Economics*, 20(12), pp. 14–34.

Larsson, A. and Viitaoja, Y., 2017. Building customer loyalty in digital banking: A study of bank staff's perspectives on the challenges of digital CRM and loyalty. *International Journal of Bank Marketing*, 35(6), pp. 858–877.

Melnikov, A.V. and Semenyuk, E.P., 2014. The information revolution and the modern printing industry. *Scientific and Technical Information Processing*, 41(1), pp. 1–11.

Nedelkoska, L. and Quintini, G., 2018. *Automation, Skills Use and Training.* OECD Social, Employment and Migration Working Papers, No. 202. Paris, France.

Repo, A.J., 1986. The dual approach to the value of information: An appraisal of use and exchange values. *Information Processing & Management*, 22(5), pp. 373–383.

Schwab, K., 2016. *The Fourth Industrial Revolution.* New York, NY: Crown Publishing Group.

Schwab, K. and Davis, N., 2018. *Shaping the Future of the Fourth Industrial Revolution: A Guide to Building a Better World.* London, UK: Penguin.

Šmihula, D., 2010. Waves of technological innovations and the end of the information revolution. *Journal of Economics and International Finance*, 2(4), pp. 58–67.

Van Dijk, J.A.G.M., 2012. *The Network Society: Social Aspects of New Media.* 3rd ed. London, UK: Sage Publications.

Vargo, S.L., Maglio, P.P. and Akaka, M.A., 2008. On value and value co-creation: A service systems and service logic perspective. *European Management Journal*, 26(3), pp. 145–152.

Veneris, Y., 1990. Modelling the transition from the industrial to the informational revolution. *Environment and Planning A: Economy and Space*, 22(3), pp. 399–416.

Vickery, B., 1999. A century of scientific and technical information. *Journal of Documentation*, 55(5), pp. 476–527.

Wachal, R., 1971. Humanities and computers: A personal view. *The North American Review*, 256(1), pp. 30–33.

Walwei, U., 2016. *Digitalization and Structural Labour Market Problems: The Case of Germany*. [online] ILO Research paper No. 17. Available at: <http://ilo.userservices.exlibrisgroup.com/view/delivery/41ILO_INST/1243479920002676> [Accessed 27 Sep. 2019].

Wisskirchen, G., 2017. *Digitalization and Automatization and their Impact on the Global Labor Market – CUE, Inc.* [online] CUE. Available at: <www.cueinc.com/digitalization-automatization-impact-global-labor-market> [Accessed 27 Sep. 2019].

Part I

Practical utilization of new technologies

2 Behind the history of labor

Technology as the driving force[1]

Alexander Bard, Jan Söderqvist and Anthony Larsson

1. Introduction

The first formations of organized work likely began even before the evolution of Homo sapiens (Kranzberg and Hannan, 2017). In fact, throughout history, labor has been a necessity for the perseverance of mankind and for the formation of society at large. At the same time, the development and organization of labor has in no small part been bolstered by improvements to the tools and equipment used. The development of technology has in turn had profound impact on the advancement of labor (and vice versa). Most prominently, it has led to an increase of production while lessening the amount of effort needed from the individual in order to carry out specific work tasks, ultimately leading to various degrees of automation in certain sectors. In its purest form, automation would entail the elimination of all manual labor through the use of automatic devices while at the same time ensuring accuracy and quality. However, automation also means eradicating various forms of labor and effectively making certain types of professions obsolete. Regardless, mankind has always turned toward improvements in technology as a means of advancing their society, even if it means the destruction of certain aspects and norms of the existing society. This raises a question of what it is that drives the organization of labor and how technology is used as a driving force, even if/when it causes society to surrender its previous norms and routines. This chapter seeks to explore this issue through a thought-provocative narrative/philosophical discussion behind the historical development of mankind's relationship with technology, its role in the development of the modern-day working society, and its integral place or role to human nature. Specifically, this chapter seeks to answer two research questions:

RQ 1: *Is there a new paradigm shift taking place in the future organization of labor?*
RQ 2: *In the event that a new paradigm shift can be expected, is it compatible with contemporary social norms or can such a paradigm shift of labor organization be expected to also transform the essence of society itself?*

2. Early days of labor

During the early stages of civilization, labor was restricted to menial tasks aimed at taking care of the bare necessities of human needs, such as food, shelter, child care, protection etc. It is likely that a division of labor emerged once certain members within the society showed some kind of proficiency or aptitude for a particular task, such as hunting, fishing, gathering crops etc. Occasionally, prehistoric humans would organize certain types of tasks such as foraging, hunting and, eventually, even agriculture. As such, division of prehistoric labor is confined to a restricted geographic area as populations were sparse and insular. Division of labor was generally organized according to differences in age and sex. Since the oldest members of a community would often lack the strength and agility to hunt or forage, they would often conduct more sedentary tasks, while the youngest members would be taught simple food-gathering techniques. While the male members of the group would take on tasks such as hunting, the female members would specialize in food gathering, cooking, and child rearing (Kranzberg and Hannan, 2017).

During this time, there was little contact with other groups of humans in other places that may have had access to different kinds of foods, and to the extent that this was so, there was little trade to be made. Since the availability of food was fickle at best, there was little surplus to be bartered with. The organization of labor would become more advanced once pottery was developed. The quality of pottery was largely contingent on the quality of clay, which in itself was not universally and equally distributed across the lands. While pottery could largely be made to some extent almost anywhere, high-quality pottery products made in some places soon became merchandise worth trading elsewhere. Naturally, the quality of these products were further enhanced by the craftsmanship of the potter and the type of tools/equipment they used, which in turn encouraged further specialization.

Similar patterns of specialization followed in other areas as well, such as for textiles. Early on, at least some 70,000 years ago, but perhaps even as early as 500,000 years ago, animal skins were used to make various forms of protective garments (Bellis, 2018; Barber, 1991). However, once agriculture started becoming more developed, the available supply of skins was reduced. This prompted a substitute material for clothing, bringing the manufacture of textiles (initially yarn, and later other materials) into fruition (Kranzberg and Hannan, 2017).

Starting in the Bronze Age, humans would eventually develop and use copper tools and weapons, effectively initializing the formation of metallurgy, This, in turn, created a new organization of labor in which individuals would devote their full time to tasks such as mining, smelting and forging. These were tasks that were so physically demanding and required full mental concentration that they likely precluded the metallurgist from other chores such as farming or hunting. Moreover, copper ore was at this time generally not found in areas with a climate and topography that was favorable toward agricultural development (copper ore was generally located in mountainous regions). This further suggests that these individuals had made metallurgy their dedicated vocation. Likewise, other professional specializations developed in similar ways.

Eventually, along with various historical paradigm shifts, such as the various agricultural revolutions as well as the industrial revolution, would lead to substantial increases in productivity. At this point, however, such increases in productivity were more contingent on the rational organization of processes rather than on individual skill. Another important development occurring throughout the seventeenth to the nineteenth century in Europe was the massive urbanization process, which also acted as an added stimulus toward bringing a more rational organization of work to pass. This would eventually set the premise for the future mass production and mass consumption, further transforming the organization of labor.

What this historical development of labor tells us, is that it is the need (and subsequent exchange) of information that has always been the quintessential *primum movens* to building and forwarding the development of society, and the human race at large. How is it that mankind has been able to utilize information sharing and the technology that is spun from it?

3. The Japanese soldier and the use of information

There is a popular story that tells how Hiroo Onoda (1922–2014), an Imperial Japanese Army intelligence officer was found in 1974. He was found alone in a Japanese holdout in an inaccessible part of an island in the Philippines in the Asian jungle, several decades after the end of the World War II, where he had remained, fighting and refusing to acknowledge that the war had since long ended (Onoda, 1974). As a result of a combination of circumstances he had been left there alone. Perhaps he had been ordered to remain at his isolated post, and had been exercising his duties to the fatherland with exemplary loyalty for all those years, or perhaps he had simply been too frightened to venture into populated areas. But time had passed and no one had told him that peace had been declared. So the Second World War was still raging inside his head.

We have no reason to laugh at this confused soldier. He may have been wrong, but then so have we been, countless times. The soldier was not particularly well-informed, but then neither are we always. We all suffer to some extent from confused perceptions of what is going on outside the small part of our immediate world that we can get a direct impression of. This does not prevent us from forming, and being forced to form, opinions about one thing after the other, even in complicated matters where our knowledge is limited to say the least. Most of what we believe that we know is precisely that: what we believe ourselves to know. Other people's actions are comprehensible to us only insofar as we actually know what they in turn believe themselves to know – which is something we seldom know. The constant inadequacy of this information means that we have to swim through an ocean of misunderstanding on a daily basis, an activity which is both demanding and costly.

Like the Japanese soldier, we form our lives inside our heads. We have to, because the world is far too large and complicated for us to open ourselves to its every aspect without protecting ourselves with a multi-layered mental filter. For this reason we create fictions for ourselves, simplified models of how we believe the world works, or how we think it ought to work. These fictions have to fill the immense vacuums between our limited areas of knowledge. It is within this world of private fictions that we think and feel, but it is outside in the collective reality that our actions have their consequences. The more complicated a situation, the higher the degree of guesswork and the greater the contribution of fiction to our perception of reality.

This dependence upon fictions often has dramatic consequences, not just for us personally but for society as a whole. Like the Japanese soldier, we are fumbling blindly through dark forests. We react to signals that we can only partially understand, the consequences of which are only partially visible to us. Important political decisions are based upon shaky foundations and often have completely different results than were foreseen; great weight is placed on diffuse expressions of opinion, most often in the form of general elections, which are in turn the result of minimal knowledge, a problem which has been discussed, amongst others, by the author and journalist Walter Lippman in a couple of perceptive and intelligent books (Lippmann, 1922, 1925; Obar, 2015). This increasing lack of an overview explains for instance why today's voters find it easier to understand the credit card fiascos and alcohol consumption of individual politicians than serious political issues. Symbolism becomes attractive when real problems are perceived as being far too complicated. The business world is constantly forced to redefine its prognoses and adjust its decisions retrospectively in order to conceal the fact that they were based upon fictional rather than factual conceptions, as a result of the perpetual and chronic lack of information.

Becoming informed is an attempt to synchronize your own head with the reality outside. There are good reasons to make the effort: It is easier to interact with your surroundings when you have a relatively correct understanding of its mechanisms. Someone who has educated themselves in the psychology of the stock market has better prospects of succeeding in the markets; someone who has educated themselves in their own and other people's inner needs has better prospects of succeeding in relationships, and so on. Every failure reveals that we were not as well-informed as we thought or had hoped. The discrepancy between our own and other people's perception of reality, and between our own fictions and actual reality, was far too great. We learn from our mistakes; we take account of our earlier failures in the future and adjust our behavior accordingly. To put it another way, we make use of information.

Fictions can be more or less truthful, more or less applicable. They come in all possible forms, from private hallucinations to scientific theories. We are constantly testing them. Our culture consists of a perpetual evaluation and combination of both seemingly promising fictions and already proven fictions. The relationship between the fictions in our heads and unaccommodating realities is a recurrent theme in literature. Don Quixote, Othello, Raskolnikov and Emma Bovary are all victims of their own feverish ignorance. They are all relatives of the Japanese soldier. In attempting to study and gain an impression of the world around us we have to learn to differentiate between our prejudices – simplified models that we make use of not because they reflect empirical evidence but because they appeal to our own personal interests – and factual analyses and prognoses – necessary and intelligent simplified models of reality which make it comprehensible to us, even if the results do not appeal to us or fit in with our cherished fictions.

Our thoughts are directed by access to information. The story of the Japanese soldier is an illustration of this: Without access to news from the outside world he lived out an imaginary war for several decades. The same thing applies to whole societies and civilizations. Available information dictates which thoughts and actions are possible. It was not a lack of raw materials that prevented the Vikings from using water skis or the Romans from videotaping their orgies – it was a lack of relevant information. Civilization, in essence, is a matter of information. This means that any technological development which dramatically alters the preconditions for actions and the dissemination of information also implies a thorough re-evaluation of old and ingrained patterns of thought. The consequences of such a technological revolution are defined as a new historical paradigm. The advent of language was one such revolution.

4. Communication

The development of linguistic communication is one of the most important aspects that would come to separate humans from other animal species (Kranzberg and Hannan, 2017). The apes, our closest relatives, are intelligent animals with fantastic learning capabilities. But we cannot teach them to speak. From a physiological perspective we can say that their upper airways cannot function as vocal organs. But apes cannot use sign language in any real sense either. Chimpanzees can learn to combine signs in order to communicate on the level of a small child; they can indicate that they want something or that they want someone else to do something, but they never exchange experiences, never speculate about the great mysteries of life. They lack the capacity to communicate their thoughts and experiences with linguistic symbols, which seriously hampers the exchange of information. Man's path diverged from that of the apes about five million years ago, but language took longer to develop. To begin with we had elementary problems with our vocal organs, and evolution is a slow process. It is difficult to specify an exact time for the advent of spoken language, but current research suggests that it occurred as recently as 150,000–200,000 years ago. Only when the development of both the brain and our anatomy was sufficiently advanced was spoken language possible. Language differentiates man from other animals. The creation of technology requires abstract thought, which in turn arises from a linguistic system of symbols. Language made it possible for man to develop socially and to gather and maintain collectives, which opened up a new world of interwoven relationships between individuals. Social life developed entirely new and rich nuances as communication became more advanced. Language offered the possibility of innovative thought, with all its countless possibilities of expression, and stimulated creativity and intelligence. It also made possible the dissemination of information to everyone who was connected to a community. The basic facts of life for a hunter-gatherer society – which plants are edible, which poisonous plants are edible after various treatments, which animals leave which tracks, and so on – became possible to communicate throughout a large group, and between generations. Other people could gain knowledge of both successes and failures, and could go on to develop further the combined experience of the collective. Mankind developed memory. Knowledge could develop, but only to a certain point. Spoken language does not permit, at least not without a tape recorder, the reliable and comprehensive storage of information.

The mathematician Douglas S. Robertson (1998) has calculated the combined amount of information that a group or tribe of linguistically capable but illiterate people can access. He takes the poem the *Iliad* as his basis, a work comprising approximately five million bits (one bit indicates a choice between two alternatives: yes or no, black or white, one or zero), and which we know it is possible for one person to memorize. If the amount of information that a human brain can store is h, then h would appear to be somewhere between one and two *Iliads*, or, in other words, somewhere between five and ten million bits. If we multiply h by the size of a prehistoric tribe, a number between 50 and 1000, we get the maximum amount of information available within a society that was not capable of writing. We ought to bear in mind that there is a sizeable amount of redundant information here. Large amounts of the total store of information – how to hunt, how to fish, and so on – can reasonably be assumed to have been shared by most members of the community, which means that the total amount of information must be adjusted downwards accordingly. The numbers themselves must, of course, be taken with a pinch of salt, but Robertson's calculations provide an excellent illustration of the impact of written language when it was developed during the third millennium BC, and of the explosion in the amount of available information this represented.

Four of the so-called cradles of civilization – Egypt, Mesopotamia, the Indus Valley and China – developed at roughly the same time, and what united them, and simultaneously differentiated them from the surrounding societies in which trade and metallurgy were also practiced, was the invention of written language. To begin with, clay tablets were used to write on. The earliest "book" consisted of several of these tablets, stored in a leather bag or case. Certain texts, laws for instance, were inscribed on large surfaces so that everyone could see them. In this way, the fundamental ideas and norms of the society were transformed from something mystical and ancient which had been communicated orally by shamans into a visible and limited number of clauses and decrees that were available to everyone. Primitive, closed societies assumed a more open and more complex character. At the same time it became clear that knowledge gave power. Early forms of writing were initially an instrument of power. The Sumerian kings and priests used scribes to work out how many sheep different people ought to pay in tax. Another use of writing was propaganda: The ruler reminded his people of who was in charge and of the glittering victories he had won for them.

It was never intended that the written word would come into the hands of every Tom, Dick and Harry. The purpose of the first writings as a means of communication was, in the words of the French anthropologist Claude Lévi-Strauss (1961, pp. 291–292) "to facilitate the enslavement of other human beings". But revolutions have their own velocity, impossible to control for any length of time, and this is particularly true of information technology. Things that occurred either long ago or far away assumed a completely different accessibility and visibility when communicated via written text. The amount of available information exploded thanks to the ingenious invention of a visual code for communication. Intellectual life became far more vital. Thanks to the phonetic alphabet – where each sign represents a sound instead of a word or concept – the ancient Greeks were able to develop philosophy and sciences that had a far firmer structure, a grammar. The replacement of the ear by the eye as the main sense of linguistic reception brought with it a radical change in mankind's way of understanding the world.

Written language looked like magic: It was entirely logical that the Egyptian god Thoth, who gave the gift of writing to mankind, was also the god of magic. Reading and writing transformed both knowledge and the world. Empires could be established and held together only when written communication had developed; only then was it possible for detailed information such as orders to be communicated across large distances. This led to the dissolution of city states. The decline in papyrus production during the reign of the last emperors is held up by many historians as one important reason for the decline and ultimate collapse of the Roman Empire. Even hand-written information had its limits. Johann Gutenberg's invention of the printing press in the middle of the fifteenth century was the start of the next epoch-making revolution in information management. The printing press was also a basic precondition of what became modern science, and of the great discoveries and technical advances that led to industrialization. Printed books were the source material of the astronomer Nicholas Copernicus, and without the printing process his manuscript may well have gathered dust on the shelves of a monastery library. Instead his *De Revolutionibus orbium coelestium* (Copernicus, 1543), the thesis proposing for the first time that the Earth moved in orbit around the sun, spread quickly across the world of learning, where nothing was ever the same again.

Once the ball had started rolling, nothing could stop it. To put it bluntly, the printing press provided gifted and innovative people with the necessary information and inspiration to a previously undreamed of extent. Christopher Columbus read Marco Polo, large numbers of manuals and other technical literature circulated in Europe, and the whole of this tidal wave of new information prompted the development of new techniques and new thinking on the management of information, methods which paved the way for the gradual development of the sciences. Among the many innovations which followed in the wake of the printing press, after a certain incubation period, and which thoroughly and comprehensively altered mankind's way of looking at itself and the world, can be counted the clock, gunpowder, the compass and the telescope. One illustrative example of the power of developed information management, provided by the physiologist Jared Diamond, is the historically decisive meeting between literate Europe and essentially illiterate America in 1532 (Diamond, 1997). In the city of Cajamarca in the Peruvian highlands, Francisco Pizarro, with 168 men, captured the Inca leader Atahualpa, who had at his command more than 80,000 troops. The event only becomes comprehensible in light of the fact that the Inca leader knew nothing about his uninvited visitors whereas the Spaniards were well-informed about their opponent. Atahualpa was completely unaware that these visitors were in the process of conquering the whole of that part of the world, and that the great Indian civilizations of Central America had already fallen to them. He was entirely dependent upon defective oral information.

Atahualpa did not take the invaders seriously, and when his troops saw troops on horseback for the first time in their lives they panicked. Pizarro himself may not have been able to read, but he was a participant in a culture of writing and printing, and therefore had access to a wealth of detailed information about foreign civilizations. He was also aware of every phase of the Spanish conquest, and based his campaign upon the tactics of Hernando Cortés when he had defeated the Aztec leader Montezuma. Pizarro's success soon became known in Europe. In 1534, a book was published describing the events of Cajamarca, written by one of his company, which was translated into several other languages and became a bestseller (MacCormack, 1989). There was a great demand for information, and its benefits were self-evident.

5. The digital technology

Today's electronic and digital media comprise the most comprehensive information revolution of all. For a long time we believed that the central purpose of the computer was to think, to produce an artificial intelligence that would far exceed our own. Many people claimed that this goal was within sight when a computer named Big Blue beat the world master Garry Kasparov at chess (Newborn, 2003; Goodman and Keene, 1997). Today we can see that technology was heading in a different direction, toward communication via networks. Increasingly powerful and fast computers are making possible infinitely complicated and time-consuming calculations and simulations which were previously impossible to perform, which is of incalculable benefit to mathematicians and other researchers. Our collective knowledge is growing exponentially. But it is the global, digital network which is the most interesting aspect of this development. A new, dominant media technology means that a new world is evolving.

The internet is something completely new: a medium in which virtually anyone, after a relatively small investment in technical equipment, and with a few simple actions, can become both a producer and consumer of text, images and sound. In this sense it is hard to think of anything more democratic; on the net we are all authors, publishers and producers, our freedom of expression is as good as total, and our potential audience limitless. There are oceans of every conceivable sort of information available at the touch of a button. The growth of this new medium has been unparalleled.

The foundations of the internet were laid as early as the 1960s with the decision of the American defense organizations to use computerized networks to decentralize their resources via a series of distant but connected terminals. The purpose of this was to protect against and limit the effects of any nuclear war with the Soviet Union. Eventually American and foreign universities were connected to the system after it had proved stunningly effective in the organization of joint research projects. This development explains why the World Wide Web, the system which later became the standard for homepages on the internet, was developed not in the USA but by researchers at CERN, the European institute for research into particle physics in Switzerland. It was not until the end on the 1980s, as a direct result of the breakthrough of the personal computer and the launch of telecommunications modems, that the internet was transformed from the ARPANET, a tool for the United States Department of Defense and the scientific communities into public property. Even in the early 1990s there were relatively few people who had heard of the internet. It was only in December 1995 that Bill Gates woke up and announced in a memo that Microsoft would be changing direction and concentrating on net traffic, one month after a prior memo stating that Microsoft had no interest in the internet (Kearns, 2002). Since then the growth of the internet has been phenomenal. It is practically meaningless to give any figures regarding the number of computers linked to the net because its development is so dizzyingly fast. Figures that were accurate when this was written will be hopelessly out of date by the time it is read.

There are various responses to this development. Critics suggest that all this talk of IT-revolutions and new economies is preposterous, or at the very least seriously exaggerated. These skeptics often point to the fact that even if IT-related shares are soaring on trend-sensitive stock markets the world over, most of these companies are posting continual losses, and that this cannot continue in the long run. The only people who have become rich from computers and IT are the various consultants and the producers of the computers and the software that make the internet possible, while consumers have invested heavily for little or no gain. Any reflected exponential growth in the economy as a whole has not materialized.

6. New paradigms?

From the point of view of the skeptic, the world is essentially the same as it was. We still manufacture and sell hammers and nails, the banks continue to devote themselves to the lending and borrowing of money, a few office routines have changed, but the significance of all of this has been exaggerated. Most people now write their own business letters on a word processor instead of using a dictaphone or a secretary, but the question is whether the state of things has been dramatically improved by this. What is known as e-commerce is just business as usual, even if we are using flashy new machines.

According to this point of view, this is largely a case of following trends, that there is a certain cachet in being first with the latest innovations, no matter what concrete benefit these may actually bring. And it matters little what technology we use to communicate: It is still the content which is important. Old and tested truths will still be just that in the future.

The contrary point of view is ecstatic. Anyone who has seen the light on their screen claims that everything will automatically turn out for the best. The internet is the solution to all our problems: The economy will blossom for everyone forever, ethnic and cultural conflicts will fade away and be replaced by a global, digital brotherhood. All the information that becomes available will make our duties as citizens more meaningful than ever, and the whole of the democratic system will be revitalized as a result. In the digital networks we shall find the social cohesion that we often lack today, and harmony will spread throughout society. Entertainment will become, thanks to the inexhaustible possibilities of this new technology, more interactive and hence more entertaining than ever.

Both the skeptic and the enthusiast are mistaken. Neither radical skepticism nor blind faith is a fruitful strategy for orientation in the accelerated process of change in which we find ourselves. Both of these points of view indicate, in essence, an unwillingness to think critically, an inability to see. They are not analyses or prognoses, but prejudices. A new, revolutionary technology for communication and information will undoubtedly change the preconditions of everything: society, economy, culture. But it will not solve all our problems. It would be naïve to believe that it could. Development means that we can approach certain problems in a dramatic way, but to balance this we will have to confront a whole raft of new problems. We can live longer and more healthily, perceive ourselves to be freer, and realize more of our dreams. But the fundamental conflicts between classes and groups of people are not going to go away, just develop into more intricate and impenetrable patterns and structures.

Change of this type is not instantaneous. The skeptic who triumphantly points out that most of the global economy is still based upon the production of physical objects like fridges, airplanes and garden furniture rather than digital services on the net is partly a little impatient – we are still in many respects only in a preliminary phase – and partly incapable of grasping the extent of the change. There is no question of the fridge disappearing, but rather that the objects around us will take on new significance and new functions in an entirely new socio-ecological system. Marketing campaigns for fridges, for example, will no longer stress their capacity to keep milk cold, because we take that for granted, but rather their capacity to communicate intelligently in a network.

It is in the nature of things that it takes a certain amount of time for changes to be absorbed. Every revolutionary technology only reveals its true colors after an unavoidable period of incubation. As far as the printing press was concerned, it took more than three hundred years before it made its definitive breakthrough, the point at which it caused a dramatic shake up of social structures and created a new paradigm: capitalism. It took time, quite simply, before literacy was sufficiently widespread for print to affect large social groups beneficially. It was not until the Enlightenment of the 1700s that thinking became sufficiently modern, the exchange of information sufficiently lively, and technical advances sufficiently explosive for there to be signs of nascent industrialism in the offing.

Literacy spread rapidly through northern Europe during the 1600s, but its growth only accelerated more noticeably during the following century, primarily as a result of Protestantism and the dissemination of Bible translations into the various national languages. The preconditions were created for a completely new sort of critical public life, whose platform was primarily the first newspapers of recognizably modern form. New publications, such as the *Spectator* in England, were aimed at (and therefore also helped to shape) an educated and cosmopolitan middle class. The aim of the newspapers was to inform about and debate the latest ideas. In France the world of the salon arose, where the aristocracy and middle classes came into contact with one another and together examined the signs of the times. This form of gathering quickly became popular and spread throughout Europe.

But even if literacy and the development of information technologies lay the basis for the changes that occurred in society, they cannot explain them fully. A whole mass of factors have to coincide and co-operate if any epoch-changing process of change is to be set in train. The French sociologist Jacques Ellul, whose interest is primarily with the internal logic of technology and its radical effects upon our lives and environment, has pointed out a number of key phenomena (Ellul, 1964). The first and possibly most self-evident precondition is that the necessary apparatus must be in place already, which in turn presupposes a longer historical process. Every innovation has its roots in a previous era. Novelty consists of what can be termed a technical complex; in other words, a series of inventions of various sorts which together form a powerful combination which is stronger than their individual parts. Innumerable innovations saw the light of day between 1000 and 1750, many of them remarkable in themselves, but they played to different tunes, they did not communicate with one another. It was only after 1750 that innovations began to work together and thereby facilitate large-scale industrialization.

Another important precondition, according to Ellul, is population growth. An increase in population means increased demands which cannot be satisfied without growth. Necessity is the mother of invention. From another, even crasser, point of view, an increase in population means greater preconditions for research and technical and economic development partly in the form of an increase in the size of the market, and partly by providing a human basis for various experiments with different types of product. A third effect is that two specific and at least partially contradictory demands are placed upon the economic environment, which has to be both stable but also in some form of dissolution. On the one hand, a stable base is required for scientific experimentation which is necessary but unprofitable in the short term, but on the other there must be a capacity for widespread and fast change, a willingness to stimulate and absorb new thought processes. The fourth precondition concerns the social climate itself, and is, according to Ellul, probably the most important of them all. There has to be a loosening of various religious or ideological taboos, and liberation from any form of social determinism. For the development of industrialism, for instance, it was vitally important that a whole raft of traditional ideas about what was "natural" were thoroughly revised. No longer were either nature itself or hierarchical social orders perceived as sacred and inviolable.

Perceptions of man and his place in the world underwent radical change. The individual gained a new position, and human freedoms and rights were spoken about, which undermined preconceptions of natural groupings and classes. Suddenly unimagined opportunities opened up, offering social advancement and an improvement in living standards. The liberation of the individual and increases in technological efficiency co-operated. An historical resonance arose, where various factors dramatically strengthened one another in an accelerating spiral. The middle classes were rewarded for their willingness to adapt and made the most of this opportunity. Hence the middle class became the dominant class of the paradigm of capitalism.

The Industrial Revolution meant that that mankind's physical power was multiplied many times over through the use of machines. "The Digital Revolution" means that the human brain will be expanded to an incomprehensible degree through its integration with electronic networks. But we are not there yet, the necessary preconditions are not yet in place. Technology may be accelerating with breathtaking speed, but we humans are slow. Once again we are hampered by all kinds of religious and ideological taboos. Once again we are on the brink of a period of necessary creative destruction. This development cannot be controlled to any great extent. History shows that every new technology worth the name has, for better or worse, "done its own thing", completely independently of what its originators had imagined. In the words of the communications expert Neil Postman (1992, p. 7), technology "plays out its own hand".

Take the clock, for example, an apparently neutral and innocent artifact, but actually an infernal little machine that creates seconds and minutes, which has retrospectively given a whole new meaning to our perception of time. When the first prototypes were developed by Benedictine monks during the twelfth and thirteenth centuries their purpose was to establish a certain stability and regularity to the routines of the monastery, principally with regard to the prescribed seven hours of prayer each day. The mechanical clock brought precision to piety. But the clock was not satisfied with this. It soon spread beyond the walls of the monasteries. It may well have kept order over the monks' prayers, but above all else the clock became an instrument which synchronized and watched over the daily lives of ordinary people. It was thanks to the clock that it became possible to imagine something like regular production during a regulated working day. It became, in other words, one of the cornerstones of capitalism. This invention, dedicated to God, "did its own thing" and became one of Mammon's most faithful servants.

The same thing happened to the printing press. The devout Catholic Gutenberg could scarcely have imagined that his invention would be used to deliver a fatal blow to the authority of the Papacy and promote Protestant heresies by making the word of God accessible to everyone, which in turn made everyone his own interpreter of the Bible. When information became generally available, the natural but no less unforeseen consequence was that various accepted "truths" were put into question. From the 1700s, modern rationalism developed alongside the notion of the educated citizen, and it was the printed word that was to do the job. The goal was the extinction of every form of superstition, principal amongst them religion and the monarchy. According to the French Enlightenment thinker, Denis Diderot, "Men will never be free until the last king is strangled with the entrails of the last priest" (Burns, 1954, p. 478).

As long as information was an exclusive rarity, confined to the privileged few, it was unthinkable that ideas like that could be widely disseminated. Instead it became, after an incubation period of two hundred years, a mass movement. Technology played out its hand. And in the process, everything was changed. When the true agenda of the printing press began to appear, there was no longer any question of the old Europe plus a nice new invention, but of a completely new Europe which thought and acted in new ways. The progression had been uncovered, the historical process began to become clearer, and common sense and science would lift mankind out of the darkness of ignorance and progressively improve standards of living. A new world view, and a new view of man, had been born.

A new, dominant information technology changes everything, not least language. This is partly because of new terminology, new words for new toys, but the most interesting and, to an extent, most problematic aspect of this is that old words assume new meanings. As language changes, so does our thinking. New technology redefines basic concepts such as knowledge and truth; it reprograms society's perceptions of what is important and unimportant, what is possible and impossible, and, above all else, what is real. Reality assumes new expressions. This is what Neil Postman means when he talks of society going through an "ecological" change (Postman, 1992, 1995, p. 192). Technology shakes up the kaleidoscope of our intellectual environment and world of ideas and shows new, unforeseen patterns. We are entering a new social, cultural and economic paradigm.

The paradigm defines which thoughts can be thought, quite literally. The paradigm is simply the set of preconceptions and values which unite the members of a specific society. To take one example: when "everyone" at a certain point in time is convinced that the world is flat, it is pointless to try to work out a way of sailing round the world. When Copernicus claimed that the Earth actually moved around the sun many people thought him mad. This is no surprise. Ridiculing his critics with the benefit of hindsight merely proves that one does not understand how a paradigm works. It is not possible to say categorically that his critics were wrong, because what they meant by the term "Earth" was precisely a fixed point in space.

The terms still carried their former meanings, the paradigm shift had not yet taken place, people were still thinking along ingrained lines. The same thing occurred with the transition from Newton's physics to Einstein's. Many people dismissed Einstein's general theory of relativity for the simple reason that it presupposed that the concept "space" stood for something which could be "bent", when the old paradigm dictated that space was constant and homogenous. This was wholly necessary – if space had not possessed just these qualities, Newtonian physics could not have functioned. And since Newtonian physics had apparently functioned well for such a long time, they could not be abandoned easily. Hence a situation arose in which two paradigms competed with one another.

But two paradigms cannot exist for one person at the same time. It is either/or. The Earth cannot be both mobile and immobile at the same time; space cannot simultaneously be both flat and curved. For this reason individual transitions from one paradigm to another must be instantaneous and complete. It is like the Japanese soldier leaving the jungle and suddenly realizing that he has been living an illusion for years: peace, not war, is the status quo, and Japan has become the driving force of the Asian economic miracle. We are speaking here qualitatively rather than quantitatively. To move from an old paradigm to a new is not merely a question of becoming informed in the sense of adding new facts to old ones with which we are already familiar, but rather in the sense that new facts, and old facts in a new light, change our worldview entirely. And once we have perceived that our old worldview is exactly that, old, and is no longer capable of explaining difficult phenomena, which it is in turn no longer possible to ignore or deny, then it is necessary to abandon large amounts of irrelevant knowledge. This is one of the sacrifices demanded by a paradigm shift.

From a narrower perspective this is an acute situation for someone trying to orientate themselves in the world which is being formed around us within and by the electronic networks. The problem is no longer a lack of information, but an incalculable excess of it. What appears to be new information and new ideas might actually be yesterday's news, or in the worst cases abject nonsense, which will direct us into time- and resource-wasting cul-de-sacs. Old recipes for success become outdated fast. It is only human to become more attached to old strategies if they have proved successful in the past, and it is therefore all the more difficult to abandon them. Someone who has built up a successful business, or who has merely managed to make his life tolerably comfortable, seldom recognizes the necessity of dropping everything and starting again from scratch.

It is here that we find the true novelty in what is happening now. Previously the point of a paradigm was that it provided us with firm ground beneath our feet after a longer or shorter period of tremors. We need to get accustomed to losing that luxury and recognize that change itself is the only thing that is permanent. Everything is fluid. The social and economic stability that has been the ideal and the norm is becoming more and more the exception and a sign of stagnation. It is not enough to think, or to think in new ways; it is now necessary to rethink constantly, and to think away old thoughts. Creative destruction never rests.

Within the world of scientific theories, where the concept of paradigms was first established, there is talk of anomalies and crises. Anomalies are phenomena which are in part unforeseen, and in part difficult to adapt to fit the current paradigm. We can see them all around us these days: in society, within our cultural life and media, and in the economy. The preconditions which underlie politics are altering at a dizzying pace. Yesterday's ideological maps have nothing to do with the reality of today. Whole branches and great empires within the media are collapsing before our eyes. Working life is undergoing a dramatic revolutionary process which is effectively destroying all our old preconceptions of secure employment, automatic promotion and hierarchical organization. Youngsters still wet behind the ears and wearing strange clothes are becoming multi-millionaires in a few short months, in businesses which few of their shareholders have any real grasp of.

When a large number of anomalies appear there are two possibilities. The first is to try to squeeze the new phenomena into the old system of explanations. This is what people have always done within science: patched up and repaired old theories, like for example the old Ptolemaic system of astronomy with the Earth in the center and all the other heavenly bodies circling around it. It holds for a while, bearably, but with time it becomes gradually more apparent that the conditions produced by the old theory are no longer of any use. And then we are confronted unavoidably with option number two: to admit that the old system has had its day, even if there is no new system ready to take its place. This precipitates a crisis. The importance of this crisis is that it signals a need for new thinking. And this is where we are at the moment, in the middle of the crisis which has arisen from the old capitalist paradigm showing that it is incapable, but before any new system has won over enough adherents to be able to function as a generally accepted explanatory model. A lot of people are still patching up and repairing the old system, and there is a noticeable lack of new thinking. Sullen skepticism as to whether the new is actually anything genuinely new, and blind faith in the new which maintains that everything is now on its way to ordering itself automatically for the best, do not count as new thinking.

7. Conclusion

Writing about the future is obviously incredibly problematic because it does not yet exist. The best we can do is to produce more or less qualified guesswork. Someone who understands how dominant information technologies have played out their hands throughout history, and who understands how the dynamism within and between digital networks functions, has the best possible preconceptions for grasping the essential points of the current revolution. This chapter set itself out to explore the factor(s) that drive the organization of labor and how technology is used as a driving force, even in those instances where it may cause society to surrender its extant norms and routines. The discovery was that information exchange has been and remains a quintessential factor in driving this development. In answering the two postulated research questions, we claim two things. The first is that a new social, cultural and economic paradigm is taking shape. The main reason is the ongoing revolution within the management of information: digitalization, and the astonishingly fast development of electronic networks. One immediate consequence of this is that our mental ecology is drastically changing, which in turn forces a whole sequence of necessary adjustments. And secondly, we suggest that the form that the new paradigm is in the process of assuming will not be concrete, but fluid. It is not merely that we are developing new social norms; it is a matter of a completely new sort of norm.

The Japanese soldier in the jungle was ill-informed, and was fighting his own private world war within his own head, but then his circumstances were hardly optimal. We, on the other hand, cannot blame anything other than laziness or stupidity if we do not manage to garner a relatively clear picture of what is going on around us and if we cannot draw the relevant conclusions from this picture. Because one thing we can say without any doubt is that it will not be the meek who inherit the earth.

Note

1 This chapter is a reworked and expanded version of a text originally published as "Technology as the Driving Force of History", chapter 1 (pp. 13–34) in the book *The Netocrats – The Futurica Trilogy, Part 1* (Stockholm, Sweden: Stockholm Text), by Bard and Söderqvist (2002). Permission for reprint has been granted by the copyright holder.

References

Barber, E.J.W., 1991. *Prehistoric Textiles: The Development of Cloth in the Neolithic and Bronze Ages with Special Reference to the Aegean*. Princeton, NJ: Princeton University Press.

Bellis, M., 2018. *The History of Clothing*. [online] Thoughtco. Available at: <www.thoughtco.com/history-of-clothing-1991476> [Accessed 18 Sep. 2019].

Burns, E.M., 1954. *Western Civilizations: Their History and Culture*. 4th ed. New York, NY: W.W. Norton & Company.

Copernicus, N., 1543. *De Revolutionibus Orbium Coelestium [On the Revolutions of the Heavenly Spheres]*. Nuremberg, Germany: Johannes Petreius.

Diamond, J., 1997. *Guns, Germs, and Steel: The Fates of Human Societies*. New York, NY: W.W. Norton & Company.

Ellul, J., 1964. *The Technological Society*. New York, NY: Vintage Books.

Goodman, D. and Keene, R., 1997. *Man Versus Machine: Kasparov Versus Deep Blue*. Cambridge, MA: H3 Publications.

Kearns, D., 2002. No lip service on security this time, Bill. *Network World*, 19(4), p. 20.

Kranzberg, M. and Hannan, M.T., 2017. *History of the Organization of Work*. [online] Encyclopædia Britannica. Available at: <www.britannica.com/topic/history-of-work-organization-648000> [Accessed 18 Sep. 2019].

Levi-Strauss, C., 1961. *Tristes Tropiques: An Anthropological Study of Primitive Societies in Brazil*. New York, NY: Atheneum.

Lippmann, W., 1922. *Public Opinion*. New York, NY: Harcourt, Brace and Company.

Lippmann, W., 1925. *The Phantom Public*. New York, NY: Harcourt, Brace and Company.

MacCormack, S., 1989. Atahualpa and the book. *Dispositio*, 14(36/38), pp. 141–168.

Newborn, M., 2003. *Deep Blue: An Artificial Intelligence Milestone*. New York, NY: Springer.

Obar, J.A., 2015. Big Data and The Phantom Public: Walter Lippmann and the fallacy of data privacy self-management. *Big Data & Society*, 2(2), p. 205395171560887, pp. 1–16.

Onoda, H., 1974. *No Surrender: My Thirty-Year War*. New York, NY: Kodansha International.

Postman, N., 1992. *Technopoly: The Surrender of Culture to Technology*. New York, NY: Alfred A. Knopf.

Postman, N., 1995. *The End of Education: Redefining the Value of School*. New York, NY: Alfred A. Knopf.

Robertson, D.S., 1998. *The New Renaissance: Computers and the Next Level of Civilization*. New York, NY: Oxford University Press.

3 The substitution of labor

From technological feasibility to other factors influencing the potential of job automation[1]

Jochem van der Zande, Karoline Teigland, Shahryar Siri and Robin Teigland

1. Introduction

This chapter, which illustrates the potential of a number of technologies to replace labor, begins with a brief overview of digitalization and automation and the three primary technologies enabling job automation – artificial intelligence (AI); machine learning (ML) – a subcategory of AI; and robotics – in order to create a solid understanding of the concepts. We then proceed to discuss the distinct human capabilities that are required in the workplace and to what degree the three primary technologies can substitute these capabilities based on their current state of development. We then turn to a categorization of job tasks based on a commonly-used framework of routine vs. non-routine and cognitive vs. manual tasks and map the human capabilities in the workplace from the previous section onto this matrix. In the next section, we discuss the resulting automation potential of tasks, jobs and industries. We then turn to discuss a set of factors beyond technological feasibility that influence the pace and scope of job automation. The chapter concludes with a brief summary of the findings that support our prospects for the future of labor.

2. Brief overview of digitalization and automation

Before one can make a proper judgment on the substitution potential of specific tasks, or even complete jobs, it is essential to first develop a solid understanding of the processes and technologies that underlie this substitution. This section aims to create the first part of this understanding by exploring the definition and history of each of the involved technologies and processes.

First, it will touch upon the process of digitalization as it is technology-led and it arguably has had, and will continue to have, a significant influence on labor. We then turn to automation, which is the overarching concept describing the substitution of human labor by machines. Subsequently, artificial intelligence and its subfield of machine learning along with robotics will be discussed as these have been identified as the three most prevalent technological areas within automation.

2.1. Digitalization

2.1.1. Definition of digitalization

Of all the topics in this chapter, digitalization is arguably the broadest concept with the most dispersed definition. Concepts such as Internet of Things (IoT), big data, mobile applications, augmented reality, social media and many others all fall within the scope of digitalization.

In business, digitalization is generally used to describe the process of improving or changing business models and processes by leveraging digital technologies and digitized resources in order to create new sources of value creation.

At the core of this process lies the rise of data-driven, networked business models (Mäenpää and Korhonen, 2015), also known as digital businesses. Digitalization is also used to describe the wider global trend of adopting digital technologies and the effects of this adoption throughout all parts of society (I-Scoop, 2017).

The term digitalization is frequently used interchangeably with digitization and digital transformation. However, it is helpful to make a clear distinction between the three. In this study, digitization will refer solely to the process of transferring analogue data (like pictures, sounds, etc.) into a digital format, i.e., binary code (Khan, 2016; Oxford Dictionaries, 2018b; I-Scoop, 2017).

With digitalization, we will refer to the business process previously described. Lastly, digital transformation is both used to describe a company's journey to become a digital company as well as the larger effects of digitalization on society at large.

Digitalization is also occasionally confused with concepts like mechanization, automation, industrialization and robotization. However, these terms usually refer to improving existing processes, such as workflows, whereas digitalization refers to the development of new sources of value creation (Moore, 2015).

2.1.2. A brief history of digitalization

The history of digitalization began with the development of the modern binary system by Gottfried Wilhelm Leibniz in 1703. However, digitalization, as we refer to it today, started with the introduction of the first digital computers in the 1940s and accelerated with the widespread adoption of the personal computer in the second half of the century (Press, 2015; Vogelsang, 2010).

Digitalization surged with the establishment and development of the World Wide Web in the 1990s, which revolutionized the access to and diffusion of information around the world. Today, with the rapid development of digital technologies like Internet of Things, big data, and AI, this transformation is happening at an unprecedented pace. Though digitalization has caught the attention of both the public and private sector, most organizations are still insufficiently prepared for a digital future, according to IBM (Berman, Marshall and Leonelli, 2013).

2.2. Automation

2.2.1. *Definition of automation*

The term automation refers to the process of introducing technologies to automatically execute a task previously performed by a human or impossible to perform by a human (Grosz et al., 2016). The field is closely related to mechanization, which refers to the replacement of human labor by machines (Groover, 2018). This is different from systems operating *autonomously*, which relates to the achievement of a goal without predefined execution rules provided by humans. The term automation therefore suggests that the system follows a fixed set of rules to complete its goal (Sklar, 2015). Automated systems are typically made up of three building blocks (Groover, 2018):

1 Power sources. Power sources, such as electricity, are necessary to execute the required action. In general, power sources are used to execute two types of actions: processing, which relates to the mutation/transformation of an entity, and transfer and positioning, which relates to the movement of an entity.
2 Feedback control systems. Feedback control systems monitor whether the required action is performed correctly or not. An example is a thermostat, which monitors the temperature in a room to match a target temperature, and adjusts the heating element's output if this is not the case.
3 Machine programming. This comprises the programs and commands that determine the system's aspired output and the required execution steps. Typical methods for machine programming are using paper/steel cards, tapes, and computer software. Automation by computer-controlled-equipment is also known as computerization (Frey and Osborne, 2013).

One of the most prevalent use cases for automation is within manufacturing. Automation in this field is also known as industrial automation (PHC, 2016). There are three types of industrial automation (Groover, 2018):

1 Fixed automation. The equipment configuration is fixed and cannot be adapted to perform another process. Hence, the sequence of processing operations is permanent.
2 Programmable automation. The equipment can be reprogrammed to perform another process, but the reconfiguration takes time and requires human interference.
3 Flexible automation. The system is controlled by a central computer and can be reprogrammed automatically and instantly. Therefore, the system can perform different processes simultaneously.

Modern, complex automated systems comprise several technologies (Robinson, 2014). Consequently, developments in the field of automation are closely related to advances in these technological sub-fields. Examples are artificial intelligence, neural networks, and robotics (Chui, Manyika and Miremadi, 2016). These will be discussed later in the chapter.

2.2.2. A brief history of automation

The term automation was coined in 1946, but its history stretches back to the dawn of humanity. As mentioned previously, automated systems usually comprise three building blocks. The history of automation can be explained by the development of these three blocks (Groover, 2018):

The first large development in automation came with the invention of tools that utilized a power source other than human muscle. This development started in the early stages of humanity with the creation of tools that magnified human muscle power, like the cart wheel and the lever.

Subsequently, devices were invented that could operate in complete absence of human power by harnessing the energy of wind, water and steam.

In the nineteenth and twentieth centuries, stronger power sources, like electricity, were incorporated into the machines, which significantly increased their power.

The growing machine power gave rise to the need for control mechanisms to regulate the output. At first, human operators were needed to control the energy input to the machine. However, the invention of the first negative feedback system removed human involvement from the process. These systems monitor whether the output of the machine corresponds to the desired level and enable a machine to self-correct its input if the output is off. Developments in this field from the seventeenth century onwards gave rise to modern feedback control systems.

The third large development in the history of automation was the introduction of programmable machines. The first was developed by Joseph-Marie Jacquard in 1801, who used steel cards with different hole patterns to determine the output of his automatic loom. Nowadays, machines are programmed by using paper cards with whole patterns and computers.

The combination of these three developments ultimately led to the rise of automation. The introduction of electrical power enabled a surge in automation at the turn of the nineteenth century. During the second half of the twentieth century and the start of the twenty-first century, the capabilities of automated systems increased significantly following several technological advancements. Firstly, automated systems became much more sophisticated and faster after the introduction and incorporation of the digital computer. This increase in power accelerated following advances in computer science, programming language and storage technology. Meanwhile, the prices of these technologies decreased exponentially. Secondly, developments in mathematical control theory and sensor technologies amplified the capabilities and power of feedback control systems, increasing the systems' versatility and ability to operate autonomously in unstructured environments.

2.3. *Artificial intelligence*

2.3.1. *Definition of artificial intelligence*

Artificial intelligence (AI) is a technological field that arguably holds considerable potential for the future. It is such a broad field that it is hard to define precisely what it really is. A famous and useful definition made by Nils J. Nilsson (2010) reads, "*Artificial intelligence is that activity devoted to making machines intelligent, and intelligence is that quality that enables an entity to function appropriately and with foresight in its environment.*" In other words, AI is computers performing tasks that normally require human intelligence (Oxford Dictionaries, 2018a). However, "intelligence" is a complex phenomenon that has been studied in several different academic fields, including psychology, economics, biology, engineering, statistics and neuroscience. Over the years, advancements within each of these fields have benefitted AI significantly. For example, artificial neural networks were inspired by discoveries within biology and neuroscience (Grosz et al., 2016).

The field of AI research has grown significantly over the past few decades and it has been used for a variety of applications, from beating professionals in board games such as chess and Go to the navigation of self-driving cars (Marr, 2016a). Terms such as big data, machine learning, robotics and deep learning all fall within the scope of AI, alluding to the breadth of the technology.

There are several ways to divide and categorize the different methods, subsets, and applications within AI. One way is to distinguish between general and applied AI. Applied AI, also known as weak or narrow AI, is more common and refers to algorithms solving specific problems and programs completing specified tasks (Aeppel, 2017). For example, a computer may excel in one specific board game that is bounded by specific rules, but outside this task it is useless (MathWorks, 2018c). General AI, or strong AI, aims to build machines that can think and perform almost any task without being specifically programmed for it (Copeland, 2018). This means that the machine has a mind of its own and can make decisions, whereas under weak AI, the machine can only simulate human behavior and appear to be intelligent (Difference Wiki, 2017).

Another way of dividing AI is into research areas that are currently "hot". This is an appropriate division as AI arguably suffers from the "AI effect", or "odd paradox", which means that once people get accustomed to an AI technology, it is no longer perceived as AI. Today, "hot" research areas include large-scale machine learning, deep learning, reinforcement learning, neural networks, robotics, computer vision, natural language processing (NLP), collaborative systems, crowdsourcing and human computation, algorithmic game theory and computational social choice, Internet of things (IoT) and neuromorphic computing (Grosz et al., 2016).

Robotics, deep learning and machine learning are all discussed further on in this chapter; however, NLP is also a subset that has made substantial progress in the last few years. NLP applications attempt to understand natural human communication, written or spoken, and to reply with natural language (Marr, 2016b). The research in this field is shifting from reactiveness and stylized requests toward developing systems that can interact with people through dialogue (Grosz et al., 2016). The other subfields will not be discussed individually.

2.3.2. A brief history of artificial intelligence

The term artificial intelligence was first used by John McCarthy in 1956 at the Dartmouth Conference, the first conference in history on artificial intelligence (Childs, 2011). The goal of the conference was to discover ways in which machines could be made to simulate aspects of intelligence. Although this was the first conference on AI, the technical ideas that characterize AI existed long before. During the eighteenth century, the study on *probability* of events was born; in the nineteenth century, *logical reasoning* could be performed systematically, which is much the same as solving a system of equations; and by the twentieth century, the field of *statistics* had emerged, enabling inferences to be drawn rigorously from data (Grosz et al., 2016).

Despite its long history, AI has only recently begun to pick up speed in research advancements. Between the 1950s and 1970s, many focal areas within AI emerged, including natural language processing, machine learning, computer vision, mobile robotics and expert systems.

However, by the 1980s, no significant practical success had been achieved and the "AI winter" had arrived; interest in AI dropped and funding dried up.

A decade later, collection and storage of large amounts of data were enabled by the internet and advances in storing devices. Moreover, cheap and more reliable hardware had stimulated the adoption of industrial robotics and advances in software allowed for systems to operate on real-world data. As a confluence of these events, AI was reborn and became a "hot" research field once again (Grosz et al., 2016).

2.4. Machine learning

2.4.1. Definition of machine learning

A plethora of papers discuss machine learning (ML), but none truly succeed in explaining what it is or what subdivisions there are. As a result, the term machine learning is often misused and confused with artificial intelligence.

According to the Oxford Dictionary, ML is a subset of artificial intelligence and is defined as "*the capacity of a computer to learn from experience, i.e., to modify its processing on the basis of newly acquired information*" (Copeland, 2018). This definition describes what machine learning is, but it does not explicitly explain what the field encompasses. The following paragraphs attempt to explain what machine learning comprises.

Machine learning has grown into a fundamental research topic with several different approaches and algorithms to be used depending on the problem. One way of dividing the field is into supervised and unsupervised learning. In supervised learning, the answer is known (found in past or completed data), whereas in unsupervised learning it is not (Libesa, 2016). Supervised learning uses a known dataset (a training dataset that is a set of labeled objects) to make predictions for datasets in the future. Unsupervised learning, on the other hand, draws inferences from datasets where input data have no labelled response (MathWorks, 2018b).

Unsupervised learning allows computers to reason and plan ahead in the future, even for situations they have not yet encountered or for which they have been trained (Bengio, 2017).

For example, both types of ML can be used for image recognition, a common machine-learning problem in which the system has to classify objects based on their shape and color. If supervised learning is used, the computer has already been taught how to identify and cluster the objects. It will know that an octagon has eight sides and will hence cluster all eight-sided objects as octagons. Under unsupervised learning, the system does not follow a predefined set of clusters or object characteristics. The system must create these clusters itself by identifying logical patterns between the objects; it will notice that several objects have eight sides and cluster them if the characteristics are deemed prevalent (MathWorks, 2018a).

Supervised learning itself has two distinct categories of algorithms: (1) classification – used to separate data into different classes, and (2) regression – used for continuous response values (MathWorks, 2018d).

Unsupervised learning can also be divided into two different categories: (1) cluster analysis – used to find hidden patterns or groupings in data based on similarities or distances between them (MathWorks, 2018b), and (2) dimensionality reduction – where smaller subsets of original data are produced by removing duplicates or unnecessary variables (Ghahramani, 2004).

Supervised learning is the less complicated of the two since the output is known, and it is therefore more universally used. Nonetheless, unsupervised learning is currently one of the key focus areas for AI (Bengio, 2017).

One of the machine-learning techniques that has been widely covered the last few years is deep learning (Deng and Yu, 2014). Deep learning is used within both supervised and unsupervised learning and teaches computers to learn by example, something that comes naturally to humans. Deep learning uses deep neural networks, a network consisting of several layers of neurons loosely shaped after the brain, to recognize very complex patterns by first detecting and combining smaller, simpler patterns.

The technology can be used to recognize patterns in sound, images and other data. Deep learning, is, among others, used to predict the outcome of legal proceedings, for precision medicine (medicine genetically tailored to an individual's genome), and to transcribe words into English text with as little as a seven percent error rate (Marr, 2016b).

2.4.2. A brief history of machine learning

Arthur Samuel coined the term machine learning in 1959, three years after AI (Puget, 2016), but, just as for AI, the technical ideas around ML were developed long before. The two major events that enabled the breakthrough of machine learning were the realization that computers could possibly teach themselves, made by Arthur Samuel in 1959, and the rise of the internet, which increased the amount of digital information being generated, stored and made available for analysis.

The focus point within machine learning has changed over time. During the 1980s, the predominant theory was knowledge engineering with basic decision logic. Between the 1990s and 2000s, research focused on probability theory and classification, while in the early to mid-2010s, focus switched to neuroscience and probability. More precise image and voice-recognition technologies had been developed which made it easier. Memory neural networks, large-scale integration and reasoning over knowledge are currently the predominant research areas. The recent discoveries within these fields are what has brought services such as Amazon Echo and Google Home into scores of households, particularly within the US market (Marr, 2016a).

2.5. Robotics

2.5.1. Definition of robotics

The field of robotics comprises the science and technology of robots and aims to develop, operate and maintain robots by researching the connection between sensing and acting (Siciliano and Khatib, 2016; Grosz et al., 2016).

Robotics is a mix between several academic fields, including computer science, mechanical engineering and electrical engineering, and is one of the primary technologies used for automation (Groover, 2018). The field is strongly related to AI (Encyclopaedia Britannica, 2018) and particularly to the fields of machine learning, computer vision and natural language processing (Grosz et al., 2016).

Developing an overall definition for robots is difficult as robots differ widely in terms of purpose, level of intelligence and form (Wilson, 2015). The Oxford Dictionary defines a robot as "*a machine capable of carrying out a complex series of actions automatically, especially one programmable by a computer*" (Oxford Dictionaries, 2018c). The International Federation of Robotics (IFR) makes a distinction between two types of robots: industrial robots and service robots.

The IFR has aligned its definition for industrial robots with the definition of the International Organization for Standardization (ISO) and refers to them as "*automatically controlled, reprogrammable, multipurpose manipulators programmable in three or more axes, which may be either fixed in place or mobile for use in industrial automation applications*" (International Federation of Robotics, 2017, p. 2).

An example of an industrial robot is a robot arm used in a car manufacturer's production process. Service robots are defined as robots "*that perform useful tasks for humans or equipment excluding industrial automation applications*". The IFR further distinguishes between personal service robots and professional service robots. The first are service robots that are not used for commercial purposes, for instance a domestic vacuum-cleaning robot, while the latter include all service robots that are used for commercial purposes, such as delivery robots in hospitals and offices (International Federation of Robotics, 2017, p. 2).

Combining the previous definitions, Wilson (2015) defines robots as "*artificially created systems designed, built, and implemented to perform tasks or services for people*". Moreover, he expands the definition of robots to include cognitive computing, which refers to automated computer programs. In other words, physicality is not a requirement and many robots solely consist of software (Horton, 2015). Examples of this are Twitterbots and IPSoft's virtual assistant, Amelia.

For the purpose of this study, the term robot will refer to all artificially created systems that perform tasks and services for people, whether they have a physical state or not. We will also adhere to the split between industrial robots and service robots. In addition, while some authors distinguish between robots and automated vehicles, for the purpose of this study they will both fall under the umbrella of robotics.

2.5.2. A brief history of robotics

From Greek mythology to da Vinci's machine designs, humans have always fantasized about creating skilled and intelligent machines, but the word robot was only introduced in 1920 by Karel Čapek, a Czech playwright (Siciliano and Khatib, 2016). The first electronic autonomous robots were created in the 1950s and the first industrial robot was developed in 1959. Nevertheless, it took two more years until the first industrial robot was acquired and installed in a manufacturing process (International Federation of Robotics, 2017). From that moment, robotics became widespread in industrial, warehousing and military applications (Boston Consulting Group, 2014; Siciliano and Khatib, 2016).

The first generations of robots consisted of large, immobile machines with a narrow skillset and limited power to adapt to their surroundings (Latxague, 2013).

Over the past decade, the field of robotics has made a gigantic leap as advances in programming, sensors, AI and robotic systems have significantly increased the intelligence, senses and dexterity of robots (Decker, Fischer and Ott, 2017; Sander and Wolfgang, 2014; Manyika et al., 2013). This has resulted in robots that are more versatile (Decker, Fischer and Ott, 2017), smaller and better connected to each other. Consequently, it is much safer for robots and humans to work closely together and the range of applications for robots has increased significantly. For example, the technological advances have enabled robots to enter the realm of services, which was previously deemed impossible (Manyika et al., 2013). In the future, technological advances are expected to further increase the capabilities of robots and prices are expected to drop. As a result, the field of robotics is expected to surge (Sander and Wolfgang, 2014).

3. The current state of the three technologies

The second step in assessing the technical feasibility of technologies posed to take over work activities is to analyze the technologies' current capabilities. In other words, what are the technologies currently able to do? To do this, we follow a framework from Manyika et al. (2017) that identifies five broader areas of capabilities: sensory perception, cognitive capabilities, natural language processing, social and emotional capabilities and physical capabilities, which enable humans to perform 18 activities in the workplace. These categories were developed based on an analysis of 2000 distinct work activities across 800 occupations. The framework is displayed in Figure 3.1.

This section discusses the current state of the technologies for each of these five broader areas of capabilities. The three technologies will be discussed simultaneously because they are closely related and are often used in combination to perform a single activity. It is important to note that many of these capabilities are still only proven in laboratories and are not yet available on the market.

3.1. Sensory perception

The area of sensory perception, or machine perception, covers the sensing and processing of external information from sensors and includes the three subfields of visual, tactile and auditory (Anderson et al., 2017). Sensory perception covers the capabilities of the sensors as well as the underlying software that processes and integrates the information. Sensory perception is essential for a variety of applications, including feedback control systems of automated systems and physical capabilities of robots (Grosz et al., 2016). Over the years, sensors and the underlying machine-learning algorithms have become increasingly sophisticated (Hardesty, 2017), and in some fields machines have even reached a capability level that is at par with the human level, according to McKinsey (Anderson et al., 2017).

Computer vision has developed significantly over the past decade, enabled by advances in sensors, deep learning and the abundance of data due to the internet. In some narrow-classification tasks, computer vision systems can outperform their human counterparts. Meanwhile, developments in sensors and algorithms for 3D object recognition, for example LIDAR (Laser-Imaging Detection and Ranging), allow for more precise distance measuring than ever before. Nonetheless, complex tasks, such as dealing with cluttered vision and fields, still present a challenge for the current technology (Manyika et al., 2013; Frey and Osborne, 2013; Robinson, 2014).

Computer vision is essential for machines to perceive and adapt to their environments and is one of the major enablers of autonomous vehicles. Advances in vision technology also enable progress in other applications, e.g., industrial and software robots.

For example, it enables robots to manage patients at the front desk of a pharmacy and to assemble customized orders in pharmaceutical settings (Qureshi and Sajjad, 2017; Manyika et al., 2013).

"Machine touch" refers to the processing of tactile/haptic information and is indispensable for a robot's ability to grasp and manipulate objects (Izatt et al., 2017; Hardesty, 2017). Though progress is being made to develop sophisticated

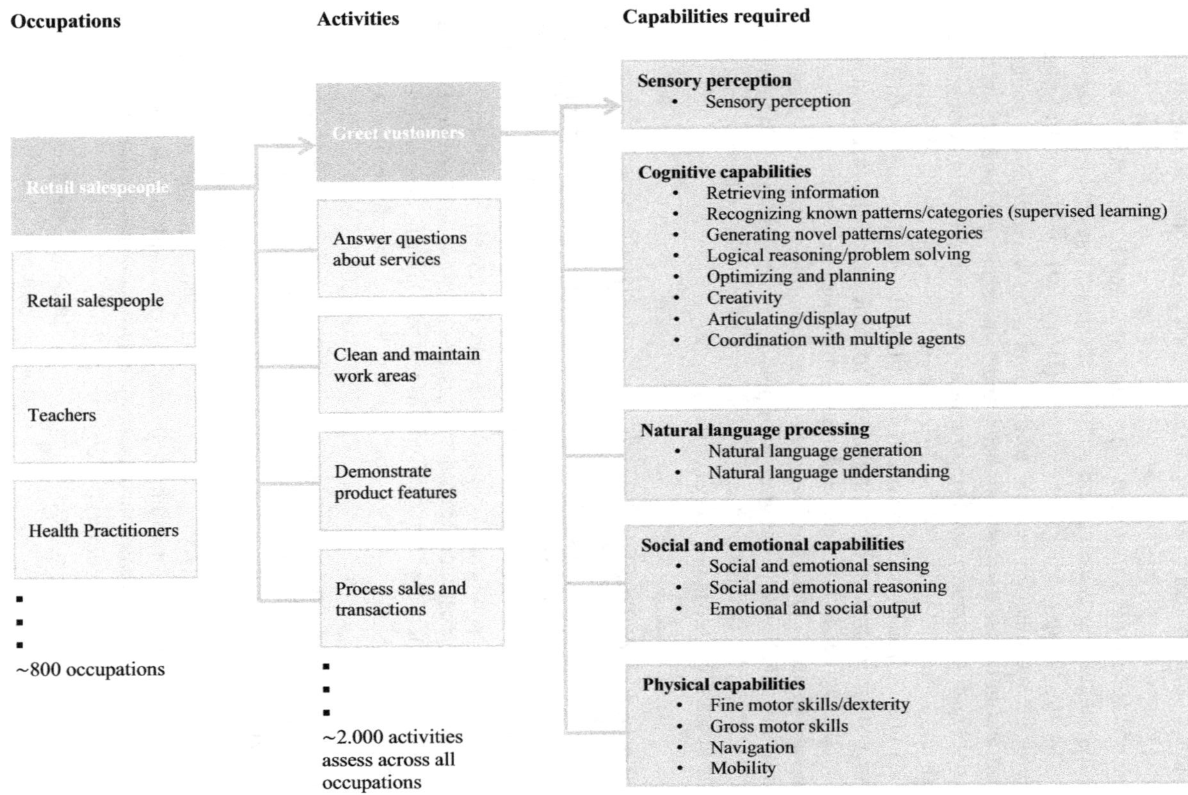

Figure 3.1 Capabilities required in the workplace (Manyika et al., 2017).

haptic sensors that mimic human capabilities, robots still struggle to obtain accurate local information. For example, it is hard to estimate how much force to apply when grabbing an object or to accurately estimate an object's position once it is in the robot's gripper and out of its camera's sight. One recent development is robot skin, a development by Georgia Tech, which gives robots the ability to feel textures (Manyika et al., 2017).

"Machine Hearing" refers to the processing of sound by computers. It is vital for natural language processing and auditory scene analyses, which is the ability to separate and group acoustic data streams (Hahn, 2017). The goal of machine hearing is for machines to be able to distinguish between different sounds, to organize and understand what they hear, and to react in real time (Lyon, 2017, pp. 131–139). For example, a serving robot in a restaurant should be able to distinguish and group the voices of the different customers at a table and accurately take their orders. Today, machine hearing is still in its infancy stage compared to machine vision. For machine-hearing models to be designed, analyzed and understood, math, engineering, physics and signal-processing are essential.

Although some subfields of sensory perception have advanced rapidly, it remains a large challenge to integrate multiple sensor streams into a single system (Hahn, 2017), and it will take several years for the technology to completely surpass the human level (Manyika et al., 2017).

3.2. Cognitive capabilities

This area covers a wide range of capabilities, including making tacit judgments, retrieving information, logical thinking, optimizing and planning, creativity, coordination with multiple agents and recognizing and generating known and novel patterns/categories. Significant developments have been made within the area, but it is also where the most technical challenges lie (Hodson, 2016; Manyika et al., 2017). As of today, there are cognitive systems that beat humans in several activities.

For example, IBM's Watson computer has a 90% success rate in diagnosing lung cancer compared to a human's 50% (Steadman, 2013). Watson also beat the reigning chess champion in 1997 and the champions in gameshow Jeopardy! in 2011 (Knight, 2016). Each individual capability will be discussed briefly.

Optimizing and planning for objective outcomes across various constraints can currently be done by a computer with the same precision as the most skilled humans in this field (Manyika et al., 2017). It includes optimizing operations and logistics in real time, for example, optimizing power plants based on energy prices, weather and other real-time data, or automating machinery to reduce errors and improve efficiency (Henke et al., 2016).

Retrieving information includes being able to search and retrieve information from a wide variety of sources. Based on this information, a computer should also be capable of writing research reports. As of today, technologies are far more skilled at retrieving information than humans (Manyika et al., 2017) because computers are much faster than humans and can go through millions of sources in the blink of an eye. For example, IBM's Watson searched through 20 million cancer research papers and diagnosed a patient with a rare form of leukemia in only ten minutes, while the doctors had missed this for months at the University of Tokyo (Ng, 2016).

Recognizing known patterns/categories is identical to the concept of supervised learning. As explained earlier, supervised learning uses known patterns to categorize and predict for datasets in the future (MathWorks, 2018d). The use and power of supervised learning has increased considerably with the growing availability of large data sets following the internet and advances in sensors. The capability of recognizing patterns is one where computers already outperform humans. For example, a deep-learning based lip-reading system, created by Google's DeepMind and the University of Oxford, trained by watching over 5000 hours of BBC programs, easily outperformed a professional human lip-reader (Frey and Osborne, 2013; Manyika et al., 2017).

Technology has not come as far in generating novel patterns/categories as it has with recognizing them; the field of unsupervised learning, which deals with this problem, is still in an early stage and the capability level of computers is below median human performance (Manyika et al., 2017). One of the difficulties is that the creation of something new requires creative intelligence, which is highly difficult to codify, as will be discussed next. For example, mathematicians perform tasks involving "*developing new principles and new relationships between existing mathematical principles to advance mathematical science*" (Frey and Osborne, 2013, p. 267). This task requires a lot of creativity and is therefore very hard to automate.

Creativity is currently one of the most difficult capabilities to automate. To be creative one must be able to make new combinations of familiar concepts, which requires a rich body of knowledge. The challenge for computers is to make combinations that "make sense" as they lack common knowledge. For this to happen, we must be able to specify our creative values precisely so that they can be codified. Another obstacle is the fact that these creative values vary between cultures and change over time. Despite the challenges, AI has already been used for some creative tasks, like creating music and staging performances (Grosz et al., 2016; Frey and Osborne, 2013).

Logical reasoning and problem-solving can be done on different levels of complexity; from limited knowledge domains with simple combinations of output to many contextual domains with multifaceted, potentially conflicting, inputs. An example of such a task is the ability to recognize the individual parts of an argument and their relationships as well as drawing well-supported conclusions (LSAC, 2018). This capability is also one of the toughest for machines to perform, and performance is still at a low level compared to humans. However, the technologies are improving. Some activities requiring judgment might even be better off being computerized because AI algorithms make unbiased decisions while humans often may not. For example, it has been shown that experienced judges are considerably more generous in their rulings after a lunch break (Manyika et al., 2017; Frey and Osborne, 2013). An algorithm would deliver the same output regardless of the time of day.

Coordination with multiple agents reflects a machine's ability to work together with other machines as well as with humans. This capability, especially human-machine collaboration, is still underdeveloped (Manyika et al., 2017). Early stages of robot collaboration have been proven, but these are largely based on laboratory research (Perry, 2014; Kolling et al., 2016). For example, researchers at Carnegie Mellon University made two different types of robots collaborate by letting a mobile robot bring work to a static robot arm that was controlled by the latter robot (Sklar, 2015).

As pointed out earlier, the general focus has shifted from substitution toward human-machine collaboration. However, the ability of machines to collaborate with humans is currently at a low level (Manyika et al., 2017), limited, for example, by the inability of AI systems to explain their decisions and actions to humans (Turek, 2017) and to understand and produce natural language.

One early example is the humanoid robot Asimo, which has a limited ability to respond to voice commands and human gestures (Boston Dynamics, 2018).

3.3. Natural language processing

Natural language processing comprises both the understanding and generation of natural language. Research within this field has shifted from reacting to clearly specified requests with a limited range of answers to developing refined and sophisticated systems that are able to have actual conversations with people. The generation of natural language is described as "*the ability to deliver spoken messages, with nuanced gestures and human interaction*" (Manyika et al., 2017). Natural language understanding is described as "*the comprehension of language and nuanced linguistic communication in all its rich complexity*" (Manyika et al., 2017). While computers' current level of generation of natural language is comparable to humans, the understanding of natural language remains at a lower level. The development within this area is one of the key factors influencing the pace and extent of automation (Manyika et al., 2017; Henke et al., 2016).

Natural language processing requires lexical, grammatical, semantic and pragmatic knowledge. Despite the fact that computers currently possess some of this knowledge, they are still less capable than humans.

Computers face difficulties in understanding multi-sentence language as well as fragments of language, while incomplete and erroneous language tends to be the norm in society (Bates, Bobrow and Weischedel, 1993). In addition, teaching computer systems and robots to detect sarcasm (Maynard, 2016), both in written and verbal conversations as well as the difference between polite and offensive speech (Steadman, 2013), currently proves to be very difficult.

In order to generate natural language, a machine must know *what to say* and *how to say it*. In order to know *what to say*, the machine must have data and should be able to determine what information from this data to include. The latter process, *how to say it*, requires a machine to know the language rules so that it can make a text (verbal or written) that makes sense. Currently, it is still very difficult for the software to produce grammatically correct and well-formed texts that have natural flows and that fit into an individual's context and needs (Coupel, 2014).

There have been some recent developments within the field, and companies such as Google, Amazon, and Apple use NLP in their products. Every time you ask Alexa, Siri or Google Home what the weather is like at your location or where to find a Japanese restaurant, NLP allows the program to understand your speech and answer in verbal language (Hunckler, 2017).

3.4. Social and emotional capabilities

This area deals with human social intelligence, which includes a machine's capability to sense and reason about social and emotional states as well as the ability to generate emotionally suitable output. These are essential capabilities for daily (human) interaction and for tasks like negotiation, persuasion, and caring. Among the five broader capability areas, social and emotional capabilities is currently the least advanced and will probably not surpass human level for at least two more decades (Manyika et al., 2017; Frey and Osborne, 2013).

Advances in machine learning and sensing have given machines a limited ability to recognize human emotions.

However, the current capabilities of these software programs are still far below human levels and face significant challenges with regards to instantaneous and accurate recognition of emotions. It is even more difficult for machines to comprehend and reason about the social and emotional states of humans.

Existing techniques analyze facial expressions, physiological factors (e.g., heart rate or blood pressure), text and spoken dialogues to detect human emotions. These techniques hold great future potential for several applications like automated call centers (Picard, 2007) and targeted advertisements based on emotional states (Doerrfeld, 2015).

Several emotion recognition software programs are already in use. Affectiva, for example, applies facial expressions analysis to adapt mobile applications to adjust to the emotional state of the user (Turcot, 2015).

To date, even the most advanced algorithms are not capable of communicating in a way that is indistinguishable from humans, and no machine has ever passed the Turing test.[2] The generation of emotionally suitable output is complicated by the existence of "common sense", which is tacit or implicit knowledge possessed by humans and ingrained in human interaction and emotions.

This knowledge is hard to define and articulate and therefore almost impossible to incorporate in algorithms (Hager et al., 2015; Frey and Osborne, 2013; Manyika et al., 2017). Communicating, in the absence of common sense, results in awkwardness or feelings of unnaturalness. There are some robots on the market that have a limited ability to mimic human emotions, like the humanoid Pepper, which can express joy, surprise, anger, doubt and sadness, but the actual creation of emotions is far away (Murphy, 2015).

3.5. Physical capabilities

This area includes fine and gross motor skills, navigation and mobility. It is closely related to the area of sensory perception, which provides the information input for physical activities (Manyika et al., 2013). Machines have already surpassed humans in terms of gross motor skills and the use of robots is widespread in industrial and warehousing settings, for example for picking and placing, welding, packaging and palletizing. Amazon has even completely automated some of its warehouses using robots.

However, on the frontier of fine motor skills and dexterity, technology is lagging behind significantly (Ritter and Haschke, 2015; Manyika et al., 2017). Manual skills are deeply integrated into the human cognitive system. Therefore, grasping and manipulation of smaller and deformable objects are still large sensorimotor challenges for the current technology. Robot dexterity is constrained by the strength of miniaturized actuators as well as visual and tactile sensors, which currently perform far below human levels (Hardesty, 2017; Ritter and Haschke, 2015; Frey and Osborne, 2013). Moreover, robots do not yet have the same degrees-of-freedom as human hands and current control systems are not yet capable of dealing with the multifaceted and unstructured nature of manual tasks. Nevertheless, there are several anthropomorphic robot hands with human-like capabilities on the market. The most advanced of these is the Shadow Dexterous Hand (Ritter and Haschke, 2015), which can perform delicate tasks such as opening a bottle cap and grabbing strawberries without crushing them.

Empowered by advances in machine vision and machine learning, navigation has already surpassed human capabilities. Advanced GPS systems, supported by vast amounts of spatial data, enable the pinpointing of exact locations and navigation toward almost every destination imaginable.

These capabilities are already widely used for example in (partly) autonomous cars and navigation apps, like Google Maps.

Despite advances in computer vision, robot mobility is still at a low level, especially autonomous mobility. Autonomous movement through static environments, e.g., specially designed warehouses, has largely been solved (Grosz et al., 2016; Manyika et al., 2017), but adapting motion to new and dynamic environments remains a substantial challenge (Heess et al., 2017).

Some of the reasons for this are technical challenges, including balance and control (Electronics Teacher, 2017), as well as insufficiently developed algorithms (Heess et al., 2017). Moreover, a lack of research on robot mobility in indoor settings has hampered progress in the area of indoor mobile robots (Grosz et al., 2016).

However, progress is being made on algorithms, as is shown by the DeepMind computer which recently taught itself to move through new, complex environments in a computer simulation (Heess et al., 2017). Real-life examples of advanced mobile robots are Boston Dynamics' Atlas, a humanoid robot which can move to various unknown terrains on two legs (Boston Dynamics, 2018), and Asimo, a humanoid robot capable of running, walking, kicking a ball and reacting to human instructions (Honda, 2018).

3.6. The overall state of current technologies

Though substantial progress is being made in all five capability areas, several capabilities currently remain out of reach for the available technologies. Most notably, technology is underdeveloped for processing and generating natural language and social/emotional output, autonomous mobility, fine motor skills and a range of cognitive capabilities. On the other hand, technology is excelling in fields such as recognizing known patterns, gross motor skills and navigation, and is largely at par with humans in the field of sensory perception. Moreover, further advances are expected in all areas, and machines will likely be at or above human levels for most capabilities within one to two decades (Chui, Manyika and Miremadi, 2015).

However, current technological progress is mainly focused on narrow, individual capabilities.

The integration of several capabilities into well-functioning holistic solutions is another significant challenge that needs to be overcome and will probably take much longer than for the individual capabilities (Frey and Osborne, 2013).

On the other hand, environmental control can mitigate the current limitations of machines. This concept refers to the alteration of the environment or the task to make it simpler and more structured, for example by breaking it down into smaller tasks or by transforming an unstructured environment into a structured one. Environmental control can obviate the need for advanced flexibility, mobility, manual dexterity and cognitive capabilities. For example, Amazon placed bar-code stickers on the floor of its warehouses to assist the robots in their warehouse navigation. They adapted the environment so it would become structured.

However, though environmental control is applied in warehouses and other local environments, countries and cities are still lagging behind in adapting their infrastructures to accommodate the new technologies (Frey and Osborne, 2013; Grosz et al., 2016).

4. The substitution of job tasks

Having discussed the current technological capabilities in the previous section, the ensuing section aims to relate these capabilities to their potential of substituting labor, focusing on the individual tasks that constitute jobs, rather than jobs in their entirety. The reason for this is that jobs include several different types of tasks, which all have a different relation to the current capabilities of technologies. Consequently, some types of tasks can already by automated while others cannot. Hence, it is essential to first understand which individual tasks can be substituted before one analyzes the effect on jobs and labor in general.

The different types of tasks are introduced in the next section, following the task model by Autor, Levy and Murnane (2003), and the substitution potential of each task category will be discussed in relation to the previously mentioned capabilities. In the next section, *The Impact on Labor*, we utilize our findings to make a judgment on the overall effect of automation on a selection of jobs and industries.

4.1. Four types of job tasks

To determine the job substitution potential of computers, Autor, Levy and Murnane (2003) conceptualized work as a series of tasks rather than complete jobs. Specifically, the paper distinguishes routine tasks from non-routine tasks and manual from cognitive tasks. This classification results in a 2 × 2 matrix, which is displayed in Figure 3.2. Routine tasks are defined as tasks that follow explicit rules, which can be exhaustively specified and, hence, translated into code. For non-routine tasks, these rules are not understood sufficiently well, which makes them much harder to codify. As a corollary of this definition, routine tasks are automatically classified as tasks that are easily substituted by technology while non-routine tasks are not.

Manual tasks are physical activities that require motor skills and mobility whereas cognitive task relate to mental processes.

In addition to the matrix in Figure 3.2, there are several other task classifications. For example, Manyika et al. (2017) have developed seven broader activity categories:

1 Predictable physical
2 Processing data
3 Collecting data
4 Unpredictable physical
5 Interfacing with stakeholders
6 Expertise
7 Managing and developing others

These seven categories fit largely within the 2 × 2 matrix of Autor, Levy and Murnane (2003). Predictable and unpredictable physical activities are aligned with the routine manual and non-routine manual task classification of Autor, Levy and Murnane (2003). Data collecting and processing largely fall under

	Cognitive	*Manual*
Routine	Explicit rules Mental processes	Explicit rules Motor skills
Non-routine	Rules difficult to codify Mental processes	Rules difficult to codify Motor skills

Figure 3.2 Four categories of job tasks (Autor, Levy and Murnane, 2003).

routine cognitive tasks, whereas interfacing with stakeholders, applying expertise and managing and developing others can be placed under non-routine cognitive tasks.

Each of the four categories is discussed in more detail in the next section.

4.1.1. Routine manual tasks

The routine manual task category includes physical activities that require systematic repetition of a consistent procedure, i.e., structured physical tasks that take place in predictable environments. The primary capabilities required to perform these types of activities are gross and fine motor skills, sensory perception and, to some extent, mobility.

Examples of activities include assembling, picking and sorting, welding and cooking. These tasks are easily translatable into computer programs and the technology to perform them is at an advanced level, especially for gross motor skills, where machines have been outperforming humans for a long time.

Consequently, this task category has the highest technological potential for substitution by machines (Manyika et al., 2017; Frey and Osborne, 2013; Autor, Levy and Murnane, 2003). Manyika et al. (2017) even predict that in the United States as much as 81% of the tasks in this category can be substituted.

The substitution of routine manual tasks has a long history and goes back to the introduction of the first machines that were capable of functioning automatically. Since then, machines have continuously pushed out humans, and a vast number of manual activities have been automated in the twentieth century (Finnigan, 2016). For example, many processes in the agriculture and car manufacturing industries are currently performed by machines. As a corollary, Autor, Levy and Murnane (2003) found that the percentage of people active in jobs with large proportions of routine manual activities declined between 1960 and 1998.

More recently, advances in sensory perception and manual dexterity have made it possible for robots to be assigned to tasks that require higher precision, e.g., slicing meat, assembling customized orders, manufacturing electronic components (Sander and Wolfgang, 2014; Sirkin, Zinser and Rose, 2015). Robots have also become safer and much more flexible to use, which allows them to quickly switch between different tasks and to safely work next to humans. Furthermore, the advances in mobility and navigation allow robots to move autonomously in static environments like warehouses.

In addition, robots are increasing their presence in the service industry. Simple service tasks, like cleaning, have been performed by robots for over a decade, the most notable example being the robot vacuum cleaner. However, with their increased dexterity and mobility, robots are increasingly able to take on complex routine manual tasks in the service industry. A prime example is the food sector where robots can be deployed to prepare and serve food and beverages (Frey and Osborne, 2013; Manyika et al., 2017).

For instance, the pizza delivery company Zume Pizza has automated its production process almost completely using sophisticated robots (TechCrunch, 2016).

Nonetheless, robot deployment is still in an early stage in this industry and the substitution potential remains limited.

Many routine manual tasks can and most likely will be performed by robots in the future and the share of repetitive, rule-based activities in jobs will decrease. With advances in sensors and increasing robot dexterity, more high-precision tasks will become candidates for substitution, such as manufacturing tasks in the electronics sector. As robots become safer, they will likely take up more positions next to their human coworkers. Further engineering advances are necessary to increase the flexibility of robotic systems by decreasing the reconfiguration time (Robotics Technology Consortium, 2013).

4.1.2. Non-routine manual tasks

Non-routine manual tasks are non-structured physical tasks that take place in unpredictable environments, often involving situational adaptability and in-person interaction. They require capabilities like sensory perception, fine and gross motor skills, social and emotional capabilities, natural language processing, navigation and mobility. The majority of these capabilities have not yet reached human-level performance and the incorporation of flexibility remains a considerable challenge (Autor, 2015; IPsoft, 2018). Consequently, the automation potential of this category is low, only 26% according to Manyika et al. (2017). Examples of tasks include operating a crane, assisting with surgery, janitorial work and making hotel beds.

Recent advances in sensory perception and physical capabilities as well as machine learning have enabled machines to take over an increasing number of manual non-routine tasks. Improvements in sensor technology and manual dexterity allow robots to perform high precision, non-standardized tasks, such as the manipulation of delicate products like fruit and vegetables. By incorporating advanced sensors, computer programs can also take over condition-monitoring tasks, such as checking the state of an aircraft engine or examining the moisture level in a field of crops. When alerted by the program, human operators can perform the required maintenance. Even some maintenance tasks are being substituted.

For example, General Electric has developed robots to climb and maintain wind turbines (Frey and Osborne, 2013).

Another well-known new application of machines for non-routine manual tasks is the autonomous vehicle. Autonomous driving was deemed impossible not so long ago as it requires activities such as parking, switching lanes and adapting to traffic lights, other vehicles and pedestrians (Autor, Levy and Murnane, 2003; Manyika et al., 2017).

However, today, facilitated by machine learning and advanced sensors, Google's autonomous car is driving the streets completely by itself and is even seen by some as safer than human-controlled cars (Frey and Osborne, 2013; Grosz et al., 2016). Autonomous mobility has also entered the warehousing industry (Autor, 2015). Here, enabled by environmental control, many warehouses, such as Amazon's warehouses, have become largely automatic.

Nonetheless, most non-routine manual tasks remain out of reach for machines for now and the near future. Despite the advances in the field of autonomous cars, autonomous mobility in general remains a significant challenge. Likewise, significant progress in perception and dexterity technologies is required before autonomous manipulation is viable in unstructured and delicate settings (Robotics Technology Consortium, 2013). Moreover, tasks that require human interaction demand further advances in language recognition, social and emotional capabilities and user interfaces. One example is walking a patient down a hospital (or nursery) hallway (Grosz et al., 2016). This requires a robot to help a patient get out of bed, which requires that the robot communicate with the person based on their emotional state, possess fine motor skills and sensory perception, to know where to hold/touch the patient and how much force to apply and to navigate through an unstructured environment. The activity is therefore not likely to be automated in a near future.

4.1.3. Routine cognitive tasks

Routine cognitive tasks include all mental (non-physical) tasks that repeat a certain procedure in a predictable environment. To a large extent, this relates to the different aspects of processing structured information, such as data collection, organization and storage (Autor, Levy and Murnane, 2003).

The required capabilities for these tasks are retrieving information, recognizing known patterns, optimizing and planning, logical reasoning/problem solving and natural language processing.

Examples of tasks are data-processing tasks such as calculating and bookkeeping but also routine customer-service activities performed by people such as cashiers, telephone operators and bank tellers. Because of their routine nature, these tasks have a high potential for machine substitution, ranging from 64% for tasks relating to data collection to 69% for tasks relating to data processing in the US, according to Manyika et al. (2017).

The automation of cognitive tasks started with the introduction of the computer (Autor, Levy and Murnane, 2003), which enabled the digitization and automatic processing of information. Subsequently, many processes, including administrative tasks, bookkeeping, invoicing, optimizing resource needs, and numerous others, have already been automated (Acemoglu and Autor, 2011).

Today, technological advances and the current focus on digitalization have brought the automation of routine cognitive tasks to an unprecedented scope and pace. Many companies have embarked on so-called "digital transformations", which refer to the simplification, standardization, and digitalization of an entire organization (Ketterer, Himmelreich and Schmid, 2016).

At the front-end, this means that large parts of customer interaction interfaces can be automated. Examples range from the automation of customer data collection for mortgage brokers to the employment of full-fledged, AI-based, virtual employees who can take over all aspects of customer interaction (IPsoft, 2018). At the back-end, the restructuring of the organization's IT landscape obviates many processes and activities (Ketterer, Himmelreich and Schmid, 2016).

In addition, for some structured processes that remain in existence, robotic process automation can be employed, which uses software robots to automate well-defined transactions/user actions normally performed by humans (Bughin et al., 2017; Ketterer, Himmelreich and Schmid, 2016). These software robots can be seen as virtual employees who work with existing applications in a similar fashion to humans (Forrester Research Inc., 2014).

The further proliferation of automated data collection and processing activities depends on the pace of digitalization. As companies progress on their digital transformations, more data and processes will be digitized and therefore likely automated. Moreover, further automation of customer service activities will depend on the machines' capability to interact with customers and thus depends on advances in natural language processing and emotional capabilities.

4.1.4. Non-routine cognitive tasks

Non-routine cognitive tasks are mental (non-physical/abstract) tasks that do not follow a structured procedure and/or take place in unpredictable environments (Autor, Levy and Murnane, 2003). These types of tasks require several cognitive capabilities, including creativity, logical reasoning, generating novel patterns and coordination with multiple agents. In addition, natural language processing and social and emotional capabilities are often of high importance (Acemoglu and Autor, 2011). These types of tasks include activities that relate to interfacing with stakeholders, applying expertise and managing and developing others. Examples of activities include legal writing, negotiations, teaching and diagnosing diseases.

Historically, these types of tasks have been the most difficult to automate (Frey and Osborne, 2013; Autor, Levy and Murnane, 2003). However, the availability of big data and recent advances in machine learning (pattern recognition in particular) have enabled machines to enter the realm of unstructured tasks. By applying unsupervised learning, a computer can create its own structure in an unstructured setting. Moreover, developments in the field of user interfaces, like language recognition, enable computers to respond directly to voice and gesture instructions (Manyika et al., 2013).

One of the tasks that can now be automated is fraud detection, a task that requires the ability to detect trends in data as well as to make decisions (Frey and Osborne, 2013). By using machine learning to build models based on historical transactions, social network information, and other external sources, the system can use pattern recognition to detect anomalies, exceptions, and outliers. This means fraudulent behavior can be spotted and fraudulent transactions can be prevented (Wellers, Elliot and Noga, 2017).

The legal domain is another area that machines are entering; nowadays, computers can analyze and order thousands of legal documents swiftly and present their findings graphically to the attorneys and paralegals (Frey and Osborne, 2013).

Yet, most of the involved capabilities remain far under human level for now. Especially tasks that require creativity, problem-solving and complex communication (a confluence of natural language processing and social and emotional capabilities) have a very low substitution potential (Manyika et al., 2017; Autor, Levy and Murnane, 2003).

Even in fields in which machines can outperform people on narrow tasks, like route planning, humans are often still required to set the target, interpret the outcomes and perform common-sense checks. Arguably there, major advances are required before machine learning and artificial intelligence become mature technologies. For instance, there are several examples of failing AI systems, like Microsoft's Tay Chatbot, who had to be shut down only 16 hours after launch because of the highly controversial messages it tweeted. Correspondingly, the three categories identified by Manyika et al. (2017), interfacing with stakeholders, applying expertise, and managing others, all have a substitution potential of below 20%.

Besides other required advances in cognitive, social and emotional capabilities, the availability of a sufficient amount of task-specific information is essential for the automation of cognitive non-routine tasks. In absence of this information, pattern recognition cannot be applied. In addition, as with the other types of tasks, environmental control, or task simplification, can be applied to mitigate engineering bottlenecks. For example, self-checkout stations in supermarkets obviate the need for advanced customer interaction (Frey and Osborne, 2013; Autor, Levy and Murnane, 2003).

4.2. The overall substitution of job tasks

As is evident from the previous discussion, technologies can take over an increasing number of activities. Routine, both manual and cognitive, tasks have been in the automation process for some time, whereas machines have only just acquired the ability to substitute for human labor in some non-routine tasks. The substitution potential for routine tasks is high and will only increase with technological advances. The substitution of non-routine tasks, on the other hand, remains largely limited to narrow applications for which human involvement is still required. A summary of the discussion for each of the job task categories is provided in Figure 3.3. To bring the automation of non-routine tasks to the next level, significant advances in all five capability areas are necessary, with natural language processing capabilities being the most important according to Manyika et al. (2017).

5. The impact on labor

Though several books and papers argue that technology will take over many jobs resulting in mass unemployment (Berg, Buffie and Zanna, 2016; OECD, 2016), as

	Cognitive	*Manual*
Routine	***Primary Required Capabilities*** Retrieving information Recognizing known patterns Optimizing and planning Logical reasoning/problem solving Natural language processing ***Sample Tasks*** Data processing tasks, e.g., calculating and bookkeeping Customer service tasks by e.g., cashiers, telephone operators, bank tellers **Predicted Substitution Rate: 64–69%***	***Primary Required Capabilities*** Gross and fine motor skills Sensory perception Mobility to some extent ***Sample Tasks*** Assembling Picking and sorting Welding Cooking **Predicted Substitution Rate: 81%***
Non-routine	***Primary Required Capabilities*** Creativity Logical reasoning/problem solving Generating novel patterns Coordinating with multiple agents Natural language processing Social and emotional capabilities ***Sample Tasks*** Legal writing Negotiating Teaching Diagnosing diseases **Predicted Substitution Rate: <20%***	***Primary Required Capabilities*** Fine and gross motor skills Sensory perception Social and emotional capabilities Natural language processing Navigation Mobility ***Sample Tasks*** Operating a crane Assisting with surgery Janitorial work Making hotel beds **Predicted Substitution Rate: 26%***

Figure 3.3 Summary of required capabilities, sample tasks and predicted substitution rate (in the USA) for each job task category.

of yet, this scenario seems unlikely to happen (Arntz, Gregory and Zierahn, 2016; Frey and Osborne, 2013; Manyika et al., 2017). Many activities can currently not be substituted by machines, and machines are not capable of performing several types of activities in an integrated way (Manyika et al., 2017; Autor, 2015). Hence, they are generally not capable of substituting labor for entire jobs, which usually include many bundled activities. Rather, to determine the substitution potential of a particular job, it is better to focus on the substitution of the individual activities within that job. A large body of research aligns with this approach and suggests that technology will take over significant parts of every job across all industries and levels of society (Manyika et al., 2017; Arntz, Gregory and Zierahn, 2016; OECD, 2016).

The following section will first analyze the automation potential of individual occupations and broader occupation categories and subsequently the nature of work and the impact of technology on industries.

5.1. The potential of job automation

Estimations of the potential of job automation differ significantly across studies. Frey and Osborne (2013) estimate that as much as 46 percent of all occupations in the United States consist of more than 70% activities that can be automated and are therefore highly automatable. By using the same methodology, but with a task approach rather than an occupation approach, Arntz, Gregory and Zierahn (2016) find that only nine percent of jobs in the US have an automation potential of more than 70%.

While Manyika et al. (2017) does not use 70% as a threshold for high automation potential, one can deduct from their study that around 25% of all jobs are more than 70% automatable in the United States.

Clearly, making an accurate estimation of automation potential is difficult and largely depends on subjective judgment of the capability of technologies and the task structure of occupations. Despite this variance, however, several high-level observations can be made.

Firstly, jobs that can be automated completely are likely to consist entirely of routine manual and routine cognitive tasks that require no human interaction or manual dexterity. Examples of these types of occupations are sewing-machine operators and order clerks.

Secondly, jobs with a high risk of automation also largely consist of routine manual and routine cognitive tasks, but will most likely include some degree of human interaction or unpredictable/high-precision physical activities. Occupation categories that include many highly automatable jobs are, for example, manufacturing and production because of their high degree of manual routine tasks, as well as sales, office and administrative support jobs because of their high dependence on information collecting and processing (World Economic Forum, 2016). Other occupation categories with large elements of routine manual activities are transportation (Frey and Osborne, 2016) and material-moving as well as food and accommodation services. According to Manyika et al. (2017), the latter even has the highest automation potential of all categories.

Lastly, the higher the proportion of non-routine tasks, the lower the automation potential of the job. This effect is enhanced if capabilities such as human interaction (requiring natural language processing and emotional and social capabilities), creativity, logical reasoning/problem solving, high-level dexterity or mobility are required. Jobs that consist entirely, or to a large extent, of these kinds of capabilities are not at all susceptible to automation (Arntz, Gregory and Zierahn, 2016; Manyika et al., 2017).

For example, the job of a choreographer primarily consists of the creative task to develop choreography and of human interaction to deal with stakeholders and train the dancers to bring the choreography to life.

A dentist, on the other hand, requires high-level dexterity and sensory perception as well as emotional and social capabilities to interact with their clients. Hence, both occupations have almost no activities that can be automated.

Still, the majority of occupation categories fall somewhere in between. This includes both routine and non-routine tasks. Therefore, they can be partly automated. For example, cognitive tasks are the core value drivers for investment bankers, yet a large proportion of their job consists of gathering and analyzing information and could thus be automated. The same holds for many legal professions. It is likely that these types of jobs will not disappear, rather, they will harness technology to improve efficiency of humans and the quality of output (Frey and Osborne, 2013).

It is important to note that this is a generalized view. The aforementioned occupation categories also include substantial proportions of jobs with low levels of automation potential, and the substitution potential of a job varies significantly across industries. For example, while supermarket cashiers and specialized software sales agents both fall under the sales occupation category, the substitution potential of the first is high while that of the latter is low because of the required technical expertise and emotional intelligence (World Economic Forum, 2016).

Furthermore, the substitution potential of similar jobs varies across different countries due to alterations in the structure of the jobs, industries and education, and previous investments in technology (Arntz, Gregory and Zierahn, 2016). For example, the automation potential in Sweden might be lower than average because Sweden sits at the forefront of technology investment. Consequently, technology will already have been included in many processes, making it difficult to automate large parts of the remaining activities. In addition, Sweden has a strong focus on high-skilled employees, who typically perform fewer tasks that are automatable. Correspondingly, Arntz, Gregory and Zierahn (2016) estimate that only seven percent of jobs in Sweden are at high risk of being substituted, compared to nine percent for all OECD countries. A discussion of other considerations such as these is provided in the next section.

5.2. The future nature of work

The large-scale substitution of individual tasks will likely change the nature of work and of all jobs (Frey and Osborne, 2013). As machines start to take over routine manual and routine cognitive tasks, human employees will be able to spend more time on complementary tasks where they hold a comparative advantage, such as activities involving creativity and human interaction (Autor, 2015; Finnigan, 2016; Arntz, Gregory and Zierahn, 2016).

Moreover, for many of these tasks, humans will be augmented by machines, and a closer collaboration between technology and humans is expected (International Federation of Robotics, 2017). For example, while a doctor is likely to remain responsible for the final diagnosis of a patient in the next decades, they will be able to base a decision partly on the automated diagnosis advice provided through AI.

As a result, jobs will require more training and a higher understanding of technology. In addition, as the incorporation of technology increases productivity, human employees might spend less of their time on work, resulting in shorter workweeks.

5.3. The effect on the labor market

The automation of activities has caused a well-documented shift in the labor market over the past decades. As part of this shift, scholars observed a polarization of the labor market in both the United States and Europe (Autor and Dorn, 2013; Autor, 2015). This polarization included a sharp decline in the share of middle-skilled jobs accompanied by increases in the share of low-skilled service jobs and high-skilled jobs (Frey and Osborne, 2013; Autor and Dorn, 2013). These middle-skilled jobs could be automated because they consisted primarily of routine manual and routine cognitive tasks, such as collecting and processing data. Tasks that could not be automated included non-routine manual and cognitive tasks. The first are usually found on the low-skill side of the spectrum while the latter are usually found on the high-skill side.

Consequently, the increase in general demand for labor following the productivity growth from automation mostly affected low-skilled jobs, e.g., hairdressers, janitors and high-skilled jobs, e.g., computer scientists, causing an overall polarization effect (Autor, 2015).

However, because of recent and future technological developments, this polarization is expected to taper off. The reason for this is threefold. Firstly, many remaining mid-level jobs require a combination of non-routine tasks and capabilities, including emotional skills, problem-solving, and flexibility, that cannot yet be performed by machines. Secondly, the rise of new technologies has created several new types of middle-skilled jobs, such as healthcare technicians and has stimulated demand for others, such as managers of eating establishments. Lastly, as discussed in this chapter, machines are increasingly able to take over low-skilled service jobs and high-skilled cognitive jobs (Holzer, 2015; Autor, 2015; World Economic Forum, 2016).

There has also been a global debate on the effect of technology on offshoring and reshoring initiatives, especially within the US manufacturing industry. Because the implementation of robotics obviates the need for cheap labor (Robotics Technology Consortium, 2013; International Federation of Robotics, 2017), many argue that it would give rise to a trend of reshoring manufacturing activities to the Western world while the offshoring trend would slow (Van den Bossche et al., 2015). However, more recently, opposing views have arisen, arguing that technology is also enabling the offshoring of many services and simplifying the management of complex global supply chains, leading to an increase in offshoring of manufacturing activities. The latter effects seem to be stronger and the reshoring trend, for example, advocated by the consultancy BCG, seems to have already ended (Boston Consulting Group, 2015). Meanwhile, offshoring is only found to increase (Van den Bossche et al., 2015).

Accurately estimating the overall effect of the previously discussed change drivers on the labor market is nigh on impossible and estimates range from mass unemployment to increases in labor demand. As large parts of jobs can be automated, fewer people will be needed to deliver the same output (Finnigan, 2016).

Consequently, automation could lead to unemployment in the short term (OECD, 2016) before gains in overall productivity raise the demand for labor again. Historically, technological progress has not significantly increased unemployment in the long run, but it remains to be seen whether this time will be the same (Autor, 2015). What is certain is that technology will cause large labor displacements, especially in high-routine occupation categories. Organizations and employees will need to increase their focus on education and training in order to be able to keep up with the increasing pace of change.

5.4. The automation potential of industries

The automation potential of work varies across industries because different industries have different job constellations and similar jobs in different industries might comprise different sets of tasks. In addition, there are also significant differences among countries regarding the job constellation of their industries. For example, an attorney in Sweden might perform very different tasks on a daily basis than an attorney in the United States.

As mentioned before, according to Manyika et al. (2017), the accommodation and food industry has the highest proportion of automatable tasks globally. These findings are supported by a study made in the US on the relation between innovation and employment (Frey and Osborne, 2015). The sector has such a high automation potential because food preparation consists of highly predictable manual tasks. For instance, tasks such as order taking and order serving do not require high levels of emotional intelligence, making them both susceptible to automation. The fast-food chain McDonald's, for example, has automated its ordering and payment processes using digital screens, and many casual-dining operators are implementing tabletop tablet systems in their restaurants.

Other industries with large proportions of automatable tasks identified by both studies are transportation and warehousing, retail trade, wholesale trade and manufacturing. For example, Amazon has already shown that robots can run entire warehouses and the technology for autonomous vehicles is largely ready, creating the opportunity to automate truck transportation.

On the low-end of the automation spectrum are industries such as educational services and the management of companies and enterprises. For many jobs in these sectors, emotional intelligence and complex communications are large and essential parts of daily activities, which substantially decreases automation potential.

The studies also disagree on the automation potential of several industries. For example, for some of the mining, real-estate rental, administrative and support services and construction industries, automation potential is estimated as average by Manyika et al. (2017) and high by Frey and Osborne (2015) while for other industries it is exactly the other way around. For example, the agriculture and information sectors are hardly automatable according to Osborne and Frey while they are averagely automatable according to Manyika et al. (2017).

Manyika et al. (2017) has also performed a study on the Swedish economy. According to the study, three industries have the highest proportions of automatable tasks. These are manufacturing, mining and transportation and warehousing. The industries with the lowest automation potential are educational services, the information sector, and the arts, entertainment and recreational sector.

In terms of the absolute number of employees who could be substituted, the manufacturing sector has by far the largest share. The study estimates that the work of as many as 420,000 people could potentially be automated. Other industries representing large numbers of people are healthcare and social assistance, administrative support and government and retail trade. Overall, Manyika et al. (2017) estimates that 46% of activities could be automated in Sweden, representing a potential redundancy of 2.1 million employees.

6. Other considerations for automation

Though it is technically feasible to substitute human labor with machines in many jobs and job tasks, there are several other factors affecting the pace and extent of automation. Five of these factors are discussed in the following sections: commercial availability, cost of implementation, economic benefits, labor-market dynamics and social, legal and ethical acceptance. We have based these factors on the five factors affecting the pace and extent of automation identified by Manyika et al. (2017). However, we renamed their first factor of technical feasibility as commercial availability in order to remove any confusion with our use of the term technological feasibility in this chapter.

6.1. Commercial availability

Although the previously discussed technologies have been proven in laboratories, the majority of them are yet to be commercialized. Many technologies are still in the early or middle stages of their development; they have not yet reached full maturity and require more scientific research. An example of this is artificial general intelligence (AGI). Despite the vast amount of research in this technology and the demonstration of some applications, much more scientific research is needed and academics estimate it might be 2050 before we can expect widespread adoption of robust AGI platforms (Vorhies, 2016).

Moreover, there is a distinct difference between technological feasibility and commercial adoption. Whereas basic (scientific) research focuses on broad generalizable cases, applied research focusses on developing engineering solutions for specific use cases. Developing viable products out of new technological concepts takes time and effort.

For example, predictive engineering for aircraft engines and predictive health care could be seen as similar scientific problems since both predict the failure of a system. However, both applications would need entirely different software, models and hardware to work and each would take years to be developed (Manyika et al., 2017).

Moreover, the ability to diagnose diseases can already be performed to some extent by computers, but computers diagnosing all types of diseases in the near future is unlikely due to technical difficulty (Bughin et al., 2017).

6.2. Cost of implementation

Besides the availability of commercially ready applications, there must be a solid business case for a company to implement automation and digitalization technologies. Hence, the development and implementation costs of new technologies are an important determinant of their adoption speed and scope. When analyzing these costs, there is a profound difference between the cost size and structure of hardware and software solutions.

6.2.1. Hardware

Hardware includes all physical components involved in a technological solution and often requires sensory perception, fine motor skills, gross motor skills and/or mobility. The capital expenditures for these components are often high and require significant upfront investments. This makes the business case more challenging and raises the need for available capital. Large companies in advanced countries, such as Sweden, are expected to have the fastest adoption rates of these solutions because they face high labor costs and are in the possession of readily available capital. Furthermore, the adoption cycles for industries facing high capital intensity are likely to be longer (Chui, Manyika and Miremadi, 2017).

The primary example of a hardware solution is an industrial robot. The cost of sophisticated robots has been declining over the past decades (Manyika et al., 2013; Frey and Osborne, 2013) and is expected to continue to decline in the future (Sirkin, Zinser and Rose, 2015).

This price drop has been enabled by significant cost decreases of advanced sensors and actuators. In addition, due to increases in production volumes of robots, economies of scale might lead to further cost reductions (Manyika et al., 2013; Grosz et al., 2016).

Despite the price drops, the cost of reliable mechanical devices remains high, and most industrial robots are still relatively expensive, ranging from several tens of thousands to hundreds of thousands of dollars. Moreover, besides the costs of the robot itself, large investments are required for engineering the robot's work cell (Robotics Technology Consortium, 2013). For example, to be able to work safely, an industrial robot often needs advanced safety equipment, and if a robot arm is to work with different tools, a tool-changing system needs to be in place. This kind of equipment is very expensive and can more than double the price of the robot's implementation (Slepov, 2016).

However, with the introduction of simpler general-purpose robots, the automation costs for simple tasks might drop significantly. Besides being cheaper themselves, these robots are more flexible and do not require extensive work cells. Likewise, they are safer for humans to work with, obviating the need for expensive safety equipment. The proliferation of this type of robots could significantly impact the adoption rate of robots. Service robots are, in general, cheaper than their industrial brothers and do not require surrounding equipment (Frey and Osborne, 2013; Manyika et al., 2013).

6.2.2. Software

For software solutions, the capital requirements are much lower, especially for solutions that are cloud-based. These low costs are enabled by increasing performance and decreasing costs of computing power, data storage and cloud computing. Often, the marginal cost of an additional software unit is negligible (Manyika et al., 2013; Autor, 2015).

However, the deployment of software can also incur highly taxing implementation costs, especially if legacy software systems are in place.

These implementation processes comprise activities such as software customization, staff training and new process architecture, and they can be more expensive than the software itself (Forrester Research Inc., 2014). Moreover, the talent required to develop, customize and implement advanced solutions is scarce and therefore extremely expensive.

For example, a study by Paysa, a career-consultancy firm, estimated that, in the United States alone, there were 10,000 open positions for AI talent in 2016, and that companies such as Alphabet and Microsoft are paying millions to acquire talented employees (Ketterer, Himmelreich and Schmid, 2016).

Robotic process automation forms a cheaper and quicker solution than the implementation of expensive new software solutions. This technology can automate workflows and substitute human labor without major investments. However, the overall benefits are limited compared to a complete system redesign (Horton, 2015).

6.3. Economic benefits

Another component in making a solid business case for the adoption of new technologies are the derived economic benefits from implementation. Companies will only be inclined to incorporate new technologies into their organizations if the benefits exceed the costs.

The first and most obvious economic benefit from the implementation of automation technologies is the reduction of labor costs, resulting from the substitution of human labor. As previously discussed, it is unlikely that many jobs will be substituted completely, but it is likely that fewer employees will be necessary to achieve the same output due to increased productivity.

The economic benefits of automation do, however, not only show in forms of saved labor costs but also in the form of new value creation. Examples include benefits such as increased throughput and productivity, improved safety, reduced waste and higher quality, all of which can increase profit in one way or another. These additional benefits can sometimes even exceed the benefits of labor substitution.

For example, implementing autonomous trucks would not only reduce labor costs but would also improve safety, fuel efficiency and productivity as there is no driver that requires stops. In turn, these improvements lead to increased profit. Google DeepMind is another example; the implementation of AI from DeepMind machine learning in Google's data centers has reduced energy consumption by 40%, resulting in increased profit (Grosz et al., 2016; Manyika et al., 2017).

Furthermore, due to the advancements in robotics, robots have become more economically viable options for tasks that were once seen as too expensive or delicate to automate, such as robotic surgery assistance.

As mentioned in the section *Definition of Digitalization*, digitalization is a means to create and capture new value within an organization. For example, it allows companies to open new digital customer channels and to develop new customer insights and products and services, leading to the creation of new value for the customer and the company. Moreover, the automation of routine processes enables employees to spend a larger amount of their time on high-value tasks. For example, within the finance sector, by letting a computer monitor existing processes and learn to recognize different situations (e.g., matching a payment with an order number), finance staff is freed from this activity and can instead focus on more valuable strategic tasks (Wellers, Elliot and Noga, 2017). Consequently, companies and industries that have digitalized to a larger extent, such as media, financial services, and technology, often show higher productivity and wage growth than industries that have digitalized to a lesser extent, such as education, retail and healthcare.

Besides increased profits for companies, society as a whole can gain substantial benefits from the implementation of technologies. Transportation is a prime example. As mentioned before, the automation of truck transportation will lead to higher productivity, higher safety and lower fuel consumption. Higher productivity means that fewer trucks will be necessary, leading to higher fuel reductions and less congested roads. As a result, the public will benefit from lower pollution, fewer traffic jams, fewer accidents and lower spending on road maintenance.

The benefits previously mentioned drive the pace of automation. However, it is important to note that most industries are still in very early stages of the adoption cycle of technologies such as AI, ML and robotics. Because of the small number of existing implementations, it is difficult to estimate what the overall benefits of these technologies will be. Moreover, it often takes years before the indirect economic benefits become visible. This time-lag between investment and benefits is especially large in capital-intensive industries where investments in hardware are required. Consequently, it is difficult for companies and regulators to understand the cost-benefit trade-offs of implementing new technologies (Grosz et al., 2016).

An example is an AI-based system. According to a survey by Bughin et al. (2017), most business leaders do not know what AI can do for them, where to use it, how to integrate it and what the benefits and costs will be.

6.4. Labor market dynamics

Since labor costs form an integral part of the business case for companies, the dynamics of the labor market are an important factor influencing the adoption rate of these technologies. These dynamics include the supply, demand and cost of human labor and are closely related to the demographics of a country and the skill-level of its citizens.

The supply and demand of labor have a large influence on the cost of labor and therefore on the economic benefits derived from the substitution of labor (Frey and Osborne, 2013). A high supply of labor in combination with low demand leads to a decrease in wages. Subsequently, low wages will decrease the economic benefits from labor substitution and thus decrease the incentive for companies to automate. For example, the food industry was identified as one of the industries in the United States with the highest automation potential based on current technologies. However, wages have historically been low in comparison to most other industries due to an oversupply of labor. Consequently, this industry has had little incentive to automate and the current level of automation is low. The opposite holds true when supply of labor is low and demand is high.

The supply of labor is a function of a country's demographics and the skill level of the working population (Manyika et al., 2017). The first influences the number of people on the labor market. In countries with a large working population, there will be an over-supply of labor in many industries and the incentive to automate will be low. On the contrary, for countries with shrinking working populations, such as Sweden and many other Western countries, the incentive to automate is larger (Manyika et al., 2013).

The skill level of the working population determines in which industries there are labor surpluses and deficits. For example, if a significant number of people have followed an education to become an English teacher, the market for English teachers will be saturated and wages will drop. Meanwhile, the market for French teachers could face a deficit of supply, increasing the wages. If activities are substituted by technology, it enables a higher level of human productivity, which would increase the labor supply. These workers can be redeployed if there is demand for activities within their skill range.

However, there often is a mismatch between the skills in demand and the skills that are in oversupply. In such a situation, people are required to reskill themselves through education and training before they can be redeployed. This takes time, money and effort. Consequently, the adoption of labor-substituting technology often leads to short-term unemployment and subsequently a period in which people need to re-educate themselves. However, as the pace of technological change and adoption is increasing, the question is whether the educational and training systems can keep pace. This is particularly difficult for people at the low-end of the skill spectrum.

A labor market polarization emerges when low-skill workers and high-skill workers represent the majority of the working population. In Sweden, technology has changed the labor market over the past 10–20 years as it has in other similar countries. Some argue that the Swedish labor market is undergoing a substitution of labor and that the Swedish regulatory and social security system is not ready for these changes. This will lead to an increased polarization and Sweden will face a difficult time redeploying employees if timely investments in training plans are not made (Breman, 2015).

Lastly, one can never really predict the future of the labor market. One year it can be steady with low unemployment and the next year it can be instable with high unemployment and a large degree of polarization. Unfortunately, the labor market is unlikely to benefit everyone equally when automation technologies are adopted. Some people will be negatively affected by either losing their job or facing wage pressure while others might see wage increases and new job openings. However, government policies, the way organizations choose to work and how individuals seek to learn new skills and jobs can all reduce the disparity in provided benefits across the labor spectrum (Grosz et al., 2016).

6.5. Social, legal and ethical acceptance

In order for the substitution of human labor to truly occur, applications of new technologies must be socially and legally accepted. This factor is one of the most central influencing the pace of automation, perhaps second only to technological feasibility. Social acceptance and legal acceptance are closely connected, and both largely depend on the related concept of ethical acceptance. Therefore, these three concepts will be discussed in combination.

Legal as well as social acceptance of new technologies are processes that take a lengthy amount of time. For example, a patient accepting a robot as a nurse or for a government to implement self-driving buses is not something that will happen overnight. It is therefore inevitable that it will take years for new technology to be completely adopted and adapted into society. Some of the requirements that must be fulfilled are decision makers realizing the potentials and benefits of AI as well as employees and workers adapting to the technologies once they are installed.

One of the major barriers for the automation process is privacy concerns. In order for new technologies and solutions to develop in the best interest of society, a large amount of data is needed. However, due to privacy concerns and regulations, data is difficult to access or anonymize. In addition, people are afraid of giving out their personal information because they do not know who will have access to it, who will use it and for what purpose (Bughin et al., 2017). It also becomes an ethical question when, for example, an employer has access to one's medical records. If someone is ill for some reason, or because they are overweight, an employer may not be interested in hiring this person.

The ethical issue also comes into consideration when technologies are, for example, used for predictive policies. It is a technical challenge to not feed the systems with biased information – e.g., racial, sexist or religious discrimination – to avoid innocent people being unjustifiably monitored and discriminated, when the real world is in fact biased (Grosz et al., 2016). However, when predictive hiring processes are performed with caution, and through careful design, testing and deployment, there is a chance that AI algorithms will make less-biased decisions than humans.

As mentioned, the extent and pace of automation rely on the social acceptance and trust for technology and AI. For example, many of the activities a nurse performs can theoretically be automated, but both coworkers and patients will likely have a difficult time to accept it initially. Arguably a majority of patients expect to be greeted by humans and have human contact when they have their meal delivered to them. In order for the activity to actually be substituted, patients and co-workers have to accept and trust the machines. This can only be accomplished if hospitals exhaustively integrate the automation technologies and make sure that the interaction between intelligent computers and humans feels natural (Manyika et al., 2017; Grosz et al., 2016).

This trust and acceptance is also important for security systems to be able to use the innovative technologies. Today, cities in North America have already deployed AI technologies in border administration and law enforcement and will heavily rely on these techniques in the future. For example, autonomous cars, drones and cameras will be used for surveillance as well as algorithms to detect financial fraud and create predictive policies. However, this is only possible if there is broad social acceptance. Furthermore, regulatory acceptance is also necessary for full-scale adoption. For example, while autonomous vehicles are fully usable they will first be adopted when regulators accept them (Manyika et al., 2017).

Furthermore, questions are raised about accountability when implementing the technologies. Issues such as who is responsible for the actions and conclusions made by robots and AI have never been dealt with before, making them difficult to tackle (Bughin et al., 2017). For example, who is responsible for a traffic accident where an autonomous vehicle is involved and maybe caused it? Is it the owner of the car, the automaker, the city, one of the many software or hardware providers or one of the many programmers who wrote some of the lines of software code?

Once the technologies are adopted, there may be consequences. For example, there is no way to know if AI would optimize the labor market without regard for nuanced social preferences or sell treasured documents about people's skills to private companies or political parties. However, it is unlikely that AI would autonomously *choose* to inflict harm on people, but there nonetheless remains a real risk that it can be *used* by people for a harmful purpose.

To summarize, the social, legal and ethical acceptance are important factors that impact the adoption of automation technologies. It is understandable that social acceptance of new technologies is difficult due to the fear that a lot of people will lose their jobs. However, as discussed earlier in this chapter, it is activities within jobs that will be substituted rather than entire jobs. Regulators must clearly state this fact and that only certain people will have access to personal information, in order for the social acceptance to increase.

7. Conclusion

This chapter aimed at investigating the substitution potential of labor by a selection of technologies. We first discussed the technological feasibility of artificial intelligence, machine learning and robotics to substitute for labor. We found that technology can perform an increasingly wider variety of job activities and that automation is no longer confined to routine tasks. Nevertheless, the automation potential for non-routine tasks remains limited, especially for tasks involving autonomous mobility, creativity, problem-solving and complex communication.

For jobs themselves, we concluded that the majority of jobs will be affected by the automation of individual activities, but that only a few have the potential to be completely substituted. The jobs most at risk are those that consist largely of routine tasks and do not rely on mobility or human interaction. Though few jobs can be substituted completely, automation could still lead to short-term unemployment, often leading to re-training and further education. In addition, we concluded that the nature of jobs will change as mundane tasks will be substituted and people will work more closely together with machines. The industries that have a large potential for activity substitution are food and accommodation services, transportation and warehousing, retail trade, wholesale trade and manufacturing.

In the last section of the chapter, we discussed five major factors that come into play before automation potential turns into actual automation: commercial availability, cost of implementation, economic benefits, labor market dynamics and social and legal acceptance. All five of these factors have a significant influence on the speed and scope of technology adoption. In particular, a lack of applied research, low wages, high costs and legal and ethical boundaries hamper the adoption of technology.

Overall, technology is advancing rapidly and the pace of change is increasing. Consequently, an increasing number of activities will have the potential to be performed by machines rather than by humans. Though the extent and speed of adoption are reduced by several factors, it is inevitable that technology will have a stronger presence in the workplace. It is unlikely that this will cause long-term unemployment, but in the short-term reskilling will be required to enable the reemployment of displaced labor. To cope with the pace of automation, an increased focus on education and training will be required – for individuals, organizations, regions and countries.

Notes

1 This chapter is a reprint of an identical report that was originally released under the same title by the same authors as "Report #5" for The Internet Foundation in Sweden (IIS), as part of the "Innovative Internet" project. The report was originally published by Stockholm School of Economics: Center for Strategy and Competitiveness, Stockholm: Sweden, in 2018 (ISBN: 978–91–86797–32–4). Permission for reprint has been granted by the copyright holder.

2 The Turing test is a computer intelligence test, requiring that a human being should be unable to distinguish the machine from another human being by using the replies to questions posed to both.

References

Acemoglu, D. and Autor, D., 2011. Skills, tasks and technologies: Implications for employment and earnings. In: D.E. Card and O. Ashenfelter, eds., *Handbook of Labor Economics, Volume 4, Part B*. Amsterdam, Netherlands: Elsevier, pp. 1043–1171.

Aeppel, T., 2017. The biggest question is not whether AI will disrupt business and society, but when? In: *AI & Machine Learning Disruption Timeline Conference, Volume 3*. Cambridge, MA: MIT IDE Conference Report, pp. 1–4.

Anderson, R.J., Corydon, B., Staun, J., Bughin, J., Lunborg, J. and Schröder, P., 2017. *A Future That Works: The Impact of Automation in Denmark.* [online] McKinsey & Company and The Tuborg Research Centre for Globalization and Firms at Aarhus University. Available at: <www.mckinsey.com/~/media/McKinsey/Locations/Europe%20and%20Middle%20East/Denmark/Our%20Insights/A%20future%20that%20works%20The%20impact%20of%20automation%20in%20Denmark/A-future-that-works-The-impact-of-automation-in-Denmark.ashx> [Accessed 22 Sep. 2019].

Arntz, M., Gregory, T. and Zierahn, U., 2016. *The Risk of Automation for Jobs in OECD Countries: A Comparative Analysis*. OECD Social, Employment and Migration Working Papers, No. 189. Paris.

Autor, D.H., 2015. Why are there still so many jobs? The History and future of workplace automation. *Journal of Economic Perspectives*, 29(3), pp. 3–30.

Autor, D.H. and Dorn, D., 2013. The growth of low-skill service jobs and the polarization of the US labor market. *American Economic Review*, 103(5), pp. 1553–1597.

Autor, D.H., Levy, F. and Murnane, R.J., 2003. The skill content of recent technological change: An empirical exploration. *The Quarterly Journal of Economics*, 118(4), pp. 1279–1333.

Bates, M., Bobrow, R.J. and Weischedel, R.M., 1993. Challenges in natural language processing. In: M. Bates and R.M. Weischedel, eds., *Challenges in Natural Language Processing*. Cambridge, UK: Cambridge University Press, pp. 3–34.

Bengio, Y., 2017. *TED Talk: The Rise of Artificial Intelligence through Deep Learning.* [online] YouTube. Available at: <www.youtube.com/watch?v=uawLjkSI7Mo> [Accessed 22 Sep. 2019].

Berg, A., Buffie, E.F. and Zanna, L.F., 2016. Robots, Growth, and Inequality. *IMF Finance and Development*, 53(3), Sep., pp. 1–5.

Berman, S.J., Marshall, A. and Leonelli, N., 2013. *Digital Reinvention: Preparing for a Very Different Tomorrow.* [online] IBM Institute for Business Value. Available at: <www.ibm.com/downloads/cas/20NX8NYB> [Accessed 22 Sep. 2019].

Boston Consulting Group, 2014. *The Evolution of Robotics*. [online] BCG Perspectives. Available at: <www.bcg.com/publications/interactives/business-unit-strategy-growth-evolution-robotics.aspx> [Accessed 22 Sep. 2019].

Boston Consulting Group, 2015. *Reshoring of Manufacturing to the US Gains Momentum.* [online] Available at: <www.bcg.com/publications/2015/reshoring-of-manufacturing-to-the-us-gains-momentum.aspx> [Accessed 22 Sep. 2019].

Boston Dynamics, 2018. *Atlas: The World's Most Dynamic Humanoid.* [online] Boston Dynamics. Available at: <www.bostondynamics.com/atlas> [Accessed 22 Sep. 2019].

Breman, A., 2015. *Underlagsrapport till analysgruppen Arbetet i framtiden [Background Report for the Analytics Group Labor in the Future].* [online] Available at: <www.regeringen.se/495191/contentassets/17116303997443d69a6692270d963b01/underlagsrapport-diginomics-och-arbetet-i-framtiden.pdf> [Accessed 22 Sep. 2019].

Bughin, J., Hazan, E., Ramaswamy, S., Chui, M., Allas, T., Dahlström, P., Henke, N. and Trench, M., 2017. *Artificial Intelligence – The Next Digital Frontier?* [online] McKinsey & Company Global Institute. Available at: <www.mckinsey.com/~/media/McKinsey/Industries/Advanced%20Electronics/Our%20Insights/How%20artificial%20intelligence%20can%20deliver%20real%20value%20to%20companies/MGI-Artificial-Intelligence-Discussion-paper.ashx> [Accessed 22 Sep. 2019].

Childs, M., 2011. *John McCarthy: Computer Scientist Known as the Father of AI.* [online] The Independent. Available at: <www.independent.co.uk/news/obituaries/john-mccarthy-computer-scientist-known-as-thefather-of-ai-6255307.html> [Accessed 28 Aug. 2019].

Chui, M., Manyika, J. and Miremadi, M., 2015. Four fundamentals of workplace automation. *McKinsey Quarterly*, 29(3), pp. 1–9.

Chui, M., Manyika, J. and Miremadi, M., 2016. *Where Machines Could Replace Humans – And Where They Can't (Yet).* [online] McKinsey Quarterly. Available at: <www.mckinsey.com/business-functions/digital-mckinsey/our-insights/where-machines-could-replace-humans-and-where-they-cant-yet> [Accessed 28 Aug. 2019].

Chui, M., Manyika, J. and Miremadi, M., 2017. *The Countries Most (and Least) Likely to be Affected by Automation.* [online] Harvard Business Review. Available at: <https://hbr.org/2017/04/the-countries-most-and-least-likely-to-be-affected-by-automation> [Accessed 28 Aug. 2019].

Copeland, B.J., 2018. *Artificial Intelligence.* [online] Encylopaedia Britannica. Available at: <www.britannica.com/technology/artificial-intelligence> [Accessed 28 Aug. 2019].

Coupel, T., 2014. *Difference Between Natural Language Understanding and Natural Language Generation.* [online] Yseop. Available at: <https://yseop.com/blog/what-is-the-difference-between-natural-language-generation-and-understanding-2> [Accessed 28 Aug. 2019].

Decker, M., Fischer, M. and Ott, I., 2017. Service robotics and human labor: A first technology assessment of substitution and cooperation. *Robotics and Autonomous Systems*, 87, pp. 348–354.

Deng, L. and Yu, D., 2014. Deep learning: Methods and applications. *Foundations and Trends® In Signal Processing*, 7(3–4), pp. 197–387.

Difference Wiki, 2017. *Strong AI vs. Weak AI: What's the Difference?* [online] Available at: <www.difference.wiki/strong-ai-vs-weak-ai/#Definition_of_Strong_AI> [Accessed 28 Aug. 2019].

Doerrfeld, B., 2015. *20+ Emotion Recognition APIs That Will Leave You Impressed, And Concerned.* [online] Nordic APIs. Available at: <https://nordicapis.com/20-emotion-recognition-apis-that-will-leave-you-impressed-and-concerned> [Accessed 28 Aug. 2019].

Electronics Teacher, 2017. *Mobility – Robotics Technology.* [online] Electronics Teacher. Available at: <www.electronicsteacher.com/robotics/robotics-technology/mobility.php> [Accessed 28 Aug. 2019].

Encyclopaedia Britannica, 2018. *Robotics.* [online] Encyclopaedia Britannica. Available at: <www.britannica.com/technology/robotics> [Accessed 28 Aug. 2019].

Finnigan, D., 2016. *Robots and Automation May Not Take Your Desk Job After All.* [online] Harvard Business Review. Available at: <https://hbr.org/2016/11/robots-and-automation-may-not-take-your-desk-job-after-all> [Accessed 28 Aug. 2019].

Forrester Research Inc., 2014. *Building A Center Of Expertise To Support Robotic Automation.* [online] Available at: <http://neoops.com/wp-content/uploads/2014/03/Forrester-RA-COE.pdf> [Accessed 28 Aug. 2019].

Frey, C.B. and Osborne, M.A., 2013. The Future of employment: How susceptible are jobs to computerisation? *Technological Forecasting and Social Change*, 114, pp. 254–280.

Frey, C.B. and Osborne, M.A., 2015. *Technology At Work – The Future of Innovation and Employment.* [online] Oxford University. Available at: <www.oxfordmartin.ox.ac.uk/downloads/reports/Citi_GPS_Technology_Work.pdf> [Accessed 28 Aug. 2019].

Frey, C.B. and Osborne, M.A., 2016. *Technology At Work 2.0, the Future Is Not What It Used to Be.* [online] Oxford University. Available at: <www.oxfordmartin.ox.ac.uk/downloads/reports/Citi_GPS_Technology_Work_2.pdf> [Accessed 28 Aug. 2019].

Ghahramani, Z., 2004. Unsupervised learning. In: O. Bousquet, U. Von Luxburg and G. Rätsch, eds., *Advanced Lectures on Machine Learning. Lecture Notes in Computer Science, Volume 3176*. Heidelberg: Springer, pp. 72–112.

Groover, M.P., 2018. *Automation*. [online] Encyclopaedia Britannica. Available at: <www.britannica.com/technology/automation> [Accessed 28 Aug. 2019].

Grosz, B.J., Altman, R., Horvitz, E., Mackworth, A., Mitchell, T., Mulligan, D. and Shoham, Y., 2016. *Artificial Intelligence and Life in 2030 – One Hundred Year Study of Artificial Intelligence*. [online] Stanford University. Available at: <https://ai100.stanford.edu/sites/default/files/ai_100_report_0831fnl.pdf> [Accessed 28 Aug. 2019].

Hager, G.D., Rus, D., Kumar, V. and Christensen, H., 2015. *Toward a Science of Autonomy for Physical Systems*. [online] Available at: <https://arxiv.org/ftp/arxiv/papers/1604/1604.02979.pdf> [Accessed 28 Aug. 2019].

Hahn, W.E., 2017. *Machine Perception & Cognitive Robotics Laboratory*. [online] Florida Atlantic University. Available at: <www.ccs.fau.edu/~hahn/mpcr> [Accessed 28 Aug. 2019].

Hardesty, L., 2017. *Giving Robots a Sense of Touch*. [online] MIT News. Available at: <http://news.mit.edu/2017/gelsight-robots-sense-touch-0605> [Accessed 28 Aug. 2019].

Heess, N., Tb, D., Sriram, S., Lemmon, J., Merel, J., Wayne, G., Tassa, Y., Erez, T., Wang, Z., Eslami, S.M.A., Riedmiller, M. and Deepmind, D.S., 2017. *Emergence of Locomotion Behaviours in Rich Environments*. [online] Available at: <https://arxiv.org/abs/1707.02286> [Accessed 28 Aug. 2019].

Henke, N., Bughin, J., Chui, M., Manyika, J., Saleh, T., Wiseman, B. and Sethupathy, G., 2016. *The Age of Analytics: Competing in a Data-driven World*. [online] McKinsey Global Institute. Available at: <www.mckinsey.com/business-functions/mckinsey-analytics/our-insights/the-age-of-analytics-competing-in-a-data-driven-world> [Accessed 28 Aug. 2019].

Hodson, J., 2016. *How to Make Your Company Machine Learning Ready*. [online] Harvard Business Review. Available at: <https://hbr.org/2016/11/how-to-make-your-company-machine-learning-ready> [Accessed 28 Aug. 2019].

Holzer, H., 2015. *Job Market Polarization and U.S. Worker Skills: A Tale of Two Middles*. [online] *Brookings Economic Studies*. Available at: <www.brookings.edu/wp-content/uploads/2016/06/polarization_jobs_policy_holzer.pdf> [Accessed 28 Aug. 2019].

Honda, 2018. *Inside ASIMO Robotics by Honda: The Technology Behind ASIMO*. [online] ASIMO. Available at: <http://asimo.honda.com/inside-asimo> [Accessed 22 Sep. 2019].

Horton, R., 2015. *The Robots are Coming*. [online] Deloitte Insights. Available at: <www2.deloitte.com/content/dam/Deloitte/uk/Documents/finance/deloitte-uk-finance-robots-are-coming.pdf> [Accessed 22 Sep. 2019].

Hunckler, M., 2017. *This Open-Source AI Voice Assistant Is Challenging Siri and Alexa for Market Superiority*. [online] Forbes. Available at: <www.forbes.com/sites/matthunckler/2017/05/15/this-open-source-ai-voice-assistant-is-challenging-siri-and-alexa-for-market-superiority/#9e3c9e63ec01> [Accessed 28 Aug. 2019].

International Federation of Robotics, 2017. *The Impact of Robots on Productivity, Employment and Jobs: A Positioning Paper by the International Federation of Robotics*. [online] Available at: <https://ifr.org/img/office/IFR_The_Impact_of_Robots_on_Employment.pdf> [Accessed 28 Aug. 2019].

IPsoft, 2018. *IPsoft*. [online] Available at: <www.ipsoft.com/amelia> [Accessed 28 Aug. 2019].

I-Scoop, 2017. *Digitization, Digitalization and Digital Transformation: The Differences*. [online] Available at: <www.i-scoop.eu/digitization-digitalization-digital-transformation-disruption> [Accessed 28 Aug. 2019].

Izatt, G., Mirano, G., Adelson, E. and Tedrake, R., 2017. *Tracking Objects with Point Clouds from Vision and Touch.* [online] Available at: <http://groups.csail.mit.edu/robotics-center/public_papers/Izatt16.pdf> [Accessed 28 Aug. 2019].

Ketterer, H., Himmelreich, H. and Schmid, C.N., 2016. *Ensuring Digital Readiness in Financial Services.* [online] BCG Perspectives. Available at: <www.bcg.com/publications/2016/technology-advantage-financial-institutions-ensuring-digital-readiness-financial-services.aspx> [Accessed 28 Aug. 2019].

Khan, S., 2016. *Leadership in the Digital Age – A Study on the Effects of Digitalisation on Top Management Leadership.* [online] Stockholm University. Available at: <https://su.diva-portal.org/smash/get/diva2:971518/FULLTEXT02.pdf> [Accessed 28 Aug. 2019].

Knight, W., 2016. *IBM's Watson is Everywhere – But What is It?* [online] MIT Technology Review. Available at: <www.technologyreview.com/s/602744/ibms-watson-is-everywhere-but-what-is-it> [Accessed 28 Aug. 2019].

Kolling, A., Walker, P., Chakraborty, N., Sycara, K. and Lewis, M., 2016. Human interaction with robot swarms: A survey. *IEEE Transactions on Human-Machine Systems*, 46(1), pp. 9–26.

Latxague, F., 2013. *A Brief History of Robots.* [online] Paris Innovation Review. Available at: <http://parisinnovationreview.com/articles-en/a-brief-history-of-robots> [Accessed 28 Aug. 2019].

Libesa, 2016. *Machine Learning: A Brief Breakdown.* [online] QuantDare. Available at: <https://quantdare.com/machine-learning-a-brief-breakdown> [Accessed 28 Aug. 2019].

LSAC, 2018. *Law School Admission Test: Logical Reasoning Questions.* [online] LSAC. Available at: <www.lsac.org/jd/lsat/prep/logical-reasoning> [Accessed 28 Aug. 2019].

Lyon, R.F., 2017. *Human and Machine Hearing.* Cambridge, UK: Cambridge University Press.

Mäenpää, R. and Korhonen, J.J., 2015. *Digitalization in Retail: The Impact on Competition. Leadership in Transition: The Impact of Digitalization on Finnish Organizations.* Greater Helsinki, Finland: Aalto University publication series SCIENCE + TECHNOLOGY. 7/2015

Manyika, J., Chui, M., Bughin, J., Dobbs, R., Bisson, P. and Marrs, 2013. *Disruptive Technologies: Advances that will Transform Life, Business, and the Global Economy.* New York, NY: McKinsey Global Institute.

Manyika, J., Chui, M., Miremadi, M., Bughin, J., George, K., Willmott, P. and Dewhurst, M., 2017. *Harnessing Automation for a Future that Works.* [online] McKinsey Global Institute. Available at: <www.mckinsey.com/featured-insights/digital-disruption/harnessing-automation-for-a-future-that-works> [Accessed 28 Aug. 2019].

Marr, B., 2016a. *What Is the Difference Between Artificial Intelligence and Machine Learning?* [online] Forbes. Available at: <www.forbes.com/sites/bernardmarr/2016/12/06/what-is-the-difference-between-artificial-intelligence-and-machine-learning/#7a8738362742> [Accessed 28 Aug. 2019].

Marr, B., 2016b. *What Is the Difference Between Deep Learning, Machine Learning and AI?* [online] Forbes. Available at: <www.forbes.com/sites/bernardmarr/2016/12/08/what-is-the-difference-between-deep-learning-machine-learning-and-ai/3/#654630005834> [Accessed 28 Aug. 2019].

MathWorks, 2018a. *Face Recognition with Computer Vision.* [online] MathWorks. Available at: <https://se.mathworks.com/discovery/face-recognition.html> [Accessed 28 Aug. 2019].

MathWorks, 2018b. *Unsupervised Learning: Machine Learning Technique for Finding Hidden Patterns or Intrinsic Structures in Data.* [online] MathWorks. Available at: <https://se.mathworks.com/discovery/unsupervised-learning.html?s_tid=srchtitle> [Accessed 28 Aug. 2019].

MathWorks, 2018c. *What is Deep Learning?: 3 Things You Need to Learn.* [online] MathWorks. Available at: <https://se.mathworks.com/discovery/deep-learning.html> [Accessed 28 Aug. 2019].

MathWorks, 2018d. *Supervised Learning – Machine Learning Technique for Building Predictive Models from Known Input and Response Data.* [online] MathWorks. Available at: <https://se.mathworks.com/discovery/supervised-learning.html> [Accessed 28 Aug. 2019].

Maynard, D., 2016. *Teaching AI How to be Sarcastic is Totally the Easiest Thing Ever.* [online] Quartz. Available at: <https://qz.com/801813/teaching-ai-how-to-be-sarcastic-is-totally-the-easiest-thing-ever> [Accessed 28 Aug. 2019].

Moore, S., 2015. *Digitalization or Automation – Is There a Difference?* [online] Gartner. Available at: <www.gartner.com/smarterwithgartner/digitalization-or-automation-is-there-a-difference> [Accessed 28 Aug. 2019].

Murphy, M., 2015. *Robots in Japan Now Have Emotions.* [online] Quartz. Available at: <https://qz.com/433877/robots-in-japan-now-have-emotions> [Accessed 28 Aug. 2019].

Ng, A., 2016. *IBM's Watson Gives Proper Diagnosis for Japanese Leukemia Patient After Doctors were Stumped for Months.* [online] NY Daily News. Available at: <www.nydailynews.com/news/world/ibm-watson-proper-diagnosis-doctors-stumped-article-1.2741857> [Accessed 28 Aug. 2019].

Nilsson, N.J., 2010. *The Quest for Artificial Intelligence: A History of Ideas and Achievements.* Cambridge, UK: Cambridge University Press.

OECD, 2016. *Automation and Independent Work in a Digital Economy: Policy Brief on the Future of Work.* Paris, France: Policy Brief on the Future of Work.

Oxford Dictionaries, 2018a. *Definition of Artificial Intelligence in English.* [online] Oxford Dictionaries. Available at: <https://en.oxforddictionaries.com/definition/artificial_intelligence> [Accessed 28 Aug. 2019].

Oxford Dictionaries, 2018b. *Definition of Digitization in English.* [online] Oxford Dictionaries. Available at: <https://en.oxforddictionaries.com/definition/digitization> [Accessed 28 Aug. 2019].

Oxford Dictionaries, 2018c. *Definition of Robot in English.* [online] Oxford Dictionaries. Available at: <https://en.oxforddictionaries.com/definition/robot> [Accessed 28 Aug. 2019].

Perry, C., 2014. *A Self-Organizing Thousand-Robot Swarm.* [online] Harvard John A. Paulson School of Engineering and Applied Sciences. Available at: <www.seas.harvard.edu/news/2014/08/self-organizing-thousand-robot-swarm> [Accessed 28 Aug. 2019].

PHC, 2016. *A Brief History of Automation.* [online] Product Handling Concepts. Available at: <www.phcfirst.com/words-in-motion/2016/9/7/a-brief-history-of-automation> [Accessed 28 Aug. 2019].

Picard, R.W., 2007. Toward machines with emotional intelligence. In: G. Matthews, M. Zeidner and R.D. Roberts, eds., *The Science of Emotional Intelligence: Knowns and Unknowns.* New York, NY: Oxford University Press, pp. 396–418.

Press, G., 2015. *A Very Short History of Digitization.* [online] Forbes. Available at: <www.forbes.com/sites/gilpress/2015/12/27/a-very-short-history-of-digitization/#1fa0730249ac> [Accessed 28 Aug. 2019].

Puget, J.F., 2016. *What Is Machine Learning?* [online] IBM DeveloperWorks. Available at: <www.ibm.com/developerworks/community/blogs/jfp/entry/What_Is_Machine_Learning> [Accessed 28 Aug. 2019].

Qureshi, M.O. and Sajjad, R., 2017. A study of integration of robotics in the hospitality sector and its emulation in the pharmaceutical sector. *Health Science Journal*, 11(1), pp. 1–6.

Ritter, H. and Haschke, R., 2015. Hands, dexterity, and the brain. In: G. Cheng, ed., *Humanoid Robotics and Neuroscience: Science, Engineering and Society*. Boca Raton, FL: CRC Press, pp. 49–78.

Robinson, A., 2014. *Industrial Automation: A Brief History of Manufacturing Application & The Current State and Future Outlook*. [online] Cerasis. Available at: <https://cerasis.com/industrial-automation/> [Accessed 28 Aug. 2019].

Robotics Technology Consortium, 2013. *A Roadmap for US Robotics: From Internet to Robotics*. [online] Available at: <http://archive2.cra.org/ccc/files/docs/2013-Robotics-Roadmap> [Accessed 28 Aug. 2019].

Sander, A. and Wolfgang, M., 2014. *The Rise of Robotics*. [online] BCG Perspectives. Available at: <www.bcg.com/publications/2014/business-unit-strategy-innovation-rise-of-robotics.aspx> [Accessed 28 Aug. 2019].

Siciliano, B. and Khatib, O., 2016. Robotics and the handbook. In: B. Siciliano and O. Khatib, eds., *Springer Handbook of Robotics*. Cham, Switzerland: Springer, pp. 1–6.

Sirkin, H., Zinser, M. and Rose, J., 2015. *The Robotics Revolution: The Next Great Leap in Manufacturing*. [online] BCG Perspectives. Available at: <www.bcg.com/publications/2015/lean-manufacturing-innovation-robotics-revolution-next-great-leap-manufacturing.aspx> [Accessed 28 Aug. 2019].

Sklar, J., 2015. *Making Robots Talk to Each Other*. [online] MIT Technology Review. Available at: <www.technologyreview.com/s/539806/making-robots-talk-to-each-other> [Accessed 28 Aug. 2019].

Slepov, D., 2016. *The Real Cost of Robotics*. [online] TechCrunch. Available at: <https://techcrunch.com/2016/03/27/the-real-cost-of-robotics> [Accessed 28 Aug. 2019].

Steadman, I., 2013. *IBM's Watson is Better At Diagnosing Cancer Than Human Doctors*. [online] Wired Magazine. Available at: <www.wired.co.uk/article/ibm-watson-medical-doctor> [Accessed 28 Aug. 2019].

TechCrunch, 2016. *Zume Delivers Made-to-Order Pizza with Robots*. [online] YouTube. Available at: <www.youtube.com/watch?v=uFSdxwRVh8A> [Accessed 28 Aug. 2019].

Turcot, J., 2015. *How to Teach a Machine to Recognize Emotions (Part 1)*. [online] Affectiva. Available at: <http://blog.affectiva.com/how-to-teach-a-machine-to-recognize-emotions-part-1> [Accessed 28 Aug. 2019].

Turek, M., 2017. *Explainable Artificial Intelligence (XAI)*. [online] DARPA. Available at: <www.darpa.mil/program/explainable-artificial-intelligence> [Accessed 28 Aug. 2019].

Van den Bossche, P., Gott, J., Levering, B. and Gutierrez, H., 2015. *U.S. Reshoring: Over Before It Began?* [online] Available at: <www.atkearney.com/documents/10192/7070019/US+Reshoring.pdf/14777afa-0c14-460e-937b-11233be340b6> [Accessed 28 Aug. 2019].

Vogelsang, M., 2010. *Digitalization in Open Economies*. Heidelberg, Germany: Physica-Verlag HD.

Vorhies, W., 2016. *Artificial General Intelligence – The Holy Grail of AI*. [online] Data Science Central. Available at: <www.datasciencecentral.com/profiles/blogs/artificial-general-intelligence-the-holy-grail-of-ai> [Accessed 28 Aug. 2019].

Wellers, D., Elliot, T. and Noga, M., 2017. *8 Ways Machine Learning Is Improving Companies' Work Processes*. [online] Harvard Business Review. Available at: <https://hbr.org/2017/05/8-ways-machine-learning-is-improving-companies-work-processes> [Accessed 28 Aug. 2019].

Wilson, H.J., 2015. *What Is a Robot, Anyway?* [online] Harvard Business Review. Available at: <https://hbr.org/2015/04/what-is-a-robot-anyway> [Accessed 22 Sep. 2019].

World Economic Forum, 2016. *The Future of Jobs: Employment, Skills and Workforce Strategy for the Fourth Industrial Revolution. Global Challenge Insight Report*. Cologny, Switzerland.

4 Minimum wages for online labor platforms?

Regulating the global gig economy

Alex J. Wood, Mark Graham and Mohammad Amir Anwar

1. Background

In 2018, over half of the world's population was connected to the internet. The rise of the so-called "gig economy" has enabled internet users to find work that they might not otherwise have been able to obtain. Over the last four years' researchers based at the Oxford Internet Institute have been at the forefront of wide-ranging research into conditions on the "online labor platforms" which constitute a global remote gig economy. Online labor platforms enable clients to access labor power potentially from anywhere in the world. According to one estimate, this has created a USD5 billion market for online work that is served by 48 million workers (Kuek et al., 2015). These platforms have been the focus of much of our research. Such platforms are global in nature, and involve the remote buying and selling of digital labor which is by its nature highly mobile and "non-geographically sticky" (also known as "crowdwork" this is work that can, in theory, be done from anywhere).

Collectively we have interviewed 250 remote gig economy workers across ten countries and four continents. We have interviewed workers in Kenya, Uganda, Ghana, Malaysia, Nigeria, the Philippines, South Africa, the United Kingdom, the United States and Vietnam as well as other stakeholders such as platform CEOs and government and trade union officials. We have also conducted a survey of 679 Asian and African workers; analyzed six months of transaction data from one of the world's largest platforms and undertaken observation at dozens of gig worker community events. It is this wealth of research which informs our following discussion (Anwar and Graham, 2017, 2018; Graham, Hjorth and Lehdonvirta, 2017; Graham et al., 2017; Graham and Anwar, 2018a, 2018b; Wood et al., 2019a, 2019b; Wood, Lehdonvirta and Graham, 2018). The platforms that we looked at were global in nature, and in this response we focus on "non-geographically sticky work" (i.e., "crowdwork" or work that can, in theory, be done from anywhere).

2. Introduction

Any discussion of platform minimum wages is worth foregrounding with a few key points. First, it is clear that pay rates are not the most important issue relating to the quality of platform work. In fact, pay rates were often significantly higher than what was available locally and were often considered to constitute decent pay. More important issues to emerge from our qualitative interviews and supported by our survey research were the limited social contact which workers experienced, that they often worked long or irregular unsocial hours at intense speeds, that many felt they had little security and some had low incomes. Nevertheless, the downward pressure on pay rates created by the individualized and competitive design of online labor platforms contributed to these outcomes. However, they were also due to an oversupply of workers relative to clients, meaning that there were inadequate earning opportunities to meet the needs of all workers and this in turn generally weakened the bargaining position of workers. Therefore, while implementing minimum wages on online labor platforms might alleviate some of these problems by increasing pay rates at the bottom, doing so might also exacerbate these problems by reducing the supply of clients (by making the platforms less attractive) while increasing the supply of workers (by making the work more attractive). Thus, any intervention to increase a platform's pay rates would require increases in the quality of the services provided in order not to reduce demand and exacerbate the weak position of labor. However, in the long run the elimination of low-productivity jobs which are unable to sustain a living wage is not necessarily bad thing. As minimum wages can force employers to invest in automation and new working methods which increase productivity and thus create new jobs which have the potential to provide decent wages (Kaufman, 2010).

Second, our empirical research highlights how the competition on many online labor platforms is international. What is more, we find that many workers perceive themselves as threatened with replacement by workers in other countries who are able to work for less due to the lower cost of living in that country. This international aspect is a key consideration in thinking about minimum wages, as any intervention is likely to unevenly affect workers living in diverse contexts. For example, a minimum wage set at North American or Western European levels would erode the comparative advantage of workers in lower income countries. This is not to suggest a race to the bottom in wages, but rather a need to make sure that minimum wages do not become an overly protectionist measure at the cost of workers in the Global South.

Third, our research has detailed that some platforms have implemented global minimum wages – mainly as an attempt to ensure quality by pricing out low-quality workers. However, a major issue with these minimum wages is that they relate only to hourly paid work when much of the work is paid on the basis of a fixed price per project. This means that the effective wage can be below the minimum hourly rate.

3. Discussion: labor market principles for online labor platforms

There is currently insufficient empirical data to fully evaluate the likely labor market consequences of online labor platform minimum wages. Instead we suggest some general labor market principles which we believe should be applied to online labor platforms.

First, all work that is done happens somewhere. Therefore, paid work undertaken through online labor platforms should fall under at least one set of national jurisdictions. There are few countries on the planet that do not have some form of regulated labor standards and minimum wage regulations. Therefore, online labor platforms must not exist as mechanisms for the avoidance of labor regulations. Just because a digital platform is used to connect a client with a worker, does not mean that the underlying economic and regulatory geography of that work should be ignored (Graham and Anwar, 2018a; Wood et al., 2019a).

We should, as a starting point, adopt the principle that we do not need to reinvent the wheel. Online labor platforms should ensure that the relevant labor laws – including the classification of workers – are being followed. This is not an unusual expectation and it is widely accepted that conventional labor market intermediaries, such as employment agencies and labor brokers, have this responsibility.

When considering this issue it is useful to draw upon the discourse surrounding what is known as "tax dodging". Both tax evasion and tax avoidance are forms of tax dodging. While only tax evasion is illegal, as only these activities break the letter of the law, both evasion and avoidance are generally seen as harmful and immoral. We argue, therefore, that what matters, when thinking about labor regulation avoidance is the spirit of the law, not the letter of law.

Online labor platforms not only have a responsibility to ensure that the letter of the law is being followed but also the spirit of those laws. This is especially important regarding employment classification as minimum wages often only apply to those classified as "workers" or "employees". In the spirit of the law, "self-employed contractors" are widely understood as being equal parties to those with whom they are entering into contracts with and thus do not require minimum wages. Conversely, "employees" are regarded as being the more vulnerable party in the relationship and in need of special protections such as minimum wages. However, in the contemporary labor market, many independent contractors are best understood as "self-employed workers" as they are in a vulnerable position due to dependence on their clients and therefore in need of protections. Therefore, the spirit of these laws dictates that self-employed workers i.e., the vulnerable self-employed should be entitled to minimum wages as well as other protections outlined in relevant labor laws.

Importantly online labor platforms tend to be based upon a business model which is premised upon creating dependency. For example, there is evidence from the local gig economy that it is impossible for Uber to make sustainable profits in a competitive marketplace (Horan, 2017). Platforms usually earn income from each transaction which takes place between workers and clients. Therefore, the success of the platform rests upon keeping the worker and client using the platform, however, workers and clients tend to develop trust and confidence which can enable repeat business to bypass the platform. In order to curb this behavior, platforms utilize a number of mechanisms, which actively seek to create worker dependency. This is not to say workers do not take their work outside the platforms but to be successful the platforms must seek to limit their ability to do so. Most platforms include exclusivity clauses in their terms of services which can hinder workers and clients doing business outside of the platforms. The control and ownership of data also acts to lock users into a platform in an attempt to prevent them from taking their platform profiles and reputations with them to another platform (see Rosenblat and Stark (2016) and Shapiro (2017) for similar findings regarding the local gig economy). Finally, platforms have monopolistic tendencies due the benefits of "network effects". A network effect is a phenomenon whereby each additional user increases the value of the platform for all users. The network effect can make it difficult for new platforms to compete with established ones, as a new platform is of little value unless everyone switches platform at the same time. However, the online gig economy seems to be oblivious to, or ignore, the problems of platform dependency and the fact that as a result labor regulations should apply to workers. An employer based in Germany who sources work from a worker based in Kenya (via a platform based in the US) rarely has any knowledge of Kenyan labor law and nor do the platforms suggest that they should.

4. Concluding analysis and future recommendations

It is also important to note that many countries' minimum wage regulations include piece work. Under these laws employers are usually required to calculate a minimum piece rate which is not less than the hourly minimum. In some countries, such as the United Kingdom, the law also ensures that a "fair" minimum piece rate is one which is achievable by workers who are less skilled or more fatigued than the average worker (Gov.uk, 2018). Platforms should use the wealth of data they collect on work tasks to calculate piece rates. These rates should be cross-checked, verified and regulated by state bodies. However, there will be some situations where the time taken to complete an average task undertaken by an average worker will be too variable for the platform to accurately or meaningfully calculate. For example, the time taken to complete some programming tasks may vary significantly due to the specific problem and whether the worker has encountered something similar before. Therefore, where average productivity cannot be adequately measured or meaningfully calculated a piece rate payment method must be recognized as unsuitable and platform workers should instead be paid on an hourly basis.

In addition, EU labor law includes a posted worker directive which stipulates that "posted workers" (someone sent by their employer to carry out a service in another EU member state on a temporary basis) should be remunerated in accordance with host countries' laws and practices. Online labor platforms enable labor to be sent digitally to the client's country and therefore the posted worker directive should apply to EU remote gig workers. This is an approach which could be adopted more widely and updated to recognize the fact that while the work is being undertaken in the client's location via the internet the costs of reproducing labor will be dependent on the worker's physical location. Therefore, minimum wages should be adjusted by purchasing power parity, perhaps with platforms adjusting their minimum rates every year (this could be perhaps verified by an independent body like the Fairwork Foundation). A benefit of doing so would be that it avoids unfairly disadvantaging workers in countries with lower labor costs.

Second, (and perhaps somewhat paradoxically), platforms should get rid of their global minimum wages. Global minimums send a message to clients that if they pay above the minimum then they are in compliance with relevant local regulations. However, it is entirely possible for workers to earn above platform minimum wages, but below their client's national/local minimum wages

Third, we acknowledge that there might be claims that any attempts to enforce minimum wages could be unenforceable given the global and dispersed geographies and networks of online work. However, our research shows that the vast majority of demand for digital work comes from just five countries. Furthermore, a small handful of platforms mediate the vast majority of that work. These two facts demonstrate that initial barriers to regulation are not due to a dispersed geography or dispersed network of work. These topological and geographical bottlenecks in the global trade of digital work offer potential sites in which regulation can be enforced (we realize that many of the other submission to this call deal with some of the specifics of "how to do regulation" and we therefore leave the details of that discussion to others).

We hope that some of these suggestions can help to bring about a fairer set of relationships between the employing class, the governing class and the working class. Online gig work has brought income and jobs to many, but that does not mean that we should expect it to function as an unregulated labor market.

References

Anwar, M.A. and Graham, M., 2017. Digital gig work in Africa: Freedom, flexibility, precarity and vulnerability among workers. In: *Reshaping Work in the Platform Economy Conference, October 19–20, 2017*. Amsterdam, Netherlands.

Anwar, M.A. and Graham, M., 2018. Platform labour at global margins: Agency and autonomy of workers in the global gig economy. In: *International Labour Process Conference, March 21–23, 2018*. Buenos Aires, Argentina.

Gov.uk, 2018. *Minimum Wage for Different Types of Work: Paid Per Task or Piece of Work Done*. [online] Available at: <www.gov.uk/minimum-wage-different-types-work/paid-per-task-or-piece-of-work-done> [Accessed 17 Sep. 2019].

Graham, M. and Anwar, M.A., 2018a. Labour. In: J. Ash, R. Kitchin and A. Leszczynski, eds., *Digital Geographies*. London, UK: Sage Publications, pp. 177–187.

Graham, M. and Anwar, M.A., 2018b. Two models for a fairer sharing economy. In: N.M. Davidson, M. Finck and J.J. Infranca, eds., *The Cambridge Handbook of the Law of the Sharing Economy*. Cambridge, UK: Cambridge University Press, pp. 328–340.

Graham, M., Hjorth, I. and Lehdonvirta, V., 2017. Digital labour and development: Impacts of global digital labour platforms and the gig economy on worker livelihoods. *Transfer: European Review of Labour and Research*, 23(2), pp. 135–162.

Graham, M., Lehdonvirta, V., Wood, A.J., Barnard, H., Hjorth, I. and Simon, D.P., 2017. *The Risks and Rewards of Online Gig Work At the Global Margins*. Oxford, UK: Oxford Internet Institute.

Horan, H., 2017. Will the growth of Uber increase economic welfare? *Transportation Law Journal*, 44(1), pp. 33–105.

Kaufman, B.E., 2010. Institutional economics and the minimum wage: Broadening the theoretical and policy debate. *ILR Review*, 63(3), pp. 427–453.

Kuek, S.C., Paradi-Guilford, C.M., Fayomi, T., Imaizumi, S. and Ipeirotis, P., 2015. *The Global Opportunity in Online Outsourcing*. Washington, DC: World Bank Group.

Rosenblat, A. and Stark, L., 2016. Algorithmic labor and information asymmetries: A case study of Uber's drivers. *International Journal of Communication*, 10, pp. 3758–3784.

Shapiro, A., 2017. Between autonomy and control: Strategies of arbitrage in the "on-demand" economy. *New Media & Society*, 20(8), pp. 2954–2971.

Wood, A.J., Graham, M., Lehdonvirta, V. and Hjorth, I., 2019a. Networked but Commodified: The (Dis)Embeddedness of Digital Labour in the Gig Economy. *Sociology*, 53(5), pp.931–950.

Wood, A.J., Graham, M., Lehdonvirta, V. and Hjorth, I., 2019b. Good gig, bad gig: Autonomy and algorithmic control in the global gig economy. *Work, Employment and Society*, 33(1), pp. 56–75.

Wood, A.J., Lehdonvirta, V. and Graham, M., 2018. Workers of the Internet unite? Online freelancer organisation in six Asian and African countries. *New Technology, Work and Employment*, 33(2), pp. 95–112.

5 The digital disruption of science

Governments and scientists toward an "Open Science"

Antoine Maire

1. Introduction: Open Science, a new era for science?

The digitalization is a key element in the evolution of public policies and governments' activities. It is an important tool to foster efficiency and a better consideration for users' needs. It is an opportunity to foster the creation of "digital governments" (OECD, 2016), characterized by a greater implication of citizens and businesses in the agenda-setting, the elaboration but also in the making of public welfare policies. It is an opportunity to foster the creation of "digital governments" (OECD, 2016), characterized by a greater implication of citizens and businesses in the agenda-setting, the elaboration but also in the making of public welfare policies. Science is at the forefront of this evolution, being recognized as a strategic issue in new knowledge-based economies.

In science, digitalization comes from the availability of new technologies and tools that enable the emergence of innovative research practices. It does not only concern the use of digital technologies in the daily work of scientists: It also provokes an important evolution in the very activity of research from agenda-setting to the publication of results. The disruption provoked by digitalization in science is often summed up with the buzzwords "science 2.0", "science in transition", "digital science", "e-Science" or more generally with the concept of "Open Science" (Millerand, 2015). If nuances exist between those concepts, they all stress the idea of a transition from one type of research to another, based on collaborative work, transparency and efficiency allowed by the digitalization of scientific activities. The concept of "Open Science" will be used in this chapter, as it is the one used by governments and European institutions when they refer to this evolution (European Commission, 2015).

1.1. Digitalization and science: what is at stake?

The digitalization refers to the use of tools coming from the web 2.0 in research activities (Teif, 2013). It is sometimes referred as the basis of a second revolution in science (Bartling and Friesike, 2014). The first revolution was based on the professionalization of research activities, and on the continuous build-up of new knowledge based on the work of other scientists. The second revolution would be based on digital technologies. It would lead to a more collaborative science, a new mode of creation of scientific knowledge; complex, non-linear, heterogeneous and transdisciplinary (Szkuta and Osimo, 2016). Thus, digitalization is commonly seen as a way to go toward a "better science" (Millerand, 2015), a potential solution to the challenges that modern science is facing, that is an increasing number of authors, of publications and of data available (Burgelman, Osimo and Bogdanowicz, 2010). They are creating new opportunities to develop more complex and more collaborative approaches, where interdisciplinary is crucial, to solve wicked problems (Bly and Ginanni, 2012).

The evolution generated by digitalization can be summed up with five main characteristics (Fecher and Friesike, 2014):

- The democratization of science: It stresses the ability of digitalization to facilitate the diffusion of the scientific results across all spheres of society, from governments to NGOs, from big businesses to small and medium enterprises. This expectation is closely associated with the "Open Access movement";
- An improvement of scientific processes: the way research is done and on how digitalization will strengthen its efficiency. It emphasizes the increasing collaboration between scientists that this movement will promote;
- The development of innovative infrastructure: The new tools available in this digital age will enable scientists to deal with an increasing amount of data and to develop a data-based culture;
- The diversification of the actors involved in the making of science: Citizens and civil society's organizations can now be involved from the funding of research, through crowdfunding for example, to the processing and analysis of data. It is sometimes summed up with the nickname "citizen science";
- The emergence of new measurement tools, new type of impact measures, sometimes called "alt-metrics". It will allow to better understand the influence and impact of science over the rest of the society.

To sum up, the disruptive effect of digitalization in science is seen as a major opportunity to foster the emergence of a new mode of production of scientific knowledge, called "Open Science". It is a tool to promote a better science and to increase its efficiency and impact.

1.2. Digitalization of science from a governmental perspective

For government, the move toward Open Science is an opportunity to foster the efficiency of public investments in science and to further increase the competitiveness of their economy (OECD, 2016). Three stakes explain governments' interest in Open Science and why they are framing new policies to foster this movement. For government, the move toward Open Science is an opportunity to foster the efficiency of public investments in science and to further increase the competitiveness of their economy (OECD, 2016). Three stakes explain governments' interest in Open Science and why they are framing new policies to foster this movement.

First, digitalization is a tool to limit the differences between heavily-funded research centers and small research centers and the polarization of the scientific field (Breivik, Hovland and From, 2010). It should allow small research centers to be more competitive, with fewer investments required, in particular if such centers do not have to devote an increasing amount of money to pay for subscriptions to research journals.

Next, digitalization is also a tool to strengthen the efficiency of national scientific communities, and to take more benefits of a huge public investment (Buhr, 2014). It asks the question of the competitiveness of a scientific community in an increasingly competitive scientific world. Moving toward Open Science is a solution to this challenge as it should increase the efficiency of the research process.

The final stake for governments is that digitalization is seen as a tool to close the gap between science, businesses and other actors, strengthening the impact of science on society (Hetland, 2011). Behind this preoccupation lies the question of how to improve the efficiency of an economy. Digitalization of science is a tool to use, as it allows scientific results to be more quickly and easily spread among businesses and among the rest of the society.

1.3. Barriers in moving toward digitalization of science

If this portrait emphasizes the advantages and hopes that digitalization is creating in science, some difficulties remain on the way. One of the characteristics of the Open Science movement is the fact that it implies the evolution of how scientists are doing research. Such evolution is already noticed with the increasing number of publications being open access or with the growing importance of research social networks (Millerand, 2015). The ability of researchers to use those new tools and the evolution of their practices are one way to move toward Open Science. It could be described as a bottom-up phenomenon, researchers experimenting directly new tools, new technic and new collaboration strategies in their daily activity (Vignoli, Kraker and Sevault, 2015). However, there are also barriers in moving toward an Open Science.

Surveys on this question show that researchers have, for an overall majority, a positive image of Open Science. When asked (Schöpfel et al., 2016), researchers mentioned several potential changes that digitalization should foster: a broader diffusion of results and publications, more collaboration, the breaking-down of discipline barriers, the ability to involve non-usual actors in the research process, but also an opportunity to better meet societal needs and public demands. Nevertheless, this positive opinion does not necessarily translate into practices. In 2015, the European Commission has commissioned a survey to better understand researchers' perceptions about what was called at the time "Science 2.0" (European Commission, 2015). When asked about the barriers in moving toward Open Science, researchers mention a limited awareness about the benefits of such a move, a lack of incentives given to Open Science in their career, a lack of infrastructure and of financial supports. All those elements, combined with a cultural resistance to change, show that a bottom-up approach is not sufficient to move toward Open Science. It shows that this evolution also needs to be encompassed and furthered by a top-down approach, in other words, by public policies (Vignoli, Kraker and Sevault, 2015; European Commission, 2015; Schöpfel et al., 2016).

The debate about the digitalization of science usually takes place within the broader debate of the digitalization of government activities and public policies, creating a discussion between researchers, experts, policymakers and citizens. However, most scientific work done on the subject is devoted to Open Science as a concept and not as a new object of public policies. Due to that, this chapter will study the evolution of emerging public policies that must foster digitalization in science and a move toward Open Science.

2. Methods

2.1. Theoretical framework

From a theoretical perspective, the problem studied is specific because the inclusion of Open Science in public policies does not come from the need to solve a specific problem. It rather comes from the perception that an opportunity should be taken. Thus, the question here is not how to solve a problem but rather how to promote a specific approach on a relatively new and unknown subject. Within this perspective, this chapter will be based on the "advocacy coalition framework" (ACF) (Sabatier and Weible, 2007). This theoretical approach assumes that policy making is so complex nowadays that public policies are elaborated in subsystems made up of experts, legislators, bureaucrats, lobbyists, researchers, experts, journalists and activists. Within such subsystems, it is argued that actors are joining and progressively establishing a "coalition of causes" to influence policy making according to their own objectives and beliefs. To understand a given policy, the ACF framework argues that it is necessary to identify a "coalition of causes" within the subsystem of public policy making. Compared to other approaches, the ACF framework has the advantage of not having to focus on whether or not to make a particular decision, but rather on issues requiring long-term policies.

2.2. Survey methodology

To apply this approach on the making of public policies toward Open Science, a comparative approach will be used with two case studies: the European policy and the French policy toward Open Science. Those two cases should lead to a better understanding of the disruptive effect of digitalization on research activities. In both cases, the ACF framework will be applied. It will lead to the identification of action subsystems of the public policy toward Open Science. It will also identify core beliefs of involved actors, to better understand why they are pushing in favor of such policies and what causes they are supporting. Finally, it will identify coordination mechanisms established within those coalitions to progressively go toward a kind of institutionalization of such gatherings.

2.3. Materials and data

The case studies will be analyzed through primary sources, such as reports, decisions and presentations made by the actors involved in this process. The documents allow for a proposal that seeks to outline various policies aimed toward promoting Open Science. However, they do not say much about the debates that such policies have faced. To know more about them, a third step in this research has used semi-directive interviews with key actors involved in the implementation of such policies.

3. Open Science in Europe and in France, assessment of two case studies

3.1. The European impetus for "Open Science"

3.1.1. The elaboration of a European policy toward Open Science

The European Union (EU) has been among the first institutions to push for Open Science. Its ambition was to strengthen the competitiveness of European research and more broadly the competitiveness of the European economy. This emphasis is relatively recent and the timeline of European actions shows an increasing globalization of European institutions' policies. A sequential approach allows the identification of four main kinds of actions. They are related, in a chronological order, to Open Access, Open Data, the development of new infrastructure and next to Open Science, an encompassing concept recently adopted by EU institutions.

As for Open Access, the EU has played a key role in fostering Open Access, seen as a strategic stake in the establishment of a European Research Area (ERA). The EU has followed a gradual approach. In 2006, the European Research Council (ERC) published a statement about the importance of Open Access for all publications receiving public funds (European Research Council – Scientific Council, 2016). This first move led to the creation of a Pilot project for Open Access in the frame of the implementation of the seventh framework program for science (European Commission, 2007). This pilot project has defined two ways to develop Open Access publishing. The "golden way" provides a reimbursement of the publications costs by European funds. The "green way" allows a publication in open archives for 6 or 12 months (for social sciences) after the first publication. This policy echoes the two strategies already mentioned for Open Access in the Budapest declaration of 2002 (Chan et al., 2002). This orientation was later confirmed by the European Commission in July 2012 with a recommendation on access and preservation of scientific information (European Commission, 2012b), updated in April 2018 to give more concrete orientations to Member States (European Commission, 2018f), and a communication entitled, "Towards better access to scientific information: boosting the benefits of public investment in research" (European Commission, 2012a). This commitment in favor of Open Access has been confirmed and strengthened in the FP8 – H2020 program (European Commission, 2014a). Since 2014, all scientific publications receiving European funding should be made available through Open Access, following either the "golden" or the "green way". Ultimately, the objective is to achieve 100% of Open Access publications in Europe from 2020.

Regarding data, following the adoption of the Reda report by the European Parliament in 2015 (European Parliament, 2015), a revision of the directives related to copyrights is under review. Its objectives are to ease the move toward Open Access and to allow the development of innovative research methods such as text and data mining. The process followed is similar to the one used for Open Access. The recommendation published by the European Commission in 2012 has encouraged Member States to adopt Open Data policies (European Commission, 2012b). The debate and policies cover both the access to research data, as well as access to public data. Thus, the FP8 – H2020 has done research data open by default for all project implemented within the program. It has also imposed the adoption of a data management plan. In the data management, the European Commission is pushing for a "FAIR" approach, meaning that data should be Findable, Accessible, Interoperable and Reusable (European Commission, 2016b).

The Open Access and Open Data objectives have led the European Commission to develop innovative infrastructure needed to support both moves. The main project is the European Open Science Cloud (EOSC). Its ambition is to build an infrastructure that would enable European researchers to stock and to exploit their data in a common infrastructure (EOSC Summit, 2017). It will also connect existing scientific data infrastructure built by Member States. A pilot has been launched in 2018 (European Commission, 2018b) and the objective is to implement the project from 2020. The other infrastructure project developed by the European Commission, called OpenAIRE for Open Access Infrastructure for Research in Europe (European Commission, 2018e), is dedicated to Open Access and to foster interactions between researchers.

The question of Open Access, Open Data and the need of affiliated infrastructure has led to the birth of a debate about the consequences of digitalization for scientific activities. It is based on the assessment that digital technologies are not only making access to research and data easier but rather that they radically change the way scientific knowledge is produced. Confronted with this challenge and based on initial reflections about the influence of digitalization in science (European Commission, 2013), the European Commission has led a public consultation on the impact of digitalization on science between July and September 2014 (European Commission, 2014b). The results have been published in February 2015 (European Commission, 2015). They show the preference of researchers for the term "Open Science" rather than "Science 2.0" previously used by European institutions, and a consensus about the benefits of Open Science as a means to foster the impact of science and to promote more collaborative and multidisciplinary projects. It has also identified a lack of awareness of the

European research community about the meaning and implications of this digital evolution.

Based on the results of this consultation, the new European commissioner, Carlos Moedas has made of Open Science one of its three strategic priorities with Open Innovation and Openness to the World (Moedas, 2015). From this new impetus, the European Commission has drafted a European Open Science agenda published in February 2016 by the Directorate General for Research and Innovation (European Commission, 2016a). Five main policy actions are identified: (1) fostering Open Science; (2) removing the barriers to Open Science; (3) developing research infrastructure for Open Science; (4) mainstreaming Open Access to research results; (5) embedding Open Science in society. Those policy actions must be translated into concrete measures to be taken by European institutions and Member States.

One of the main measures is the creation in 2014 of an advisory body in charge of co-designing and co-developing policies implemented to foster Open Science (European Commission, 2018c). Named the Open Science Policy Platform (OSPP), it is placed under the authority of the Directorate General for Research and Innovation. Apart from this assignment, the OSPP is also in charge of the education and training of the European research community to Open Science. The European Commission has thus financed a project under the nickname FOSTER, which stands for "Facilitate Open Science Training for European Researcher" (European Commission, 2018c) in February 2014. It includes a website offering free training courses, toolkits, and an Open Science training book. The ambition of this project is to tackle one of the main difficulties in the evolution toward Open Science, the need of a change in researchers' behavior.

The creation of the OSPP represents an important evolution because it should play a key role in the elaboration of a new European policy toward Open Science. It gives the EU the ability to encompass all the aspects of Open Science. The work already done by the platform is proof. Working groups have been established and have produced reports (European Commission, 2018c) on alt-metrics, citizen science, on how to provide researchers with the skills they need for Open Science, on the evaluation of research careers fully acknowledging Open Science practices or on the awards, incentives and recognition for researchers practicing Open Science. This work comes in parallel with the ignition of the process leading to the elaboration of the next Framework Program, FP9, which will follow H2020. The advance report already published under the leadership of Pascal Lamy strongly emphasized the need to go further on Open Science. It argues that Europe now needs to "*embrace the transformative power of Open Science*" (Lamy, 2017, p. 8).

If actions were taken by the Commission, it was also the case at the intergovernmental level. In particular, Open Science has been promoted under the Dutch presidency of the European Union Council in the first half of 2016. The Dutch government, jointly with the European Commission, organized an important international conference on Open Science in Amsterdam in April 2016 entitled "Open Science, from vision to action." It was based on the assessment that the transition toward an Open Science needs to be accompanied by governmental actions. It has led to the adoption of the Amsterdam Call for Action on Open Science (Ministry of Education, Culture and Science, 2016) that identified five main areas of actions: (1) removing the barriers to Open Science; (2) developing research infrastructure; (3) fostering and creating incentives for Open Science; (4) mainstreaming and further promoting Open Science policies; and (5) stimulating and embedding Open Science in science and society. This call was later used to push for the adoption of a common position between the Member States regarding Open Science during the competitiveness council of May 2016 (European Union Council, 2016). Its adoption has not faced any opposition from Members States, all being aware of the importance of the topic. This common position has validated the orientation of the Commission on Open Science. It is also a testimony of the desire of all Member States to work for the emergence of an Open Science.

3.1.2. Analysis of the elaboration of the European policy

The analysis of the emergence of a European policy according to the ACF framework allows for a better understanding of the dynamics and logic behind the increasing role played by European institutions in the field of Open Science. The subsystem is the one in which the science policy of the European Union is framed. The central actor within this move is the European Commission. Compared to other subjects, the Commission has been a frontrunner on the topic of Open Access and Open Data, but also regarding the financing of the infrastructure needed to support the move. The action of the Commission precedes the ones implemented by Member States, even the most advanced ones such as the Netherlands and Germany. The Commission can be considered a frontrunner in this evolution because its commitment was a deliberate one. It comes first from the attention given to Open Access and Open Data and later to the formalization of a more comprehensive policy toward Open Science, which followed the survey commissioned in 2014. It was implemented under the orientation of the commissioner and the responsibility of this policy is divided between two directorates: A and B. The first directorate is in charge of the policy development and coordination and the second one of Open Innovation and Open Science.

The European Commission is convinced of the necessity to deal with digitalization as it changes the whole life cycle of the research project. This conviction is supported by the results of the 2014 public consultation that have shown a strong appetence of the research community for Open Science. Thus, one of the ambitions of the European Commission is to give a voice to the research community and to progressively translate its aspiration into policy actions. This should lead Member States to take into account this perspective and this new orientation. Besides this aspiration, the core beliefs of the initiators of such policies, in particular of European commissioners, is to strengthen the efficiency of the European investment in public research. Therefore, the Commission considers that moving toward Open Science is critical to maintain a dominant position for Europe in the field of Science. Moreover, the scientific policy of the European Union is often criticized due to its lack of ambition, in part because it is difficult for the EU to bypass the specificity and isolation of each Member State's scientific policies, many of them being reluctant to give up parts of their sovereignty in this strategic field. The adoption of a European policy for Open Science allows the Commission to go beyond this compartmentalization. It can promote an encompassing concept that could foster the efficiency of scientific research and that offer the occasion to promote the connectivity and the interoperability of Member States' scientific policies. The mutual learning exercise organized by the Commission in 2017 in an example of this ambition (Miedema et al., 2018).

This ambition is also related to one of the most important objectives of the European Commission: the creation of a common European Research Area (ERA), a key objective of the former Lisbon Strategy (European Council, 2000). The objective is for the ERA to do for scientific research what the Common Market did for economic exchanges. In this perspective, Open Science is a tool to further the integration of national research initiatives. This objective clearly appears in the directorate general's organizational chart as the unit in charge of Open Science is also in charge of the ERA (European Commission, 2018a). The European Open Science Cloud is another symbol of how Open Science is used by the Commission to promote a unified European Research Area.

Within this perspective, the European Commission appears to be at the center of the advocacy coalition pushing toward a more active policy for Open Science. The creation of a European Open Science Platform strengthens the coordination within the coalition and is a tool to gather and exchange for supportive actors. The European Open Science Platform actually recommends the creation of similar administrative structures in each Member States to enable them to develop a comprehensive policy on the subject (Open Science Policy Platform, 2018d). The key element for this coalition is to convince the Member States to support this orientation and more importantly to translate it at a national level. The common position adopted by the Member States regarding Open Science during the competitiveness council of May 2016 is thus an important milestone. It has given legitimacy to the European Commission to go further on this topic. It has also marked the commitment of Member States to engage in this endeavor.

3.2. Open Science in the French context

3.2.1. The elaboration of a French policy toward Open Science

The study of the French case shows several differences with the European case studied before. The appropriation of Open Science, as a concept, is relatively recent. Because of that, key documents regarding the orientation of French science policy forgot to mention the concept as an objective or a reference. For example, the French national strategy for research, published in 2015, never mentioned Open Science (Ministry of National Education Teaching and Research, 2015). More recently, the French white paper for research and higher education mentions the concept of Open Science but only on a vague perspective (Ministry of National Education Teaching and Research, 2017), in a sidebar saying that some debates exist at the European and national level about Open Science. It also mentions that the French ministry for higher education and research is working on new International Strategy for Research, Innovation and Higher Education. Its purpose is to provide strategic guidance for French stakeholders, ensuring that France establishes relevant and efficient partnerships and endowing France with a strong decision-making capacity in international bodies (Ministry of National Education Teaching and Research, 2017). This new strategy is supposed to take into account the European dimension of research and in particular the European impetus regarding Open Science. This approach tends to resume Open Science to an EU initiative, an element that the French international science policy should take into account, and not to a bottom-up phenomenon associated with the disruption of scientific activities created by the digitalization. Nonetheless, if Open Science is not directly mentioned, the question of the digitalization of scientific and teaching activities is a major concern in all those texts. The white paper insists on the disruptive effect of digitalization and on the need for research to better understand the consequences of such evolution on scientific activities. The concept of Open Science is also supposed to appear more clearly in the next French national strategy for research.

Apart from this assessment, the formalization of the debate has followed in France approximately the same process as at the European level. It has started with Open Access, moving to Open Data before leading to the progressive elaboration of an encompassing policy toward Open Science. The necessity of Open Access has been pushed in France by activists since the beginning of the 2000s. They have built upon the global movement for Open Science with the three declarations of Budapest in 2002 (Chan et al., 2002) and Bethesda (Brown et al., 2003) and Berlin in 2003 (Chan et al., 2003). Local initiatives have been implemented through COUPERIN, Open Edition or through the establishment of an archive which allows researchers to safely store their works, *Hyper Articles en Ligne* [Eng: Hyper Article Online] (HAL). Those initiatives show the progressive emergence and structuration of a coalition of actors pushing in favor of Open Access. For them, science is a global common to which everyone should have access (Chartron, 2016).

As for Open Data, the French initiative has also to be understood within a broader context, international and European. The Sebastopol meeting held in California in 2007 has affirmed the principles related to Open Data (Malamud et al., 2007). In France, the implementation of an open data policy has followed an initiative pushed by the new French President, François Hollande, for the opening of administrative data. It has led to the reform of "Etalab". Initially designed to build a public portal of administrative data (Government Bill, 2011), it is now described as an "administrative start-up" in charge of supervising the opening of public data. Besides this mission, Etatab is also in charge of fostering scientific practices based on a culture of data. It has also pushed for the French adhesion to the Open Government Partnership in 2014, later used to elaborate a first French policy toward Open Science.

The different aspects of Open Science have been highly debated in France on the occasion of the adoption of the new law, "Towards a Digital Republic", in 2015–2016 (Government Bill, 2016). On Open Access, the law allows researchers to make their articles freely accessible 6 months after the publication and 12 months for humanities and social sciences, following the European recommendations. On Open Data, the law is also making public data open by default. It is opening the way for text and data mining, even though the application decree still needs to be adopted. The debate surrounding this application decree is symptomatic of the controversy that opposed the research sphere and the publishing sphere. Publishers are usual opponents of Open Access, arguing that it threatens their economic model and the viability of their industry. In the French context, this debate is important because it indirectly touches the "cultural exception" that France is claiming. Transposed to the European debate surrounding copyright, it explains the difficulties faced by the Commission to change the directives related to copyright to allow the development of text- and data-mining practices.

Moreover, the innovative and collaborative method used to shape the law has allowed many actors, individual and institutional to publish recommendations in the frame of the elaboration of the law. For example, the French National Council for Scientific Research (CNRS) has published a white paper (CNRS, 2016) and strategic guide (CNRS, 2017) arguing in favor of moving toward Open Science in the broader context of the construction of the so-called new "digital republic". This debate is structuring because it has allowed the progressive acceptance and recognition of Open Science as a concept in France.[1]

This growing interest for Open Science has later led to the definition of a French embryonic strategy for Open Science. Practical decisions have been made, such as the creation of a special position within the Ministry of Higher Education, Research and Innovation dedicated to Open Science. It has also led to the adoption of a full engagement on Open Science in the frame of the Open Government Partnership (State Secretariat in charge of Digital Affairs, 2018). This international initiative was supposed to focus on the opening and the transparency of governmental activities. However, France has pushed to include an engagement bearing the objective of building an "Open Science ecosystem" in France. This engagement, "number 18", can be considered the first French embryonic strategy toward Open Science. It has been translated into a national plan for Open Science in July 2018 (Ministry of Higher Education, Research and Innovation, 2018b).

Apart from further investments in existing infrastructure such as the archive HAL, the ScanR search engine, or the Isidore platform, the main evolution proposed by this strategy is the creation of a special committee dedicated to Open Science (Ministry of Higher Education, Research and Innovation, 2018a). Under the leadership of the Directorate General to research and innovation of the Ministry of Higher Education, Research and Innovation, it will gather experts who will recommend policy orientations and concrete actions to foster Open Science. Its main mission will be to define a French policy for Open Science and to oversee its implementation. It will also play a role in the diffusion of knowledge related to Open Science in the scientific communities and in the gathering of potential remarks and ideas in a bottom-up approach.

3.2.2. Analysis of the French policy toward Open Science

The application of the ACF framework allows for better understanding of the progressive elaboration of a French policy toward Open Science. The sequential approach that led to the formulation of a French policy toward Open Science is similar to the phenomenon observed at the European level. It started with Open Access and Open Data, with a support to needed infrastructure and it has later ended up with the formalization of a comprehensive policy toward Open Science.

Compared to the European case, the French subsystem of action appears to be mainly dominated by actors involved in the technical aspects of digitalization. Two of the milestones in the elaboration of a French policy toward Open Science have been promoted directly and indirectly by the Secretariat of State in charge of Digital Affairs: the adoption of the law, "Towards a Digital Republic", and the French commitment to the Open Government Partnership. It is also true for the National Center for Scientific Research, since the white paper on Open Science published in 2016 was produced by the directorate for scientific and technical information (CNRS, 2016). On the contrary, the Ministry of Higher Education, Research and Innovation plays a secondary role in this evolution as the absence of Open Science in strategic documents is a testimony. The creation of an advisory position in charge of Open Science within the ministry is, however, changing the dynamic and should allow the ministry to better take into consideration Open Science in the future.

As for the advocacy coalition framework, the French case shows an interesting convergence of two different advocacy coalitions. The first one structured itself around the promotion of Open Access to scientific publications. It is mainly composed of activists who believed that science is a common good for which no one should have to pay. Their core beliefs were a bit different than the one previously mentioned for the European case. Access to science results is seen as a principle when it was mainly seen by European institutions as a tool to reinforce the efficiency of the money invested in research. The main challenge this coalition has to face is an advocacy coalition organized around publishers. The recommendations published in 2012 by the European Commission have thus played a key role in strengthening the position of the Open Access advocacy coalition within the French system. It has demonstrated that Open Access is not only a utopian concept but also a matter of economic efficiency and scientific competitiveness.

The second advocacy coalition which merged in this trend toward Open Science is related to the opening of data. It was mainly structured around entrepreneurs. They have pushed in favor of the opening of public data arguing that the private sector has now access to enough calculus power to offer innovative services to the public. Its theoretical basis is "Government as a platform" described by O'Reilly (O'Reilly, 2011) and translated in the French context by Colin and Verdier (Colin and Verdier, 2015). In this perspective, opening public data should allow citizens and businesses to do more with those data and to develop innovative services, science not being an exception. The coalition has strongly pushed in favor of the opening of public data and has succeeded in making the "Government as a platform" a new objective for the French government (State Secretariat in charge of the Modernization of Public Action, 2018).

The collaboration between those two advocacy coalitions has played a crucial role in the elaboration of a new French policy toward Open Science. It was the symbol, in a way, of the alliance between a scientific and technological perspective. The modernization of government has offered an occasion to those two coalitions to meet and collaborate through inter-ministerial consultations to frame those new policies. The law, "Towards a Digital Republic", and the commitment in the Open Government Partnership are good examples of such consultations. The success of these advocacy coalitions is linked to the ability of some of their members to access public positions, enabling them to act directly within the administrative structure. For the Open Data coalition, Henri Verdier has been appointed in 2013 director of Etalab and later inter-ministerial director in charge of digital, information and communication systems as well as administrator general in charge of data. For the Open Access, Marin Dacos has been appointed in 2017 as an advisor in charge of Open Science in the French Ministry of Higher Education, Research and Innovation.

However, despite the convergence between those two advocacy coalitions, the emergence of a comprehensive coalition arguing in favor of Open Science is still in the making. The call of Jussieu signed by many actors of the sector in October 2017 is a good illustration, both of the composition of this advocacy coalition and of the central role it played within the French debate. Entitled "Jussieu Call for Open Science and Bibliodiversity" (Bauin et al., 2017), it illustrates the central role played by the Open Access movement in the elaboration and structuration of a new advocacy coalition in favor of Open Science. The creation of the future Committee for Open Science should also play a key role in the coordination of this coalition of actors to push for Open Science. It presents a lot of similarities with the European Open Science Platform previously mentioned.

4. Discussion of the results

This study shows that the digitalization of science is still in an emerging phase. The process that led to the formulation of new policies toward Open Science is reflecting a convergence between different dynamics, Open Access and Open Data being the most prominent ones. Both the European and the French cases are confirming this tendency, even if it appears that they come from different origins. However, both cases also confirmed that Open Science policies are still fragmented and that a comprehensive policy in charge of fostering the development of an Open Science is still in the making. The adoption of the "Open Science" concept is thus playing a key role that should allow the elaboration of a more comprehensive policy. The creation of governing bodies is also a key element in this process. Those new structures, specifically dedicated to Open Science, have the ability to deal with all the aspects created by the digitalization in science. The transition toward an Open Science has such a great impact that it is difficult for a traditional administration to deal with all those elements and to develop a comprehensive policy on this subject. At the EU level, those new structures ease the cooperation between European institutions and Members States with the emergence of a single administration.

The study of both cases, however, confirms the difficulties of going toward a real digitalization in scientific activities and not only toward a simple digitization, that is to say a digitization of a paper-based system. This evolution is complex because, to be fully achieved, it needs to come not only from the top-down but to adopt a bottom-up approach. A more comprehensive approach would be necessary to better understand the challenges and difficulties that such a bottom-up approach must face.

The analysis of the elaboration of the public policies to favor the digitalization of scientific activities also shows that it paradoxically comes across two movements. Frontrunners have already created and implemented a lot of local initiatives that directly improve their daily scientific activities and enable them to develop innovative research approaches. The case of the elaboration of HAL, the French scientific archive platform, is an example of a local initiative, later taken over by the government to develop a more global policy. The elaboration of public policies toward those initiatives comes late and tries to encompass them in a more general and coherent framework. This step, a governmental intervention in this transitory period, appears to be a necessity in the development of a coherent scientific community organized around the Open Science concept.

At the same time, it also appears that in the broader scientific community, digital technologies lead to digitization rather than to digitalization. Due to that public intervention regarding the digitalization of science is facing a strong inertia within the scientific communities. The most important challenge regarding a true evolution toward the digitalization of scientific activities is related to the lack of knowledge of scientists about the movement itself, its implication and to the lack of incentives for a change in habits that could be seen positively in a career perspective. Studies about the subject are currently underway and should later play a key role in the revision of the career logic of scientists to foster the development of Open Science habits. The role of government in this movement will also be fundamental.

The assessment that digitalization is just emerging, through a transition from innovative practices implemented by frontrunners toward a much more general diffusion, means that the role of government is still central in its evolution. The digitalization of science cannot come only from a bottom-up approach, i.e., from an inherent and progressive change of researcher behaviors. The transition process is just beginning, but it is characterized by a tendency of governments to transform local initiatives in general behaviors. The Open Science and the Open Data movements are examples of such ambitions. Going further is also requesting a strong implication of governments to ease and soften the conditions of the transition. This accompanying role comes out with the creation of the needed infrastructure (archive, data center, innovative tool) but also from the progressive creation of the conditions needed for the emergence of new research practices. The movement is here just in the making. It will go through an evolution in the career system of researchers, in the evaluation of their works, of the methods used to analyze the impact of their work and through a lot of pedagogy toward this movement. The role of government is central because it has created the legal conditions to foster such evolution. It is also because it can use funding requirements to promote certain habits and behavior, as it was done by the EU regarding Open Data and Open Access for example. It is, however, too early to really identify the consequences of this evolution on the relation between governments and science when this transition process is over. Even though some tendencies can be mentioned such as the horizontalization of the research process, the diversification of research funding possibilities or the increasing importance of multidisciplinary and multi-stakeholder's research, they remain tendencies that will need to be confirmed in the long term.

This study of the elaboration of public policies toward Open Science is also opening new research perspectives. Among them, several could be mentioned:

- Research about the structures of local initiatives implemented by so-called frontrunners to favor the full digitalization of science. Such an approach could start with the cartography of these initiatives. It should also lead to a better understanding of how these initiatives are later taken over by governments;
- The other research perspective that could be pursued should be related to the implementation of the policies previously described: the creation of innovative governing structures, the way they are working and interacting with the research community but also with the policy-making process is a theme of research per se;
- The other aspect that deserves more attention is the connection between the so-called Open Science and Open Innovation. Open Science is seen as a major tool to close the gap between academics, businesses and government spheres. This connection clearly appears within the administrative structure of the European Commission because Directorate B is in charge both of Open Innovation and Open Science;
- More broadly, digitalization of science questions how knowledge is produced and used in a given society, finally asking the question of collective intelligence.

5. Conclusion

The study of the European and French policy toward Open Science has demonstrated that the way Open Science has been put on the agenda was the result of a progressive evolution that encompassed the debate around Open Access, then about Open Data and later the question of the infrastructure. Open Science later appeared as a convenient concept to encompass all the dynamics, changes and opportunities generated by the digitalization of science, thus all the policies implemented regarding this change.

Nonetheless, much remains to be done regarding the emergence of this potential new scientific revolution. The movement toward Open Science comes mainly from local and individual initiatives that have pushed in favor of Open Access, Open Data but also for the development of innovative research habits. The proliferation of such initiatives in the French context is the best example of the vitality and the expectation that this digitalization movement is creating. However, the spread of this bottom-up evolution is also facing barriers. Because of them and because of the importance of the stakes involved, governments have started to implement top-down approaches. Their objective is to encompass existing initiatives and to encourage a broader implication of the whole scientific community.

Regarding the evolution of the relation of government toward science, the transition toward the digitalization of research activities is ongoing. Thus, the role of governments is still central, as they must accompany this transition and create the progressive conditions needed for a generalization of innovative practices so far implemented by frontrunners.

Acknowledgements

My sincere appreciation goes to the editors of this book, Professor. Robin Teigland and Dr. Anthony Larsson for their help and commitment but also for their trust. I would like to thank them for giving me this opportunity to work with them on this fascinating project. My gratitude also goes to Mr. Marin Dacos, Mr. Serge Bauin, Mr. Henri Verdier, Mr. Pierre Mounier and Mr. Jean-Claude Burgelman for their availability, help and support in the making of this research project. I am also deeply grateful to Ms. Celia Leclerc for her continuous help and support in the writing of this chapter.

Note

1 This evolution can be traced back to some various symbolic decisions. Two of which are the following: (1) the 2018 amendment of the title of the annual conference organized by COUPERIN from the "Open Access Days" to the "Open Science Days". (2) The transformation of the Digital Scientific Library into a new Committee for Open Science. This evolving transformation reflects the broader French interest in regards to the issue of Open Science.

References

Bartling, S. and Friesike, S., 2014. Towards another scientific revolution. In: S. Bartling and S. Friesike, eds., *Opening Science: The Evolving Guide on How the Internet is Changing Research, Collaboration and Scholarly Publishing*. Cham, Switzerland: Springer, pp. 3–15.

Bauin, S., Barthonnat, C., Berthaud, C., Bouche, T., Cavalier, F., Colcanap, G., Contat, O., Fargier, N., Fournier, T., Gremillet, A., Helein, F., Hologne, O., Jannes-Ober, E., Lafait, J., Le Blanc, A., Lutz, J., Malotaux, S., Millet, J., Mounier, P., Okret-Manville, C., Ollendorff, C., Respingue-Perrin, S., Roche, J., Romary, L. Roux, D., Schopfel, J., Teissier, B., Thomas, A. and Vautrin, C., 2017. *Jussieu Call for Open Science and Bibliodiversity*. [online] Available at: <http://jussieucall.org/index.html> [Accessed 22 Sep. 2019].

Bly, A. and Ginanni, K., 2012. Science Re-Imagined. *Serials Librarian*, 62(1–4), pp. 33–46.

Breivik, M., Hovland, G. and From, P.J., 2010. Trends in research and publication: Science 2.0 and open access. *Modeling, Identification and Control: A Norwegian Research Bulletin*, 30(3), pp. 181–190.

Brown, P., Cabell, D., Chakravarti, A., Cohen, B., Delamothe, T., Eisen, M., Grivell, L., Guédon, J., Hawley, S., Johnson, R., Kirschner, M., Lipman, D., Lutzker, A., Marincola, E., Roberts, R., Rubin, G., Schloegl, R., Siegel, V., So, A., Suber, P., Varmus, H., Velterop, J., Walport, M. and Watson, L., 2003. *Bethesda Statement on Open Access Publishing*. [online] Access. Available at: <http://legacy.earlham.edu/~peters/fos/bethesda.htm> [Accessed 27 Aug. 2019].

Buhr, C.C., 2014. *Open, Digital Science in Europe* (16.04.2014). [online] Science 2.0 Conference. Available at: <www.youtube.com/watch?v=kF8bWnkMfyw> [Accessed 27 Aug. 2019].

Burgelman, J.C., Osimo, D. and Bogdanowicz, M., 2010. Science 2.0 (change will happen . . .). *First Monday*, 15(7), pp. 1–12.

Chan, L., Cuplinskas, D., Eisen, M., Friend, F., Genova, Y., Guédon, J.C., Hagemann, M., Harnad, S., Johnson, R., Kupryte, R., La Manna, M., Rév, I., Segbert, M., de Souza, S., Suber, P. and Velterop, J., 2002. *Budapest Open Access Initiative (BOAI)*. [online] Available at: <www.budapestopenaccessinitiative.org/read> [Accessed 27 Aug. 2019].

Chan, L., Cuplinskas, D., Eisen, M., Friend, F., Geisselmann, F., Einhäupl, K., Gaehtgens, P., Gruss, P., Henkel, H., Kröll, W., Winnacker, E., Larrouturou, B., Schirmbacher, P., Galluzzi, P., Bullinger, H., Elkana, Y., Guédon, J., Roth, M., Bréchot, C., Ruano Leon, J., Simon, D., Braarvig, J. and Mittelstraß, J., 2003, *Berlin Declaration on Open Access to Knowledge in the Sciences and Humanities*. [online] Available at: <https://openaccess.mpg.de/Berlin-Declaration> [Accessed 27 Aug. 2019].

Chartron, G., 2016. Stratégie, politique et reformulation de l'open access [Strategy, Politics and the Reforumlation of Open Access]. *Revue française des sciences de l'information et de la communication*, 8(April), pp. 1–20.

CNRS, 2016. *Livre blanc – Une Science ouverte dans une République numérique [White Paper – Open Science in a Digital Republic]*. [online] Direction de l'Information Scientifique et Technique. Available at: <www.cnrs.fr/dist/z-outils/documents/2016%20 03%2024%20Livre%20blanc%20Open%20Science.pdf> [Accessed 22 Sep. 2019].

CNRS, 2017. *Une Science ouverte dans une République numérique – Guide stratégique [Open Science in a Digital Republic – A Strategic Guide]*. [online] Direction de l'Information Scientifique et Technique. Available at: <www.cnrs.fr/dist/z-outils/documents/guide-strategique-applications.pdf> [Accessed 27 Aug. 2019].

Colin, N. and Verdier, H., 2015. *L'âge de la multitude – Entreprendre et gouverner après la révolution numérique [The Age of Multitude]*. Paris, France: Armand Colin.

EOSC Summit, 2017. *EOSC Declaration*. [online] (July). Available at: <http://ec.europa.eu/research/openscience/pdf/eosc_declaration.pdf#view=fit&pagemode=none> [Accessed 27 Aug. 2019].

European Commission, 2007. *FP7 in Brief*. [online] Available at: <https://ec.europa.eu/research/fp7/pdf/fp7-inbrief_en.pdf> [Accessed 27 Aug. 2019].

European Commission, 2012a. *Communication from the Commission to the European Parliament, the Council, the European Economic and Social Committee and the Committee of the Regions Towards Better Access to Scientific Information: Boosting the Benefits of Public Investments in Research*. [online] Communication. Available at: <https://eur-lex.europa.eu/legal-content/EN/TXT/HTML/?uri=CELEX:52012DC0401> [Accessed 22 Sep. 2019].

European Commission, 2012b. *Recommandation de la Commission du 17.7.2012 relative à l'accès aux informations scientifiques et à leur conservation [Recommendation of the Commission of 17.7.2012 on Access to and Preservation of Scientific Information].* [online] Available at: <https://ec.europa.eu/research/science-society/document_library/pdf_06/recommendation-access-and-preservation-scientific-information_fr.pdf> [Accessed 22 Sep. 2019].

European Commission, 2013. *Digital Science in Horizon 2020.* [online] Available at: <http://ec.europa.eu/information_society/newsroom/cf/dae/document.cfm?doc_id=2124> [Accessed 27 Aug. 2019].

European Commission, 2014a. *Horizon 2020 in Brief – The EU Framework Programme for Research and Innovation.* [online] Available at: <http://ec.europa.eu/programmes/horizon2020/sites/horizon2020/files/H2020_inBrief_EN_FinalBAT.pdf> [Accessed 27 Aug. 2019].

European Commission, 2014b. *Public Consultation 'Science 2.0': Science in Transition.* [online] Available at: <https://ec.europa.eu/research/consultations/science-2.0/background.pdf> [Accessed 27 Aug. 2019].

European Commission, 2015. *Validation of the Results of the Public Consultation on Science 2.0: Science in Transition.* [online] Available at: <www.eesc.europa.eu/resources/docs/validation-of-the-results-of-the-public-consultation-on-science-20.pdf> [Accessed 27 Aug. 2019].

European Commission, 2016a. *Draft European Open Science Agenda.* [online] Available at: <https://ec.europa.eu/research/openscience/pdf/draft_european_open_science_agenda.pdf> [Accessed 22 Sep. 2019].

European Commission, 2016b. *Guidelines on FAIR Data Management in Horizon 2020 History of Changes.* [online] Available at: <http://ec.europa.eu/research/participants/data/ref/h2020/grants_manual/hi/oa_pilot/h2020-hi-oa-data-mgt_en.pdf> [Accessed 27 Aug. 2019].

European Commission, 2018a. *Directorate-General for Research and Innovation.* [online] Available at: <https://ec.europa.eu/info/sites/info/files/organisation_charts/organisation-chart_dg-rtd_en.pdf> [Accessed 27 Aug. 2019].

European Commission, 2018b. *EOSCpilot.* [online] Available at: <https://eoscpilot.eu> [Accessed 27 Aug. 2019].

European Commission, 2018c. *FOSTER.* [online] Available at: <www.fosteropenscience.eu> [Accessed 27 Aug. 2019].

European Commission, 2018d. *Open Science Policy Platform.* [online] Available at: https://ec.europa.eu/research/openscience/index.cfm?pg=open-science-policy-platform [Accessed 27 Aug. 2019].

European Commission, 2018e. *OpenAIRE.* [online] Available at: <www.openaire.eu> [Accessed 27 Aug. 2019].

European Commission, 2018f. *Recommendation of 25 April 2018 on Access to and Preservation of Scientific Information.* [online] Available at: <https://ec.europa.eu/digital-single-market/en/news/recommendation-access-and-preservation-scientific-information> [Accessed 27 Aug. 2019].

European Council, 2000. *Presidency Conclusions – Lisbon European Council (23 and 24 March 2000).* [online] Available at: <www.consilium.europa.eu/uedocs/cms_data/docs/pressdata/en/ec/00100-r1.en0.htm> [Accessed 27 Aug. 2019].

European Parliament, 2015. *Harmonisation of Certain Aspects of Copyright and Related Rights.* [online] Available at: <www.europarl.europa.eu/sides/getDoc.do?pubRef=-//EP//NONSGML+TA+P8-TA-2015-0273+0+DOC+PDF+V0//EN> [Accessed 27 Aug. 2019].

European Research Council – Scientific Council, 2016. *ERC Scientific Council Statement on Open Access*. [online] Available at: <https://erc.europa.eu/sites/default/files/press_release/files/erc_scc_statement_2006_open_access_0.pdf> [Accessed 22 Sep. 2019].

European Union Council, 2016. *La transition vers un système de science ouverte – Conclusions du Conseil (adoptées le 27 mai 2016) [Transition Towards an Open Science System – Conclusions of the Council 27 May 2016]*. [online] Available at: <http://data.consilium.europa.eu/doc/document/st-9526-2016-init/fr/pdf> [Accessed 22 Sep. 2019].

Fecher, B. and Friesike, S., 2014. Open science: One term, five schools of thought. In: S. Bartling and S. Friesike, eds., *Opening Science: The Evolving Guide on How the Internet is Changing Research, Collaboration and Scholarly Publishing*. Cham, Switzerland: Springer, pp. 17–47.

Government Bill, 2011. *Décret n° 2011–194 du 21 février 2011 portant création d'une mission "Etalab" chargée de la création d'un portail unique interministériel des données publiques [Decree n°2011–194 Related to the Creation of the 'Etalab' Mission]*. Paris: Government Offices of France [online] Available at: <www.legifrance.gouv.fr/affichTexte.do?cidTexte=JORFTEXT000023619063&dateTexte=&categorieLien=id> [Accessed 27 Aug. 2019].

Government Bill, 2016. *Loi n° 2016–1321 du 7 octobre 2016 pour une République numérique [Law n°2016–1321 for a Digital Republic]*. Paris, France: Government Offices of France [online] Available at: <www.legifrance.gouv.fr/affichTexte.do?cidTexte=JORFTEXT000033202746&categorieLien=id> [Accessed 27 Aug. 2019].

Hetland, P., 2011. Science 2.0: Bridging Science and the Public. *Nordic Journal of Digital Literacy*, 6, pp. 326–339.

Lamy, P., 2017. *Investing in the European Future We Want: Report of the Independent High Level Group on Maximising the Impact of EU Research & Innovation Programmes*. [online] European Commission. Available at: <http://ec.europa.eu/research/evaluations/pdf/archive/other_reports_studies_and_documents/hlg_2017_report.pdf> [Accessed 27 Aug. 2019].

Malamud, C., O'Reilly, T., Elin, G., Sifry, M., Holovaty, A., O'Neil, D., Migurski, M., Allen, S., Tauberer, J., Lessig, L., Newman, D., Geraci, D., Bender, E., Steinberg, T., Moore, D., Shaw, D., Needham, J., Hardi, J., Zuckerman, E., Palmer, G., Taylor, J., Horowitz, B., Exley, Z., Fogel, K., Dale, M., Hall, J., Hofmann, J., Orban, D., Fitzpatrick, W. and Swartz, A., 2007. *Open Government Data Principles*. [online] Available at: <https://public.resource.org/8_principles.html> [Accessed 27 Aug. 2019].

Miedema, F., Mayer, K., Holmberg, K. and Leonelli, S., 2018. *Mutual Learning Exercise: Open Science – Altmetrics and Rewards*. [online] Available at: <https://rio.jrc.ec.europa.eu/en/policy-support-facility/mle-open-science-altmetrics-and-rewards> [Accessed 27 Aug. 2019].

Millerand, F., 2015. Les imaginaires de la "science 2.0" [The Imaginations of 'Science 2.0']. *Communication*, 33(2), pp. 1–26.

Ministry of Education, Culture and Science, 2016. *Amsterdam Call for Action on Open Science*. Government Offices of the Netherlands [online] Available at: <www.government.nl/documents/reports/2016/04/04/amsterdam-call-for-action-on-open-science> [Accessed 27 Aug. 2019].

Ministry of Higher Education, Research and Innovation, 2018a. *Appel à Manifestation d'Intérêt (AMI) pour la constitution du Comité pour la Science Ouverte (CoSO) [Call for Expression of Interest for the Establishment of an Open Science Committee]*. [online] Available at: <www.enseignementsup-recherche.gouv.fr/cid128239/appel-a-manifestation-d-interet-ami-pour-la-constitution-du-comite-pour-la-science-ouverte-coso.html> [Accessed 27 Aug. 2019].

Ministry of Higher Education, Research and Innovation, 2018b, *National Plan for Open Science*, [online] Available at: <http://cache.media.enseignementsup-recherche.gouv.fr/file/Recherche/50/1/SO_A4_2018_EN_01_leger_982501.pdf> [Accessed 27 Aug. 2019].

Ministry of National Education Teaching and Research, 2015. *Stratégie nationale de recherche France-Europe 2020.* [online] Available at: <http://cache.media.enseignementsup-recherche.gouv.fr/file/Strategie_Recherche/26/9/strategie_nationale_recherche_397269.pdf> [Accessed 27 Aug. 2019].

Ministry of National Education Teaching and Research, 2017. *Livre Blanc de l'Enseignement Supérieur et de la Recherche 2017 [Research and Higher Education White Paper 2017].* [online] Available at: <http://cache.media.enseignementsup-recherche.gouv.fr/file/Actus/04/1/ESR_Livre_Blanc_707041.pdf> [Accessed 27 Aug. 2019].

Moedas, C., 2015. Open innovation, open science, open to the world. In: *Speech.* [online] Available at: <http://europa.eu/rapid/press-release_SPEECH-15-5243_fr.htm> [Accessed 27 Aug. 2019].

O'Reilly, T., 2011. Government as a Platform. *Innovations*, 6(1), pp. 13–40.

OECD, 2016. *Digital Government Strategies for Transforming Public Services in the Welfare Areas.* [online] Available at: <www.oecd.org/gov/digital-government/Digital-Government-Strategies-Welfare-Service.pdf> [Accessed 27 Aug. 2019].

Open Science Policy Platform, 2018. *Integrated Advice of the Open Science Policy Platform Recommendations.* [online] Available at: <https://ec.europa.eu/research/openscience/pdf/integrated_advice_opspp_recommendations.pdf> [Accessed 22 Sep. 2019].

Sabatier, P.A. and Weible, C.M., 2007. The Advocacy Coalition Framework. Innovations and Clarifications. In: P.A. Sabatier, ed., *Theories of the Policy Process*, 2nd ed. Boulder, CO: Westview Press, pp. 189–220.

Schöpfel, J., Ferrant, C., André, F. and Fabre, R., 2016. Ready for the future? A survey on open access with scientists from the French National Research Center (CNRS). *Interlending & Document Supply*, 44(4), pp. 141–149.

State Secretariat in charge of Digital Affairs, 2018. *Pour une Action Publique Transparente et Collaborative – Plan d'Action National pour la France 2018–2020 [For a Transparent and Collaborative Public Action – National Action Plan for France 2018–2020].* [online] Available at: <https://www.etalab.gouv.fr/wp-content/uploads/2018/04/PlanOGP-FR-2018-2020-VF-FR.pdf> [Accessed 22 Sep. 2019].

State Secretariat in charge of the Modernization of Public Action, 2018. *État plateforme [Platform Government].* [online] Available at: <http://etatplateforme.modernisation.gouv.fr> [Accessed 22 Sep. 2019].

Szkuta, K. and Osimo, D., 2016. Rebooting science? Implications of science 2.0 main trends for scientific method and research institutions. *Foresight*, 18(3), pp. 204–223.

Teif, V.B., 2013. Science 3.0: Corrections to the Science 2.0 Paradigm. [online] Available at: <http://arxiv.org/abs/1301.2522> [Accessed 22 Sep. 2019].

Vignoli, M., Kraker, P. and Sevault, A., 2015. Paving the way for Science 2.0: Top-down and bottom-up approaches. *International Conference for E-Democracy and Open Government (CEDEM'15)*, pp. 119–130.

6 Black boxes of cognitive computers and the impact on labor markets

Victor Erik Bernhardtz

1. Introduction

This chapter discusses the impact of digitalization on labor and labor markets. More specifically, it deliberates on the introduction of Internet of Things (IoT) and cognitive computing in labor and how it changes, to some extent, all types of labor. At the heart of this discussion is a duality: Digitalization of labor markets has the potential of bringing significant advances in productivity, work quality and work safety. On the other hand, the early stages of digitalization of labor markets has produced concerning outcomes, such as precarious work, ethical concerns and low pay, to mention a few. In other words, we *can* harness the potentials of digitalization of labor, but we are currently not always doing so.

The connecting of work tools (analog and digital) to the internet opens up new potentials for productivity. More types of work can be organized without concern to physical space, enabling remote control over devices as large as mining trucks or as delicate as surgery equipment. Connected tools have the added potential of generating data on how they are used. This means that not only products, but also how work is carried out, can be extensively evaluated without taxing human resources. Moreover, a connected work tool can also function as a location device, provided the worker is mobile.

The potential of connected tools can be unleashed with help of cognitive computing. Cognitive computing is the introduction of computer systems[1] designed to mimic the human brain, with the power of modern computers. This allows for analysis of masses of unorganized data much greater than a human brain could grasp. One application is evaluations of productivity, based on data generated by work tools. What cognitive systems *could* do is in theory all but limitless.

Even though recent research suggests that the number of jobs that will be lost to automation are significantly fewer than predictions from just a few years ago, the jobs lost still count in the many millions (Nedelkoska and Quintini, 2018). Equally important is the alteration of jobs that will remain, but in new forms. In order for workers to retain such jobs, massive retraining efforts needs to be implemented, lest employers lose human capital that cannot be replaced by software.

Cognitive computer systems will be able to replace humans in carrying out work tasks previously thought to be impossible to automate. One such task, equally alluring as a cause for caution, is management. Today, management executed by computer programs is most visible, albeit often in a rudimentary form, in so-called platform work. In platform work, tasks are distributed from the buyer of the task to workers via computer systems. Proponents of platform work emphasize the increased opportunities for correctly matching task and worker, minimizing transaction costs and friction. Critics cite risks in one-sided flexibility for platform firms, while workers' flexibility will be all but lip service.

The automated management in platform work is worth special attention, as the experience from such organization of work is likely to influence a wider implementation of cognitive computer management in traditional employment. It raises key questions, of which the most important center on the transparency of the computer systems. Computer programs designed to mimic the human brain are intimidating. In order to implement them in the daily work life they need to be demystified and less opaque, if workers are to accept them as tools for management, as well as other tasks.

Employer's organizations and unions play a crucial role in this. If they in cooperation and through negotiations can find solutions both parties can accept, the long-term success of digitalization of labor is more likely.

1.1. Method

This text is a literature study of, and a theoretical discussion on, the impact of digitalization on the labor market. In the discussion, the Nordic Social Partner Approach will be applied as a suggested model in addressing the labor market challenges posed by digitalization.

2. Evolution of digitalization of labor and labor markets

The Roman *cursus publicus* used carriages pulled by horses and oxen as well as attaching messages to pigeons. In time, mail would be sent over open sea in ships and eventually the electric telegraph made its appearance. The telautograph, the fax machine, telex and other electronic devices for sending messages, followed. Today we use email and various applications for instant messaging. In all likelihood, we will invent even more methods for sending mail.

What is here very briefly described is the evolution of sending mail. The same exercise can be carried out with other phenomena, changed by digitalization, which we in casual conversation might characterize as "revolutionary", while we should probably refer to them a "evolutionary". Understanding digitalization as an evolution paves way for the insight that we are able to adapt to and harness digitalization within existing societal structures.

2.1. Industrie 4.0

Computer programs are today processing and analyzing data of larger scales and with deeper complexity than ever before (Brynjolfsson and McAfee, 2014). This is a truism, but it should be read in the context that things will never be as *slow* again as they are today.

The work of Intel's co-founder Gordon E. Moore has given name to *Moore's law*, the observation that the number of transistors in a dense integrated circuit doubles every two years (Moore, 1965). Moore's law has proven to hold relatively true, despite being challenged academically as well as practically, for some odd 50 years (Khan, Hounshell and Fuchs, 2018). This is essential to our discussion as the integrated circuit is a prime example of a general-purpose technology, which has made possible a wide array of inventions that together have brought us to our current state of digitalization. The integrated circuit is a technology that has changed the conditions for human life.

Currently, almost two decades into the twenty-first century, the labor market is at the dawn of the so-called *Industrie 4.0*. Industrie 4.0 (or "Industry 4.0" for the purpose of this chapter) was first launched as a concept by the German government in 2013. "4.0" connotes a fourth industrial revolution (European Commission, 2017).[2] In this shift in production the "smart factory" comes to realization. The robots and digital tools in the smart factory are augmented by computer systems that largely automate the management of (increasingly complex) production processes, in collaboration with humans, throughout the value chain. In the smart factory, robots and digital tools are also connected to the *Internet of Things (IoT)* (Kramp, van Kranenburg and Lange, 2013).

2.2. Internet of Things (IoT)

The Internet of Things (IoT) encompasses the expansion of the internet beyond the computer screen to devices that traditionally have been analogue, such as (but not limited to) manufacturing robots, trucks or soccer balls. A driving force for IoT is the ambition to increase work productivity. Tech giant Cisco projects that by 2030 IoT will consist of some 500 million devices.

Whether Industry 4.0 constitute a "revolution" as such, or should perhaps be characterized as evolution, will be for historians to decide. Regardless, two component of Industry 4.0 stands out as both new and at the same time highly relevant for the labor market in a scope far beyond manufacturing: connectivity with the world outside the workplace and cognitive computing.

The potentials in connecting work devices to IoT will bring an increasing proportion of devices used wherever work takes place online. Connecting work tools to the internet allows both for the smart factory previously described to operate, but also allows for worker and tool to exist in different physical spaces. Robots controlled from a computer screen can operate in places and under conditions that are impractical or outright dangerous for humans. Equipment used in delicate surgery can be operated by a doctor hundreds of miles away. The space for innovation is vast. Public administration stands out as one sphere where the potential is particularly high, as increased productivity could benefit citizens both in their everyday lives as well as in tax money being spent more efficiently.

2.3. The advent of cognitive computing

The other key characteristic of Industry 4.0 that will have impact well beyond the smart factory is *cognitive computing*. While the term has no widely agreed upon definition, it refers to a computer system that – more or less – has the function of mimicking the human brain (Kelly III, 2016). Such systems have the ability to reason, interact with humans and adapt and learn from such interactions as well as to reflect on the work they carry out. They are not programmed in the traditional sense. Kelly III (2016, para.5) formulates the difference between cognitive systems and their predecessors in terms of output:

> Those [previous] systems were deterministic; cognitive systems are probabilistic. They generate not just answers to numerical problems, but hypotheses, reasoned arguments and recommendations about more complex – and meaningful – bodies of data.

The most common public understanding of cognitive computing is artificial intelligence (AI), which in itself has a public image quite far from reality. AI as a term was coined in 1955 (Kelly III, 2016). It has since inspired cultural works such as William Gibson's pioneering cyberpunk novel *Neuromancer* (Gibson, 1984) and the iconic action-science fiction movie *The Terminator* (The Terminator, 1984), directed by James Cameron. Both works were incidentally released in 1984 and both paint a dark future of AI in society. However, the reality of cognitive computing is quite different from those of Gibson's hacker Case and Cameron's waitress Sarah Connor.

Cognitive computer systems can make sense of data of greater volume and complexity compared to previous generations of computer systems. The value in such applications is, similar to the gains of IoT, that we are able to make machines do things that humans cannot do. Cognitive computing is not about imitating humans, the purpose is to augment what humans are able to do. The OECD (2016) provides the example of diagnosing rare diseases with the help of cognitive software extracting conclusion from data shared between thousands of health professionals, data far too vast for humans to comprehend. Kelly III (2016), a vice president of IBM, points out that cognitive computing has nothing to do with sentience or autonomy on the part of robots or digital tools and that human qualities, such as common sense and ethics, are needed to maximize the potentials of cognitive computing.

The potential of cognitive computing is, according to some, gigantic. Google CEO Sundar Pichai stated that the invention of AI is perhaps more profound than the invention of electricity or the discovery of how to make fire (Goode, 2018). Kelly III (2016) describes cognitive computing as a "new era of technology, business and society". Andrus Ansip, vice-president of the European Commission, compared the impact of AI to that of the steam engine and electricity, while calling for investments of at least €20 billion by the end of 2020 (European Commission,

2018). To provide a counter argument in order to avoid hyperbole, we would do well to remember that cognitive systems can be vastly superior to humans in carrying out specific tasks but at the same time rarely can perform the simple task of drawing a circle.

Nonetheless, the strong vocal support from both government and business leaders suggest that significant political and financial capital will be invested in cognitive computing. IBM has invested USD15 billion alone in Watson, a cognitive computing system sometimes attributed with bringing the AI industry out of a longer stretch of relatively low attention (Thomas, 2017). Moreover the market value for systems applying AI systems in office settings is projected to reach USD 48.5 billion by 2022 (Waters, 2018).

3. Impact of digitalization on labor and labor markets

The devices we use in our work will be increasingly connected to the internet and cognitive computing will gradually become part of labor and labor markets. While the introduction of new technology is something that has always been present in the life of a manufacturing worker, workers in other sectors will be acting within a, at least partly, new context.

3.1. Will robots make the worker obsolete?

Office workers may have seen the introduction of more efficient ways of sending mail (although the email inbox constitutes a serious mental health concern for some) or more sophisticated ways of writing said letter (digitization such as the evolution from typewriter to word-processing computer software). But a shift to higher penetration of IoT and cognitive computing constitutes a more profound change.

In an oft-cited study by Frey and Osborne (2017) initiated 2013, they suggest that 47% of total US employment is at risk of being computerized. The figure of 47% has since been put in headlines of newspaper articles worldwide and the data in the working paper has been extrapolated over data for other countries, with the result of slightly alarmist messages. Furthermore, in the Frey and Osborne (2017) study, there are some estimates that spark skepticism, such as models (the job, not the scientific activity) being at a 98% risk of being replaced by computers – the category of employment at highest risk.

While one can be tempted to smirk at such a prediction half a decade later, there is something to be said on modeling that has bearing on digitalization and labor. Instagram, a freeware photo- and video-sharing social-networking service owned by Facebook, has grown from 150 million active users in September 2013, when Frey and Osborne initiated their study, to over one billion monthly users in June 2018 (Kuchler, 2018; Hernandez, 2013; Chaffey, 2018). Models and so-called influencers have a very high presence on Instagram and other social media. Instagram posts that are more likely to generate engagement typically show faces (Bakhshi, Shamma and Gilbert, 2014). Hence, Instagram is an ideal work tool for a model.

Models are yet to see themselves computerized, but their work has changed to a large degree. The model's own camera is part of IoT and computer software is used in retouching their persona both by professional art directors and – with easy-to-use applications such as Instagram – models themselves. Models in the twenty-first century are hence potentially less dependent on gatekeepers. This transformation of the job "model" requires a different skill set from people who work as models, compared to what a model active in the twentieth century needed. In that sense, modelling is a good example of how impacts of digitalization on labor and labor markets can manifest. The job remains but a significant portion of the job has changed radically, as has the product.

3.2. Risks related to automation vary between countries

Since 2013, several studies have arrived at more conservative estimates than Frey and Osborne (2017). Bakhshi et al. (2017) predict that one-fifth of the workforce in the United Kingdom and the US are in jobs that are likely to be less in demand in the future as an effect of automation (Bakhshi et al., 2017). Conversely, one-tenth of the workforce are in jobs that are likely to increase in demand. Bakhshi et al. (2017) also stress the levels of uncertainty associated with such predictions. The OECD, in a working paper studying all its member states, found that about one in two jobs are likely to be significantly affected by automation (Nedelkoska and Quintini, 2018). The amount of jobs that are highly automatable to a degree that humans are no longer needed is estimated to 14%, across OECD member states. The OECD points out that this lower estimate compared to Frey and Osborne still translates into 66 million workers at a high risk of losing their jobs.

The risks are not evenly distributed. Workers in Anglo-Saxon countries, the Nordic countries and the Netherlands are at a considerably lower risk, compared to workers in Eastern European countries, Japan, Chile, Germany and South European countries. This is mainly due to two reasons. Firstly, countries differ a lot in economic structure and secondly, the way work is organized within the same industry in different countries can be quite different. According to the OECD, the latter seems to represent the bulk of jobs at risk of automation. This can be explained partly by the fact that some countries are lagging behind in automation processes that other countries have already transitioned through (Nedelkoska and Quintini, 2018).

The OECD further finds, in contrast to some previous studies, that cognitive computing will not have a higher impact on jobs requiring a higher skill set. On the contrary, the OECD concludes that the jobs with the highest risk of automation are among jobs with lower education requirements and skill levels. However, as mentioned previously, a hefty proportion of the jobs that will continue to exist are predicted to change to a large degree, as a result of automation (Nedelkoska and Quintini, 2018).

Another important finding of the OECD is that when controlling for industry and occupation, women are at a higher risk of being affected by automation. It seems that while women are over-represented in sectors where jobs are at a comparably lower risk, women have more automatable tasks than their male colleagues within these sectors (Nedelkoska and Quintini, 2018).

To summarize, while the number of workers made obsolete by smart robots in labor markets influenced by Industry 4.0 are lower than what has earlier been estimated, the number is still high, particularly in certain countries and sectors. Higher still is the number of workers that will need substantial retraining, as the jobs they hold remain, but change. Ensuring workers are equipped, in terms of education, to meet the growing demands for high-skilled workers in some sectors, is also essential (Nedelkoska and Quintini, 2018).

3.3. New methods for evaluating labor

In what way the characteristics of any specific job are altered by what is previously described is a task too daunting for this chapter and its author to stipulate. However, there are some characteristics that could be expected to be fairly similar. One area relates to labor management.

If the tools used to carry out work are connected to the internet, it means that working with the tools in and of itself can be a data-generating activity, regardless of what a worker or a robot is creating with said tool. The data generated can then be shared both with other similar tools, but more broadly, with the entirety of the internet (subject to restrictions set up by company policy and the licensor of the software). Depending on the quality of programming of the software, the evaluation of what a worker/robot produces can be measured not only by looking at a *product* (be it a car component, a salad or a quarterly budget) and the time and resources the worker/robot used in creating said product, but also how the *tools* were used.

The upside of this is that computer systems can replace humans in gathering and organizing data about how work is carried out. If the computer system analyzing the data generated in turn is cognitive as previously described, larger and more complex sets of data can be examined and compared with other analysis. In the end, humans in managerial positions could potentially have access to deeper analysis about a broader scope of production, while using far fewer human resources and not having to take the limitations and risks for measurement errors that come with humans, into account.

On the other hand, the risks of leaning on cognitive computing for evaluating labor are several. Firstly, any computer system that aims at measuring something will yield less useful outcomes if the programming quality is not sophisticated enough. Secondly, software engineers will be dependent on receiving all the needed input data from the ordering part – data that might not always be possible to provide. Thirdly, the ordering part and the software engineer must be on the same level of understanding of what outcomes are expected as well as understanding of their respective limitations.

As a consequence of the previously mentioned challenges, workers risk ending up in a situation where poorly designed computer systems evaluate their work performance. In addition, there is a risk of management providing feedback based on information generated by poorly designed systems. The effects could be even less favorable for workers if outcomes from poorly designed systems influence decisions on wage increases and promotion opportunities. We will elaborate this discussion in the following section.

4. Ethical and social dimensions of digitalization of labor

Aside from significantly altering how we work, cognitive computing has the potential of affecting the ways in which work is structured. Cognitive computing opens up for management that is informed, guided or even organized by machines. While this opens up for a number of potential productivity gains, there are also major challenges. How implementation of cognitive computing in labor markets is organized will determine its success. This is particularly true for computer-system management.

4.1. Management by app

In a world of digitalized labor, a probable scenario is that the data generated by humans carrying out work tasks will be copied, disseminated, analyzed and then applied to improve the future work tasks carried out by the same humans, as well as other humans in the same sector (and beyond). In short, cognitive computing will play a key role in management.

A scenario where the boss in practice is an app can be either utopia or dystopia, or both. Such management by app is currently best observable within the so-called "platform economy", also referred to as "gig economy" or "sharing economy". In this chapter, platform economy is used, as it best describes the condition of an intermediate platform to facilitate the transaction of labor that occurs (Söderqvist, 2017). In the platform economy, workers find short-term employment, often simpler tasks such as "deliver a pizza from address A to address B" or "man this reception for two hours", through online platforms, often downloaded as apps to their smartphones.

The potential comparative advantage for workers, companies and in some cases customers (as in the case of pizza delivery) is that platforms can be superior in finding the right worker for the right task. Who the right worker is could be dependent on a specific skill set, the physical distance between a worker and the location where the task is to be carried out, or anything else that could be deemed relevant depending on the task. The platform gains this advantage through designing its computer systems in a manner that increases the likelihood of successful matching. There are obvious benefits for utilizing cognitive computing here. Indeed, one US-based but globally operating platform refers to their matching process as "data science magic".

What is previously described is a situation where software plays a central role in management of work. The software programming will strongly influence which worker is matched with any specific task, or which three to four workers a company posting a task will have to choose from. Few, if any, human interactions occur in the time span starting when the task is posted and ending when the worker receives payment. In addition, platforms usually have built-in rating systems. Some gives companies/private individuals the opportunity to rate the worker, others allow for companies/private individuals and worker to rate each other. Through rating systems, the systems do not only organize work in terms of optimal matching, but also produce data that are in effect an evaluation of the quality of said work. Lee et al. (2015) have named this practice algorithmic management in a study of ride-sharing services Uber and Lyft. Lee et al. (2015) argue that algorithmic management should be seen one of the core innovations that enables the business models of platform firms.

4.2. Low pay and gray areas

In reality, the platform economy is not functioning as well as it potentially could. A study of 2676 workers performing 3.8 million tasks on the platform Amazon Mechanical Turk points to very low pay (Hara et al., 2018). The study showed an hourly wage (adjusted for the time invested in searching for tasks, work on rejected tasks and unsubmitted tasks) of ~USD2/h. Four percent of workers earned above USD7.5/h. As there are no long-term quantitative wage studies on platform work, it is difficult to judge if this represents conditions on a broader scale. However, it surely represents working conditions that can be said to be undesirable. In recent years, mass media reports on the platform economy have painted a rather grim picture of the impact on both workers and sectors of the labor market.

In addition to poor working conditions, there are growing concerns over the unclear employment relations in the platform economy (De Stefano, 2015). As platform work sometimes constitutes neither independent freelance work nor traditional employment, there is a lack of clarity in terms of responsibilities. Who is responsible for making sure work conditions are decent and pay is properly provided, if workers interact only with computer software, as opposed to human management? In cases where a clear chain of responsibilities can be established, who ensures that rights of workers are respected in cases where platform owners do not see themselves as employer, but merely as an intermediate? Some platforms, perhaps consciously, operate in legal gray areas that circumvent labor market laws and/or standards (Söderqvist, 2017).

As suggested previously, the platform economy is not yet delivering on its potential. Nevertheless, the platform economy poses a challenge to the political system: Regulate the platform economy rapidly, lest you find yourselves in a situation in which you have to regulate in accordance with (bad) practices that have become industry standard. But then again, regulation with haste risks putting legislation in place that is not sustainable in the long run, as well as potentially hampering genuine entrepreneurship. We will return to the issue of regulation later in this chapter.

Implementation of algorithmic management will not be reserved to new phenomena in the labor market, such as the platform economy. The deeper penetration will likely occur within traditional firms and will take various shapes. More workers are likely to have fewer interactions with humans in the organizing of their work. As discussed previously, measurement of data generated by the work itself and IoT push such developments forward. Whether an increase in the use of algorithmic management is a positive or negative development for workers is not deterministic.

4.3. Black boxes, trust and accountability

While a welcome scenario could be that humanity embraces and interacts with cognitive computing in the same way that we seamlessly have integrated the internet into our lives, such an acceptance will not come by itself. On the contrary, in order for algorithmic management to be accepted and trusted, transparency is of key importance.

Cognitive computing is to most people as incomprehensible as the human brain. Input data is fed to the computer system/brain, something occurs, resulting in output data (be it speech, a structured analysis of a large dataset, etc.). How the human brain carries out this process is a knowledge reserved for a few. However, we do not demand of neuroscientists to explain the process to us in order for us to accept human-to-human interaction regarding a new task at work. This is because we accept receiving instructions on a new work method or task (albeit sometimes grudgingly) from beings we recognize and understand, and therefore trust.

Algorithmic management on the other hand is characterized by opacity. What input data is used to organize and evaluate work? Who put that data there? What are their biases? Was that data personal information? In whose interests did they design the system? How is this data shared with other licensors of the software? And so on.

Today, few developers of cognitive computer software are willing to answer these questions, at least not without a *quid pro quo* that is likely to involve a financial transaction. Such secrecy has stimulated a development toward a critical classification of cognitive computing as something taking place inside "black boxes". We know that something is going on inside these systems, but we have no way of knowing what that is or how it functions.

This is an ethical problem that is likely to (if unanswered) push people in the direction of the more dystopic popular culture portrayals of cognitive computing. Suspicion is a logical reaction to a black box that is to heavily influence your daily work, if no one is willing to tell you how that box is designed or how it reaches its conclusions. If there is reason to believe that the box is biased, who is to be held accountable if the box stays opaque? If workers are to accept and embrace algorithmic management, or at least accept that algorithmic management will influence human decision making, workers need to be able to understand and influence how algorithmic management is designed and how it functions. Transparency is crucial.

4.4. Black boxes and democracy

We now arrive at a situation where algorithms must come out of their black boxes, if the cognitive computing influencing the organization of work is to have any legitimacy. At the same time, we should not require of people to learn programming or data science in order to be able to decipher how algorithmic management functions. On the contrary, the vision should be equal access to the ability to understand and interact with cognitive computing, as opposed to it being an activity reserved for society elites, data scientists and engineers. In some circumstances, this will require some form of intermediary that can interpret the system for the layperson. Then again, a wiser course of action here would probably be to avoid discussion being about the code, in favor of discussion being about principles and what outcomes that are desired. Also desirable is a development of cognitive computer systems that can explain to the user what the system is doing.

Black boxes constitute a fundamental challenge for democracy in the twenty-first century. Indeed, the societal aspects of cognitive computing are probably much more difficult to solve, compared to technical aspects.

4.5. Data ownership

A concern similar to the challenges put forward by black boxes is ownership over data. If labor is characterized by a high penetration of IoT, a growing number of work tools (physical and digital) will be data generating. In many occupations, workers will generate data through their physical presence and how their bodies interact with their surroundings. This constitutes the generation of a new type of value for the employer. Data on labor can be a valuable commodity, if cognitive computer software can organize such data from a multitude of workers.

The question of ownership of such data is almost philosophical, yet it presents very real problems. If the ownership of data lies with the person inhabiting the corporal presence generating the data, will that person be compensated when that data is shared by an employer? If so, what does such compensation look like? Should data generation motivate a pay rise, or should perhaps workers license data generated by them to employers, while employed? What happens with data that transcends company boundaries and exists after a worker leaves a job? Will the worker have the right to offer data generated by them to their next employer, as part of a skill set? These are concepts that deserve to be analyzed further. The guiding principle of such analysis should probably be that some form of compensation is reasonable.

5. The Nordic social partner approach and digitalization of labor

Digitalization has a profound impact on labor. It challenges existing labor laws and practices and has the ability to largely transform how labor is managed and organized. As with previous technological shifts in labor, digitalization is a harbinger of different scenarios and outcomes for different actors in the labor market. All have in common that there is no technocratic, value-neutral method with which to approach them. We shall be analyzing them through the lens of the Nordic social partner approach.

5.1. Sweden: business as usual

> *In Sweden, if you ask a union leader, "Are you afraid of new technology?" they will answer, "No, I'm afraid of old technology."*

These words emanated from the Swedish Minister for Employment and Integration Ylva Johansson, succinctly describing the Swedish response to the advancement in automation, in a 2017 *New York Times* article (Goodman, 2017, para.9). This article contends that 80% of all Swedes hold positive views on robots and AI. The article elaborates by drawing a connection between Nordic countries' comparatively high investments in labor-market policies, and workers' optimism regarding new technologies. The article can, in short, be summarized as follows: If the government promises to assist you in adapting to a new labor market, you will not fear that labor market.

Embracing that which is new while simultaneously ensuring that citizens can engage in it, is one of the key components of the Nordic social partner approach. In the Nordic approach, employers' organizations and trade unions represent the interests of capital and labor in negotiations that ultimately regulate the labor market – wages, work conditions, et cetera – through collective bargaining (Lundh, 2010). The state ensures that citizens have access to education, health care, day care for children and other social services.

In addition, if job loss comes knocking, the state provides unemployment benefits, often topped up by the union. Further, wages of workers in companies filing for bankruptcy are guaranteed by the state for a certain period. In the Nordic approach, entrepreneurship is crucial, as the model cannot harbor companies that are not making a profit. Incidentally, Swedish workers have enjoyed real wage increases for decades.

With such characteristics, the Nordic approach is an organization model of the labor market that with no small benefit can be applied in meeting the challenges of digitalization. Digitalization brings new issues and conundrums to the table, such as the aforementioned issues of black-box transparency and data ownership. Within the framework of the Nordic approach, all partners have a strong interest in maintaining productivity gains while keeping the peace on the labor market. High levels of conflict and/or discontent are simply poor for business.

This means that employers and unions have a joint interest in solving problems emerging in new contexts, be it changes brought on by technological leaps or sharp turns in the global economy. Within the Nordic approach, both employers and unions have strong incentives to reach agreements and work in concert. Such a mutual and interdependent situation is fertile ground for innovative solutions. When negotiations in the end produce a way forward, it is a direction anchored in both employers and unions. Lastly, there is a clear advantage of regulating the new components of the labor market through collective bargaining, compared to regulation through legislation.

Collective bargaining agreements can be tailored to the specific context of each sector of the labor market, whereas legislation applies in the same way to all different contexts. Collective bargaining is flexible and has its point of departure with the stakeholders, whereas legislation by default needs to be "popular". Regulating through legislation hence risks a situation where the public opinion wants to deal with a problem specific to one sector of the labor market, not taking into consideration that perfectly healthy sectors might be disadvantaged. In addition, agreements can be renegotiated. Changing legislation on the other hand is (rightly so) quite difficult to amend.

5.2. Outsmarting cognitive computing and opening black boxes?

What then, is the Swedish way of dealing with cognitive computing and black boxes? The short answer is that it depends on which sector of the labor market, with some 650-odd collective bargaining agreements, one refers to. The upside of avoiding a one-size-fits-all approach, such as legislation, is that the incorporation of cognitive computing can be tailored to very specific opportunities and limitations of a sector of the labor market. Employers and unions responsible for that sector can cooperate on how new technology should be implemented in a way that benefits growth, while not used to the disadvantage of workers.

What does this mean for cognitive computing and black boxes? As Swedish workers can be expected to have a generally positive view on technology, Sweden has a bit of a head start in the implementation of Industry 4.0-esque ways of organizing work. However, quite a few situations will likely occur where cognitive computer systems do not function as intended. Limitations in system design is one factor, another is subpar input data. It is therefore crucial that employers and unions mutually allow for unintended consequences to occur, with the understanding that they will be corrected.

Then again, there will be cases where employers will try to use management by app as a method for keeping workers in check and/or pressure workers to perform beyond healthy limits. Strong unions (69% of Swedish workers are unionized, 90% of workers are covered by a collective bargaining agreement) will play a key role in opposing such activity (Kjellberg, 2010). But unions will need the support and cooperation from employers. The incentive for the latter being that companies not adhering to good working conditions are using social dumping as a business model, which is bad for employers both in terms of unfair competition and in the undermining of the Nordic approach.

Transparency of black boxes is a potentially harder nut to crack, as it does not only involve the motivations of employer and worker, but also that of computer system developer. Companies that develop and license cognitive computing systems have no obvious incentive for opening up their software to scrutiny. Employers have a role to play here as a safeguard, refusing to license and implement systems unless employers and representative bodies in unions are able to fully understand as well as exercise real influence over the system. This is crucial for levelling the playing field between system and user. Keeping the algorithms in a non-transparent black box will feed mistrust among workers, resulting in non-cooperation, strikes or sabotage – as seen throughout the history of industrial revolutions. This is ideally where the interest for the developer to be transparent manifests – it is the only way to stay in business.

6. Conclusion

As digitalization has such high prospects, it is likely that actors on the Swedish labor market will dive in quite happily. This has potentially high effects for a large group of people while it at certain stages might be a bumpy ride. The Nordic approach does provide quite a few methods for smoothing out the road, but at the same time, evolution of the Nordic approach is (as always) necessary. Problems will not solve themselves. Currently, the partners within the Nordic Approach can be said to be asking the right questions. There is, for example, a consensus on the great need for worker retraining, as some jobs will disappear, while far more will change in nature. On the other hand, there is yet no consensus on how such retraining should be organized or funded.

Asking the right questions is often a good way to start and digitalization is not the first time the Nordic social partner approach faces a great challenge. But as challenges go, digitalization can be said to be a rather complex one. It is worth stressing that the challenges of highest importance are not necessary the technical aspects. The hardest part will probably be making the mental transition of the conceptualization of cognitive computing. Trust will be a crucial issue.

The robots, digital tools and computer systems that are increasingly more entwined in the daily routine of work, do not constitute threats in and of themselves. Of greatest importance is to keep in mind that the impact from digitalization of labor markets will be different, depending on what values and policies inform key decisions. Different responses will have correspondingly different outcomes for workers, management, shareholders and customers. These responses will be guided by values as well as policies. There are choices to be made, by states, enterprise, unions and citizens.

Regardless of exactly how many percentage points of jobs that will disappear or radically change in nature, we should expect that many millions of workers will be out of a job or at least struggle to retain their job. It is imperative that societies not allow that to transform into neo-luddism, social unrest or a fertile ground for populist sentiments. Digitalization, in order to be successful, must benefit the vast majority of people it will affect. The way to achieving that will be a political endeavor that will require profound deliberations, compromise and action, from all who in one way or the other have an interest in the future labor market.

Acknowledgements

The author would like to thank Carl Fredrik Söderqvist for research insights.

Notes

1 A computer system is a functional computer, including all necessary hardware and software to make it functional. Computer systems vary in size from a single device to large multiuser systems.

2 The first three being (1) mechanization powered by water and steam; (2) mass production and assembly lines powered by electricity; and (3) the digitization and automation of production, respectively.

References

Bakhshi, H., Downing, J.M., Osborne, M.A. and Schneirder, P., 2017. *The Future of Skills: Employment in 2030*. [online] Available at: <www.nesta.org.uk/report/the-future-of-skills-employment-in-2030> [Accessed 16 Sep. 2019].

Bakhshi, S., Shamma, D.A. and Gilbert, E., 2014. Faces engage us: Photos with faces attract more likes and comments on Instagram. In: *Proceedings of the 32nd Annual ACM Conference on Human Factors in Computing Systems – CHI '14, Toronto, Canada, 2014*. Toronto, Canada, pp. 965–974.

Brynjolfsson, E. and McAfee, A., 2014. *The Second Machine Age: Work, Progress, and Prosperity in a Time of Brilliant Technologies*. New York, NY: W.W. Norton & Company.

Chaffey, D., 2018. *Global Social Media Research Summary 2018*. [online] Smart Insights. Available at: <www.smartinsights.com/social-media-marketing/social-media-strategy/new-global-social-media-research> [Accessed 16 Sep. 2019].

De Stefano, V., 2015. *The Rise of the 'Just-In-Time Workforce': On-demand Work, Crowdwork and Labour Protection in the 'Gig-economy'*. [online] International Labour Office, Geneva. Available at: <www.ilo.org/wcmsp5/groups/public/---ed_protect/---protrav/---travail/documents/publication/wcms_443267.pdf> [Accessed 16 Sep. 2019].

European Commission, 2017. *Germany: Industrie 4.0*. [online] Digital Transformation Monitor. Available at: <https://ec.europa.eu/growth/tools-databases/dem/monitor/sites/default/files/DTM_Industrie%204.0.pdf> [Accessed 16 Sep. 2019].

European Commission, 2018. *Artificial Intelligence: Commission Outlines a European Approach to Boost Investment and Set Ethical Guidelines*. [online] Press Release. Available at: <http://europa.eu/rapid/press-release_IP-18-3362_en.htm> [Accessed 16 Sep. 2019].

Frey, C.B. and Osborne, M.A., 2017. The future of employment: How susceptible are jobs to computerisation? *Technological Forecasting and Social Change*, 114, pp. 254–280.

Gibson, W.F., 1984. *Neuromancer*. New York, NY: Ace.

Goode, L., 2018. *Google CEO Says AI Will Be More Important to Humanity Than Electricity or Fire*. [online] The Verge. Available at: <www.theverge.com/2018/1/19/16911354/google-ceo-sundar-pichai-ai-artificial-intelligence-fire-electricity-jobs-cancer> [Accessed 16 Sep. 2019].

Goodman, P.S., 2017. *The Robots Are Coming, And Sweden Is Fine*. [online] *The New York Times*. Available at: <www.nytimes.com/2017/12/27/business/the-robots-are-coming-and-sweden-is-fine.html> [Accessed 16 Sep. 2019].

Hara, K., Adams, A., Milland, K., Savage, S., Callison-Burch, C. and Bigham, J.P., 2018. A Data-Driven analysis of workers' earnings on Amazon mechanical Turk. In: R. Mandryk and M. Hancock, eds., *CHI '18 Proceedings of the 2018 CHI Conference on Human Factors in Computing Systems Paper No. 449, Montreal QC, Canada – April 21–26, 2018*. Montreal, Canada: ACM.

Hernandez, B.A., 2013. *Instagram Reaches 150 Million Monthly Active Users.* [online] Mashable. Available at: <https://mashable.com/2013/09/08/instagram-150-million-monthly-active-users> [Accessed 16 Sep. 2019].

Kelly III, J.E., 2016. *Computing, Cognition, And the Future of Knowing: How Humans and Machines Are Forging a New Age of Understanding.* [online] Computing Research News. Available at: <https://cra.org/crn/2016/09/computing-cognition-future-knowing-humans-machines-forging-new-age-understanding> [Accessed 16 Sep. 2019].

Khan, H.N., Hounshell, D.A. and Fuchs, E.R.H., 2018. Science and research policy at the end of Moore's law. *Nature Electronics*, 1(1), pp. 14–21.

Kjellberg, A., 2010. Kollektivavtalens täckningsgrad samt organisationsgraden hos arbetsgivarförbund och fackförbund [The collective agreement's coverage and the degree of organization of employer associations and unions]. *Studies in Social Policy, Industrial Relations, Working Life and Mobility*, 1(2013:1), pp. 1–93.

Kramp, T., van Kranenburg, R. and Lange, S., 2013. Introduction to the internet of things. In: A. Bassi, M. Bauer, M. Fiedler, T. Kramp, R. Van Kranenburg, S. Lange and S. Meissner, eds., *Enabling Things to Talk: Designing IoT solutions with the IoT Architectural Reference Model.* Heidelberg, Germany: Springer, pp. 1–10.

Kuchler, H., 2018. *Instagram Targets YouTube With Standalone Video App.* [online] Financial Times. Available at: <www.ft.com/content/a4541cb0-74bd-11e8-aa31-31da4279a601> [Accessed 16 Sep. 2019].

Lee, M.K., Kusbit, D., Metsky, E. and Dabbish, L., 2015. Working with machines: The impact of algorithmic and data-driven management on human workers. In: *Proceedings of the 33rd Annual ACM Conference on Human Factors in Computing Systems – CHI '15, Seoul, Republic of Korea, April 18–23.* New York, NY: ACM Press, pp. 1603–1612.

Lundh, C., 2010. *Spelets regler: Institutioner och lönebildning på den svenska arbetsmarknaden 1850–2010 [Rules of the Game: Institutions and Wage Formations in the Swedish Labor Market 1850–2010].* 2nd ed. Stockholm, Sweden: SNS Förlag.

Moore, G.E., 1965. Cramming more components onto integrated circuits. *Electronics*, 38(8), pp. 114–117.

Nedelkoska, L. and Quintini, G., 2018. *Automation, Skills Use and Training.* OECD Social, Employment and Migration Working Papers, No. 202. Paris, France.

OECD, 2016. *Digital Government Strategies for Transforming Public Services in the Welfare Areas.* [online] OECD Comparative Study. Available at: <www.oecd.org/gov/digital-government/Digital-Government-Strategies-Welfare-Service.pdf> [Accessed 16 Sep. 2019].

Söderqvist, F., 2017. A Nordic approach to regulating intermediary online labour platforms. *Transfer: European Review of Labour and Research*, 23(3), pp. 349–352.

The Terminator. 1984. Directed by J.F. Cameron. USA: Orion Pictures.

Thomas, D., 2017. *Artificial Intelligence Investing Gets Ready for Prime Time.* [online] Forbes. Available at: <www.forbes.com/sites/greatspeculations/2017/10/25/getting-ready-for-prime-time-of-artificial-intelligence-investing> [Accessed 16 Sep. 2019].

Waters, R., 2018. *Businesses Turn to Software Robots for Office Work.* [online] Financial Times. Available at: <www.ft.com/content/4580f43a-2191-11e8-9efc-0cd3483b8b80> [Accessed 16 Sep. 2019].

7 AI leadership and the future of corporate governance

Changing demands for board competence

Fernanda Torre, Robin Teigland and Liselotte Engstam

1. Introduction

When discussing digitalization and its impact on the future of labor, much of the practitioner and academic literature tends to focus on labor in general. However, this chapter takes a different approach and focuses on one subset of labor that to date has attracted considerably less attention in the literature: corporate boards. Corporate boards may be the organizational unit that has the most influence on firm performance and behavior as they influence decision-making and are involved throughout the different phases of a firm's strategic process (Huse, 2007). Leblanc and Gillies (2005, p. 6) even argued, "Nothing is more important to the well-being of a corporation than its board of directors". This should not be too surprising as corporate boards and executives are responsible for major strategic decisions such as mergers and acquisitions, new product launches, and digital transformation (Libert, Beck and Bonchek, 2017). Today, however, corporate boards are increasingly struggling with taking the right decisions. For example, a 2015 McKinsey study found that only 16% of board directors said they fully understood how technological advances were changing their company's trajectory and how the dynamics of their industry were changing (Sarrazin and Willmott, 2016).

Due to this increasing complexity of board tasks, it is expected then that digitalization will not lead to the automation or obsolescence of board directors within the foreseeable future (Bankewitz, Åberg and Teuchert, 2016; Libert, Beck and Bonchek, 2017). Rather, research and industry both point to the need to continuously develop the competence of boards to successfully tackle the many challenges brought by digitalization, especially as the external environment continues to become more volatile and uncertain due to digital technologies. For example, a recent study by MIT found that firms whose boards of directors were *digitally savvy*, i.e., members had "an understanding, developed through experience and education, of the impact that emerging technologies will have on businesses' success over the next decade" (Weill et al., 2019, p. 17), significantly outperformed other firms on key metrics – such as revenue growth, return on assets and market cap growth (Weill et al., 2019).

Of all the various digital technologies, artificial intelligence (AI) has been predicted by global leaders across industries to have a greater impact on the world than the internet (PWC, 2019). Indeed, it has even been predicted that AI will become the basis of essential competitive advantage when employed for strategic and operational decision-making, similar to electricity in the Industrial Revolution and enterprise resource planning software (ERP) in the information age (Libert, Beck and Bonchek, 2017). However, AI is still poorly understood by firms and their leaders, and the majority are still unsure as to when and how AI should be implemented (EY, 2018).

To date, the majority of activities by researchers and practitioners alike have focused on the implementation of AI at the operational level of firms (Acemoglu and Restrepo, 2019). Few are investigating what impact AI will have on the governance of organizations and how corporate boards may need to develop their competence to successfully lead their organization in this new evolving AI-based era. This seems surprising as the governance of AI, and the "big data" on which AI is based, is predicted to become one of the greatest board issues in the next ten years (Featherstone, 2017).

In order to address this research gap, we embarked on a two-year study investigating how boards will govern and leverage AI. This chapter presents some of the preliminary results from this study based on a literature review and a series of interviews with leading global experts in corporate governance and AI as well as with chairmen, board directors and top management in some of Sweden's largest multinationals. In particular, we limit our discussion in this chapter primarily to two competence areas that we propose corporate boards need to develop in order to successfully govern in a world where AI is increasingly the basis of competitive advantage: (1) guiding AI operational capability and (2) supervising AI governance capability. We also present the *Boards 4 AI Leadership Matrix*, a tool that we suggest boards may use to facilitate the development of these competence areas. In addition, we touch briefly on how AI may change the future of board work such as new board processes and augmenting board tasks (Libert, Beck and Bonchek, 2017).

The structure of this chapter is as follows. The next section presents the background for this chapter. Section 3 presents the first competence area – Guiding AI operational capability – while Section 4 presents the second competence area – Supervising AI governance capability. Section 5 then introduces our *Boards 4 AI Leadership Matrix*. Section 6 presents a reflection on how AI may influence board work in the future, followed by our concluding remarks in Section 7. Again, it is important to note that the aim of this book chapter is not to focus on the future of labor in general, even though boards will themselves have a major impact on organizations and the future of work. Rather, the primary focus is on labor at the board level and how the implementation of AI in industry will require boards to develop new competence areas to successfully govern. Furthermore, we should note that while there are different models of corporate governance based on a number of factors, e.g., ownership models, development stages, jurisdictions, this study takes a more generalist approach and applies a broad view on the competence areas that boards need to develop to ensure AI leadership.

2. Background

Corporate boards and top management are ultimately responsible for a firm's success as they are the ones taking strategic decisions and thereby putting the firm at risk. Today's competitive environment is becoming increasingly more volatile and uncertain, leading boards to take on even more strategic risk. The challenge arises when board members lack the necessary competence to take such strategic decisions, in what has been labeled "ungoverned incompetence" (Cebon, 2017). In other words, ungoverned incompetence occurs when the board tries to make the right decision, yet it ends up making the wrong one due to a lack of competence by board members (Cebon, 2017). One of the most well-known examples is Lehman Brothers during the global financial crisis when the board took the decision to invest in a product that it did not understand (Cebon, 2017). Since then many boards have failed to take the right decisions for their firms, especially when it comes to digitalization and new digital business models. For example, more than 50% of the firms that were on the Fortune 500 list in the year 2000 have disappeared from this list due to digital disruption (Nanterme, 2016). Furthermore, MIT research in 2018 on more than 1000 multinationals with over USD1 billion in revenues showed that firms with boards with a relatively low level of digital competence had significantly lower revenue growth, lower ROA, and lower market growth than those firms with digitally competent boards (Weill et al., 2019).

One of the authors of this report, under the auspices of Digoshen AB, investigated further how firms and their boards are meeting changing competitive environments due to digitalization. Building on research on digital transformation by organizations such as MIT, Institute of the Future, and the Centre for Creative Leadership and Altimeter as well as their own research and work with clients, Digoshen AB found that those firms that are digital leaders in their industry have relatively high capabilities in two areas: "digital *business* capability" and "digital *leadership* capability" (Engstam and Caroan, 2016; Pagano, 2017). In other words, as the risks continue to rise due to an increasingly complex and uncertain environment, it is not enough for a firm to merely have a high level of *digital business capability*, i.e., the use of digital technologies in areas such as local and global marketing efforts as a means to enable collaboration across firm boundaries, as the basis for a new customer value proposition or business model, and as a driver of rethinking the firm. Rather, the firm must also have a strong *digital leadership capability* to ensure successful digital transformation, i.e., digital competence at the board level, participation by the board in the identification of digital opportunities, the board's monitoring of risks related to digital transformation, and the board's use of social media and other digital technologies to share knowledge, listen to customers and increase visibility of their company.

To learn more about firms' digital business and digital leadership capabilities, Digoshen administered a survey to board directors within the European Confederation of Directors Association (ecoDa) and the Swedish Academy of Board Directors during 2016. Approximately 400 board members answered the survey with the majority of respondents from ten European countries and others from the US, Australia, China and Africa. A second survey was then conducted with approximately 400 board members from 2017 to 2019, including participants from the Swedish Academy of Board Directors Chairman Program and members of the INSEAD Directors Network, a global network.

Comparing the Digital Business Capabilities results from the surveys revealed that digital transformation was predominantly at the functional level, with only some firms starting to have their strategy influenced by digital trends. While only 30% initially had a digitally influenced vision, this more than doubled to 73% in the second survey. As for using digital technologies to understand customers better, this increased from 40% to 58%. Another interesting change was that the percent of firms launching new business models rose from 40% to 58%, with 47% starting to sacrifice existing revenue – up from 25%. While only 14% had started to look into the next wave of digital opportunities, such as AI, robotics and 3D printing, this only increased to 19% in the second.

Looking at digital leadership capabilities, more than 50% responded in 2016 that their CEOs had been leading key strategic digital business initiatives, and this number climbed to 73% in the second survey. While initially 25% had clarity in roles and responsibilities in governing digital initiatives, this only improved to 36% in the second round. One area that surfaced as critical for success was the monitoring by boards of the risks related to digital technologies and digital transformation. However, 60% of the companies were not clear about, nor did they monitor, their digitally related risks, and this number remained the same in the second set of results, even though the digital risks for most organizations had increased. A notable change was that 66% of board members claimed in the second survey to be listening via social media to customers, employees, partners, competitors and industry experts, up from 50% in 2016.

This research and these surveys revealed that digital transformation is affecting not only firms but also the work of boards. Boards are adapting their focus, changing their behavior and increasing their competence. However, the focus by boards has been primarily on understanding digitalization's influence on a firm's operations and less on how to lead digital transformation.

In our current research project, 4boards.ai, we built on the previously mentioned findings to further investigate the impact of digitalization on corporate governance by narrowing our focus to AI. We have chosen this specific focus since AI is the digital technology that is expected to have the greatest impact on firm competitiveness, and as previously noted, AI governance, and the "big data" on which AI is based, is predicted to become one of the greatest board issues in the next ten years (Featherstone, 2017). Thus, one of our underlying research aims is to examine the competence that board directors need to develop in order to successfully govern their firms in a world where AI is increasingly the basis of competitive advantage.[1]

To fulfill this aim, we present the preliminary results from our research based on extensive firsthand board work experience by one of the team members, a systematic review of academic and practitioner literature on corporate governance and AI implementation in industry, and a series of interviews with board members of leading multinationals and with global AI experts. In short, we found that while boards are aware of the importance of AI implementation as a key competitive advantage, they do not yet have sufficient competence in two key areas to best steward their companies within AI Leadership: (1) guiding AI operational capability and (2) supervising AI governance capability. Next we discuss each of these competence areas in depth, basing our discussion on our preliminary findings from our research.

3. Guiding AI operational capability

As representatives of shareholders and stakeholders, boards cannot ignore the extraordinary value-creation opportunities that AI is enabling in today's digital era characterized by a constantly changing strategic context, short-term strategizing, availability of large amounts of data and crowd-approaches to knowledge sharing (Bankewitz, Åberg and Teuchert, 2016). Through applications such as recommendations, live translations, facial recognition, autonomous vehicles and smart cities, AI offers tremendous opportunities and already is changing how value is created by firms and delivered to end users. By 2025, some 75.4 billion devices will be connected globally, compared with 26.6 in 2019 (Statista, 2019). This hyper-connectedness will generate unique innovation opportunities as well as completely new relationships between customers, suppliers, stakeholders, regulators and the greater ecosystem. Looking into the future, these relations will manifest themselves in the dissolving borders of traditional pipeline-based firms toward multi-sided business models and collaborative platforms, which will, in turn, enable new business structures in the form of networked ecosystems (Ringel et al., 2019).

Digitalization in general, and AI in particular, creates a unique context for sensing and seizing new opportunities, i.e., both the process of identifying opportunities before they arise and the process of responding to these same opportunities (Bankewitz, Åberg and Teuchert, 2016). In order to both sense and seize opportunities, boards need to have sufficient competence to guide AI operational capability, which we discuss next.

3.1. Guiding the gathering, harvesting and analysis of big data

Data are a new type of asset that organizations need to consider since our digitized society has brought to light a key aspect of technology: the connectedness between different nodes in the system. The 2018 New Vantage Partners annual executive survey shows that today, for the first time, large corporations report that they have direct "access to meaningful volumes and sources of data which are providing AI solutions with sufficient meaningful data to detect patterns and understand behaviors" (NewVantage Partners, 2018, p. 7). This is probably because it is now widely accepted that the size of available data sets represents a competitive advantage (Lauterbach and Bonime-Blanc, 2018). Indeed, data sources are numerous and include publicly available open data sets (external), data created by a company's customers, suppliers and other partners (and collected by the company within the ecosystem) and data created by the company itself (internal). Additionally, firms are beginning to take note of "alternative data", an expression developed by investment companies to label data from non-financial and non-traditional sources to improve investment decisions (Kolanovic and Krishnamachari, 2017). Big and alternative data can come from individuals (e.g., social media, news, reviews, web searches/personal data), business processes (e.g., transaction data, corporate data, government agency data), and sensors (e.g., satellites geolocation, other sensors) (Kolanovic and Krishnamachari, 2017).

Gathering quality data and building a reliable data-lake to train algorithms is no easy task. In fact, one of the most challenging tasks of building an AI program is the cleaning, preparing and labeling (tagging) of data (Lauterbach and Bonime-Blanc, 2018). Accenture suggested in a report that the firm's reluctance of investing in AI is largely driven by data concerns, as 48% of surveyed companies reported data quality issues, while 36% reported a lack of sufficient data for training and 35% reported data existing in silos (Sinclair, Brashear and Shacklady, 2018). Thus, boards need to develop an understanding of not only the gathering but also of the harvesting and analyzing of data. To address this challenge, boards can learn from the *Data Management Life Cycle* as proposed by the World Economic Forum/Accenture (WEF, 2018) that consists of four steps:

1. Data origination: Strong data infrastructure to enable data harvesting
2. Data storage: Robust data warehousing to enable storage (combination of on-premise, cloud and hybrid models)
3. Data structure and analysis: Capabilities to structure and analyze data (data quality over data quantity)
4. Communication and action: Tools and assets to communicate and take action on insights

Furthermore, our research revealed that the timely collection and harvesting of data will become increasingly critical as boards will need faster and more transparent indicators of the status of the business and industry in order to gain insights relative to strategic decisions.

Finally, our research finds that there will be a need for boards to acquire a deeper understanding of the complexities of data ownership and data access rights issues. Boards need to make balanced decision regarding their company's usage of data and who they should protect among their stakeholders.

3.2. Guiding AI-driven innovation

AI presents organizations with the opportunity to innovate their businesses in a multitude of ways, ranging from incremental improvement to complete reinvention (McWaters, 2018). Table 7.1 provides an overview of the innovation that AI enables as well as some examples provided by the World Economic Forum:

Table 7.1 From core to radical innovations with AI.

Leaner, faster operations	AI allows operational enhancements, such as improving efficiency, decreasing costs and freeing capacity. Example: Using automation and pattern detection to improve core business processes.
Tailored services, products and advice	AI resolves traditional trade-offs between cost and customization, enabling tailored products at near-zero marginal cost. Example: Big data analytics for personalization.
Ubiquitous presence	AI expands reach by enabling better self-serve applications that allow more services to be delivered digitally. Example: A suite of offerings that capture new market share by using AI to offer a seamless experience automating the purchasing process.
Smarter decision-making	AI enhances decision-making capabilities, unlocking novel insights that drive improved performance. Example: Identification of unexplored patterns to outperform markets.
New value propositions	AI redefines core offerings, unlocking untapped segments and revenue opportunities through new products and services. Example: Big data analytics to identify new areas of customer demand.

Source: (Adapted from McWaters [2018] and WEF [2018])

In order for firms to take advantage of AI opportunities, corporate boards need to be able to implement a portfolio approach addressing a range of AI opportunities. From exploiting AI for leaner, faster operations to exploring AI for new value propositions, a portfolio approach is important since some projects will generate quick wins while others will focus on transforming end-to-end workflows (Lauterbach and Bonime-Blanc, 2018). It is important not to mistake the mere launch of a few isolated use cases as complete AI deployment. McKinsey & Company has noted that if an AI strategy is not implemented beyond a few use cases, then this is a warning signal of AI program failure (Fleming et al., 2018). Additionally, corporate boards need to understand the strong relationship between successful innovation management in general and AI innovation capabilities, i.e., a firm that is successful at innovation generally is successful at AI deployment (Ringel et al., 2019).

Some of the capabilities common to innovation management and AI implementation are the following:

- Cross-functional, diverse teams working on AI and algorithmic development are a "must-have in the adoption of safe and beneficial technology" (Lauterbach and Bonime-Blanc, 2018, p. 145).
- Strong feedback loops in an iterative development process in close connection with business development are required because "the best algorithms will

not succeed in delivering results if they do not improve a product or a service experience for a customer" (Lauterbach and Bonime-Blanc, 2018, p. 145).

- Clear top-management buy-in since if the "executive leadership team is not ready to redesign business models and end-to-end processes across the whole organization, a company may never benefit from the full potential of AI" (Lauterbach and Bonime-Blanc, 2018, p. 147).
- An innovation culture that embraces both a "succeed fast" approach to innovation and that focuses on finding unmet real needs (Main, McCormak and Lamm, 2018).
- Training and hiring programs with innovation at the core is a key enabler for digital transformation: "whichever strategy it pursues, an organization must offer its workforce an engaging work environment that enhances the employee experience, incubates ideas and encourages creative thinking" (WEF, 2018, p. 15).

Of note is that some of the most urgent opportunities for AI-driven innovation are related to the 2030 Agenda for Sustainable Development adopted by all UN Member States in 2015 (Rolnick et al., 2019). Due to the complexity of social-ecological systems, AI presents specific opportunities within big data analysis and the management and optimization of the global technological infrastructure that extracts and develops natural resources such as minerals, food, fossil fuels and living marine resources. Furthermore, algorithms facilitate global trade flows that form the basis of environmental monitoring technologies (Galaz and Moberg, 2015).

Even though there are strong arguments to use AI in the context of enabling innovation, our research shows, however, that the level of AI implementation varies greatly across organizations. This represents an additional area for board consideration since looking into the future, we see that while organizations struggle to invest in their dynamic capabilities for innovation and AI implementation, the "first-mover" advantage might be of key importance in this innovation game. AI is a technology that lends itself to a "winner-takes-all" strategy due to either potential networks effects of the solutions presented or due to the nature of the technology itself. AI does not allow a "plug-and-play" approach, which generates a performance gap between AI "pioneers" that appear to be "pulling further away" from organizations that are still lagging behind (Ringel et al., 2019, p. 8). The performance gap between AI performers and non-performers might be of particular concern for corporate boards since it might require a more ambitious AI deployment strategy, which also increases the pressure for leadership contributing to a distinct "bandwagon" effect. This effect can be described as "a psychological phenomenon in which people do something primarily because other people are doing it, regardless of their own beliefs, which they may ignore or override (Kenton, 2018). The "bandwagon" effect is known to contribute to speculative bubbles; therefore, corporate boards should be considerate of this bias and aim toward a meaningful implementation of AI according to best practices.

To better guide AI-driven innovation in the firm, boards will, however, need to strike the right balance between development and control activities. Our research has revealed to date that currently, most companies spend the majority of their board meetings discussing control issues, thereby greatly limiting time spent on innovation. A better practice would be to develop processes for control outside the larger board meetings either in committees or through online fora and instead devote more board time to discussing development activities while considering the right KPIs (key performance indicators) to reflect this balance. As a consequence, a bigger focus by boards on innovation will require the development of new competences in the board and capabilities in the firm. In order to develop their companies' businesses, boards will need to better understand innovation, technology and sustainability, and their impact on opportunities, threats and new business models.

3.3. Guiding the growth of a digital business ecosystem

As mentioned previously, there is a strong correlation between companies that consider themselves strong innovators and those that see themselves as being strong at AI (Ringel et al., 2019). One would expect this to be primarily true for technology firms; however, a closer look at a BCG report reveals that the most innovative firms are not all technology firms. Rather the most innovative firms are those that develop not only AI but also platforms and ecosystems across their industry regardless of industry (Ringel et al., 2019). While a platform structure is nothing new per se, for example, newspapers have connected subscribers and advertisers for many years, the enhanced ability to capture, analyze and exchange huge amounts of data will increase a platform's value to all (Van Alstyne, Parker and Choudary, 2016). Furthermore, the usage of digital platforms, APIs, IoT technology and new tools for data collection and analysis will allow for new products and services that go beyond the boundaries of traditional business (Fuller, Jacobides and Reeves, 2019). This is a shift that is predicted to have a significant impact in the near future. For example, a McKinsey study showed that an emerging set of digital ecosystems could account for more than USD60 trillion in revenues by 2025, or more than 30% of global corporate revenues (Bughin et al., 2018).

A platform is a specific kind of ecosystem, i.e., all platforms are ecosystems but not all ecosystems are platforms. A platform leverages "networked technologies to facilitate economic exchange, transfer information, connect people, and make predictions . . . thus a platform drives value from its role as an intermediary" (Fenwick, McCahery and Vermeulen, 2019, p. 3). Currently, seven of the 12 largest companies by market capitalization – Alibaba, Alphabet (Google), Amazon, Apple, Facebook, Microsoft and Tencent – are ecosystem orchestrators (Bughin et al., 2018) that use platforms to create value by facilitating exchanges between different yet interdependent groups (Fenwick, McCahery and Vermeulen, 2019). For example, developers of voice-recognition-based smart-home platforms, such as Amazon's Alexa or Google's Home, make it easy for others to create new consumer services that use their AI-enabled platforms – and in the process to attract the critical mass of applications needed to make their platform and thus their ecosystem a clear leader (Ringel et al., 2019).

For most firms, the relation between the implementation of AI and operating in a business ecosystem becomes increasingly relevant as it will be extremely difficult for a firm to implement an advanced AI program completely alone. Currently, the costs of implementing AI and finding the appropriate data scientists are extremely high, especially as AI technology is becoming increasingly relevant for all business sectors, not only within tech firms. As a result, the search for recruiting and retaining AI talent is also becoming more competitive (Perisic, 2018), in what has been called a war on AI talent (Kelnar, 2019). Boards need to develop an understanding of how organizations collaborate in digital business ecosystems to hasten the pace of implementation of an AI program, reduce costs and to potentially tap into value from ecosystem partners. As identified in Accenture Technology Vision 2017, "The competitive advantage of tomorrow won't be determined by one company alone, but by the strength of the ecosystems chosen, and the company's plans to help the ecosystems grow" (Accenture, 2017, p. 39).

Moving forward, corporate boards must develop their understanding of complex adaptive systems. To implement AI through a platform and even to orchestrate a digital ecosystem can be described as the management of a complex adaptive system, i.e., an understanding of the individual parts does not automatically convey an understanding of the whole system's behavior. The management of a complex adaptive system requires what could be called "competing on the edge" that requires "adaptation to current change and evolution over time, resilience in the face of setbacks, and the ability to locate the constantly changing sources of advantage . . . engaging in continual reinvention" (Brown and Eisenhardt, 1998, p. 19). The goal is flexibility, requiring the board to be able to shape strategy where the organization both influences and is influenced by ecosystem stakeholders, while evolving the ecosystem for mutual benefit (Fuller, Jacobides and Reeves, 2019). In other words, corporate boards will need to support the development of an organization's adaptive capabilities so they can provide real-time responses to strategic issues and opportunities provided by AI (Bankewitz, Åberg and Teuchert, 2016).

An implication of digital business ecosystem participation is that boards will need to be more dynamic in their work. Today many companies have four to seven board meetings a year, which, given the complexities of governing a digital business ecosystem, will probably need to be complemented with more flexible options. For example, board work can be complemented with full or temporary committees for areas such as innovation and technology. Additionally, board meetings could be both physical and virtual, thereby enabling the ability to react more quickly to changing conditions. For example, virtual meetings could occur in between the traditional face-to-face board meetings to discuss upcoming opportunities or threats.

4. Supervising AI governance capability

While AI offers considerable innovation opportunities in both strategic and operational areas (Vinnova, 2018), we have found that corporate boards must also develop the competence to supervise AI governance capability in order to successfully govern and mitigate the risks that go hand-in-hand with implementing AI within an organization. Next we discuss three areas in which boards need to develop their competence in order to supervise AI governance capability.

4.1. Supervising data management, ethics and black box decision-making

Currently, few companies perceive data as a valuable asset, and thus they do not devote sufficient attention to how they manage their data. As a result, they lag behind in implementing clear rules and policies to ensure data are trustworthy, clean and usable (Protiviti, 2019). Accenture found that 79% of executives responded that their organizations were basing their most critical systems and strategies on data, yet many had not invested in the capabilities to verify the truth within (Accenture, 2018). This is a basis for concern for boards because if an AI system is based on incomplete or poor data quality, it could lead to the wrong training of the algorithms, opening concerns for the trustworthiness of the AI decisions.

Besides data quality, algorithms that are programmed by humans may be subject to bias, leading to ethical conflicts. Programmers might inject their judgments into the code and train algorithms with biased data, leading to machines being even more untrustworthy or incapable of delivering neutral results. A recent report by Microsoft identified five areas for potential bias: dataset bias, associations bias, automation bias, interaction bias and confirmation bias (Chou, Murillo and Ibars, 2017). Dataset bias occurs when algorithms are trained on data with low diversity, leading to a generalization that will underrepresent certain elements. Association bias takes place when the data used to train an AI model reinforces and multiplies a cultural bias. In the same way, the automation of decisions might override social and cultural considerations and automate goals that go against human diversity. Interaction bias takes place when the bias comes from humans that have distinctively tampered with the AI in order to make it biased. And finally, confirmation bias takes place when AI algorithms serve up content that matches what other people have already chosen, thereby confirming preconceptions.

The best way to address possible biases is to have algorithms developed in a context of diversity, in terms of disciplines, demographics, experience and knowledge, as this will be the best way to anticipate ethical failures and minimize the risks of unintended AI harm (Pauwels, 2018). Regrettably, we live in what can be called a diversity crisis. For example, it took a group of engineers who call themselves "black in AI" to uncover the scandal of how facial recognition technologies failed to trace the features of individuals with darker skin tones (Snow, 2018).

Furthermore, there are also concerns with what is called AI black box decision-making, which can create a liability minefield. Black box decision-making refers to machine learning and the fact that it might not be possible to trace back to why certain decisions were taken by a firm's AI system, making it nearly impossible for a firm to explain its AI actions to stakeholders, the general public or regulators. AI's black box decision-making can result in considerable accountability challenges since responsibility for a decision or action taken may be difficult to pinpoint – was it the programmer who wrote the initial algorithm, the machine that learned the wrong thing due to improper data, or perhaps the company's processes that led to a failure to update the algorithm? Moving forward, as AI applications are highly complex and many managers do not completely understand how they work, regulators may be reluctant to approve AI systems if they cannot be thoroughly explained in how and why decisions are made (Protiviti, 2019). To avoid this, boards need to ensure that the firm strives for the "explainability" of its AI systems in order to be transparent and provide an explanation for decisions and actions made (AI HLEG, 2019).

Furthermore, in our interviews we have found that boards tend ***not*** to be aware of the current applications of AI in their organization, especially when it comes to off-the-shelf solutions. For appropriate AI governance, corporate boards will need to stay informed of the individual AI application uses in their companies as well as the model reviews done for those algorithms.

In a context of clouded accountability and relatively low expert understanding, AI raises risks of reputational damage and ethical concerns. Major AI leaders, such as Microsoft, Intel, Alphabet Inc./Google and IBM, have recently published social responsibility principles, showing an interest in self-regulation and taking on real-world problems. These documents provide a look into potential foreseeable troubles. For example, Microsoft, in its annual Securities and Exchange Commission (SEC) report filed in June 2018 and referent to the previous year, has put it very clearly:

> Issues in the use of artificial intelligence in our offerings may result in reputational harm or liability . . . AI algorithms may be flawed. Datasets may be insufficient or contain biased information. Inappropriate or controversial data practices by Microsoft or others could impair the acceptance of AI solutions. These deficiencies could undermine the decisions, predictions, or analysis AI applications produce, subjecting us to competitive harm, legal liability, and brand or reputational harm.
>
> (Microsoft, 2018, p. 28)

Following suit also Alphabet Inc. (holding company of Google) has also reported the risks of AI:

> New products and services, including those that incorporate or utilize artificial intelligence and machine learning, can raise new or exacerbate existing ethical, technological, legal, and other challenges, which may negatively

> affect our brands and demand for our products and services and adversely affect our revenues and operating results.
>
> (Alphabet, 2018, p. 7)

Even if some IT multinationals today show concern by showing interest in self-regulation and publishing social responsibility principles, it is unclear as to how regulators will act in the future. Perhaps a set of principles may materialize into standard practices within industry that are sufficient for regulators or perhaps strict regulations may be developed. Looking into the future, even prior to potential regulation, corporate boards will have to take a stance on the ethical implementation and regulation of AI. For example, Microsoft recently announced that it had decided to decline the sale of its facial recognition technology to both a California law enforcement agency and to an unnamed capital city because of human rights concerns (Menn, 2019).

In this context, it seems clear that boards should raise their competence in this area. For example, boards should be able to supervise the creation and monitoring of a data governance framework for the firm. This framework should focus on ensuring that the firm's data and processes are developed with a clear purpose and fulfilling ethical obligations. This is distinguishable from the current practices of many firms that aim to merely fulfill legal obligations. Currently, there is not an established and mature model that is consensual among industry, policy makers and academics, although several models are being tested (Micheli et al., 2018). This is relevant for boards because, in the words of Anastassia Lauterbach, "A visionary board should ask how the company thinks about data to solve strategic and operational problems, whether there is a solid data governance framework in place, and if and when the business considers providing wide access to data, allowing as many people as possible to find valuable insights" (Lauterbach, 2018, para.7). Additionally, the data governance framework should be linked to the firm's regulatory actions and cybersecurity activities, the subject of the next section.

4.2. Supervising AI security

Not only must boards develop the competence to ensure they can adequately supervise data governance, but they must also develop significant competence related to understanding how to best ensure data and AI system security and protection from hackers and similar ransomware activities (Else and Pileggi, 2019).

There are different kinds of cybersecurity threats, and one of the most commonly discussed is hacks, i.e., an unauthorized intrusion into a computer or a network, such as malware, phishing, man-in-the-middle attack, denial-of-service attack, SQL injection, among others (Cisco, 2018). This action can be perpetrated with different intentions, from stealing corporate secrets to executing ransomware attacks such as the 2017 WannaCry that led to losses estimated to reach USD4 billion (Berr, 2017). Robert Mueller, during his time as Director of the FBI, explained this increasing threat in an RSA Cyber Security Conference, "I am convinced that there are only two types of companies: those that have been hacked and those that will be. And even they are converging into one category: companies that have been hacked and will be hacked again" (Mueller, 2012, para.63).

Hacks are not always conducted by external malicious software. They can also be conducted through social engineering, which relates to the action of using psychological manipulation to trick targeted users into making security mistakes or giving away sensitive information. As corporations devote more resources to IT departments and vamp up firewalls, hackers are increasing their social engineering efforts to bypass these defenses, by going further beyond technology and targeting the aspect of a corporation where security has been the weakest – its employees.

An example of this practice is *phishing*, which can be defined as email scams that use social engineering to attempt to trick the recipient into providing confidential information or unintentionally installing malware through the use of links or attachments (Proofpoint, 2019). According to the enterprise security company, Proofpoint, 83% of global info-security respondents experienced phishing attacks in 2018, which is up from 76% the previous year (Proofpoint, 2019).

To ensure AI security, boards should understand the relevant talent issues. Questions arise such as which talent should be outsourced, when, and how versus whether talent should be employed in house? While conventional security principles are about keeping the bad guys out, social engineering raises another type of question: what to do when the "bad" guys are already inside (Gregersen, 2018)? Thus, one area of discussion for boards is how to ensure employee education as employees who feel they have sufficient training and support to deal with technology at work will be better at their jobs and save the company from hacking attacks along the way. In fact, according to Proofpoint's 2018 report, security awareness training had a significant impact on preventing attacks, and nearly 60% of organizations saw an increase in employee detection once their staff was better trained to identify possible attacks (Spadafora, 2019). Looking forward, one suggestion is that boards understand how to drive AI security implementation by applying the same friendly customer-centric experience that companies have with clients on their own employees (Gregersen, 2018), addressing both cybersecurity and talent retention.

Moreover, AI systems are particularly susceptible to attacks (Mitchell, 2019) for two main reasons: 1) machines are being used to train other machines – which scales the exposure of compromised pieces of code, and 2) machines can be fooled by adversarial examples, i.e., inputs optimized by an adversary to produce an incorrect model classification (Elsayed et al., 2018; Lauterbach, 2018). Image classification systems could be attacked by adding a layer of noise distortion, e.g., fool an algorithm to identify a school bus as an ostrich (Szegedy et al., 2013). Autonomous driving systems could be attacked by, for example, placing stickers on a STOP sign to fool the self-driving car to interpret the sign as a "Speed Limit 80" sign (Eykholt et al., 2018). Finally, speech recognition systems could be attacked by, for example, an audio signal changed so that it is white noise to a human but is, in fact, a command to a machine (Carlini et al., 2016). Thus, corporate boards should be extra vigilant and cognizant that such attacks on their AI systems can occur. Corporate boards need to develop the competence to develop and reevaluate a routine to foresee where such attacks may occur and how to both monitor and sufficiently respond if and when an attack occurs.

Finally, besides training employees to avoid attacks, boards need to be ready to handle worst-case-scenario situations that might happen anyway. The board should have a clear process on how to deal with AI security breaches, such as how to handle reputation issues in the media or even how to run offline since "pen-and-paper" operations might be necessary in the case of extensive attacks.

All these different aspects related to governing AI as a black box that is susceptible to cyberattacks will require firms to take an intelligent, proactive and multi-layered attitude toward cyberattacks (Grasso, 2019). The implication for how board work may change is that in the future boards will need to better balance the company's focus on long-term strategies that will have to be clearly communicated with all stakeholders, shifting away from more traditional short-termism.

4.3. Supervising business ecosystem leadership

As AI businesses move into ecosystem configurations and platform models, boards will need to learn to "govern" all the stakeholders and the organization's relation to them. Traditionally, as firms grew, they would develop increasingly hierarchical structures as a way to manage the complexities of scale. Although this system might have been useful in the past, in today's dynamic and uncertain business reality, it raises challenges related to the bureaucratization of firm culture (Fenwick, McCahery and Vermeulen, 2019). Today's reality demands the creation of flat, open and inclusive organizations that take advantage of stakeholder talent. Together with live data drawn from the ecosystem, a flexible organization raises the opportunity to automate decisions in what, for example, Alibaba calls the "self-tuning enterprise" (Fuller, Jacobides and Reeves, 2019). As such, AI ecosystems and platforms should be built around the idea of delivering constant innovation via open and inclusive processes of collaboration and co-creation (Fenwick, McCahery and Vermeulen, 2019). For boards this means a flexible and holistic approach to stakeholder governance, which boards can develop following the three-step strategy (Fenwick, McCahery and Vermeulen, 2019):

(A) Leveraging current and near-future digital technologies to create more "community-driven" forms of organization
(B) Building an "open and accessible platform culture"
(C) Facilitating the creation, curation and consumption of meaningful "content"

Besides governing stakeholders to harvest their talents, boards need to learn to govern specific aspects related to data usage and data rights throughout the ecosystem, similar to what many firms have implemented when it comes to sustainability and supply-chain management. Boards will need to ensure that all participants conform to local regulations for the jurisdictions in which the organization exists (WEF, 2018). This represents a big shift in boards' focus. It will no longer be enough to guarantee a firm's own governance, but it will be increasingly relevant to apply all aspects of governance and risk management to the different partners and stakeholders of the ecosystem.

When addressing stakeholder governance, an extra point for boards to understand is the asymmetry of power between the tech-leaders and the tech-takers. Together with the powerful network effects from digital platforms, this lends itself to a "winner-takes-all" scenario (Lauterbach and Bonime-Blanc, 2018), as addressed in Section 3.2. In this context, boards should be sure to evaluate the benefits and drawbacks when choosing or joining an ecosystem or choosing an AI technology vendor.

Finally, boards should also develop the competence required to enable complex ecosystems. As mentioned previously, the management of a complex adaptive system requires adaptation and indirect shaping in what is called a *shaping strategy*. This still feels counterintuitive to many boards and leadership teams more used to a traditional "plan and execute" controlling approach. A BCG Henderson Institute study found in a strategy simulation game that only 18% of managers succeeded in ecosystem strategy versus an AI opponent, while they would succeed 71% in a classical strategy simulation (Fuller, Jacobides and Reeves, 2019). As a consequence, boards will need to increase their focus on stakeholders from a primary focus on shareholders.

5. Boards 4 AI leadership matrix – a tool for developing board competence

As previously discussed, AI warrants the close attention of the board because firms that successfully implement and govern AI can disrupt the market, drive growth and manage their risk. To support boards to develop the two competence areas necessary to successfully steward the firm to leverage AI, we have developed the tool presented in Figure 7.1, based on the preliminary findings from our research.

To apply this tool, we suggest that a firm's board members should *individually* evaluate where the board is in terms of its competence in the two areas: (1) guiding AI operational capability and (2) supervising AI governance capability. The board can then use the results as a basis for discussion on how the board can improve its two sets of AI capabilities in the firm. For example, a board may not know where or how AI is being implemented in their firm. If such is the case, the board could use this opportunity to address this gap and develop a critical opinion about how the board should develop its competence in order to guide the firm's AI operations and supervise its AI governance. It is also important for boards to relate themselves to others in their industry as there may be differences across industries. For example, the boards of dominant companies such as Facebook, Amazon, Alibaba, Tencent and Google seem to be building their guiding AI operational capability faster than their supervising AI governance capability. However, in the medical service industry with strict regulations concerning patient data, boards may have a very strong supervising AI governance capability but still a rather low

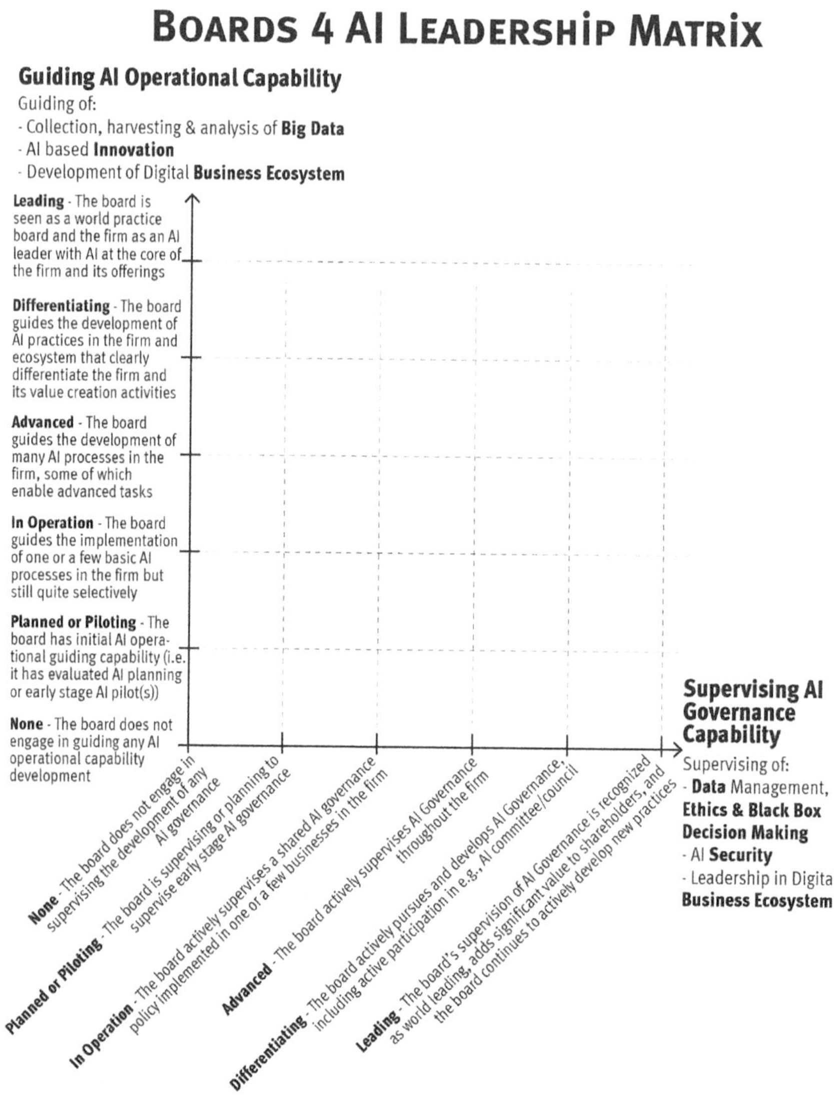

Figure 7.1 Boards for AI leadership matrix.

guiding AI operational capability. Firms in either of these categories would need to leapfrog if they want to become complete AI leaders and need to consider how to fast forward implementation to become the AI benchmark. In this context, as explained in the ecosystem section, it may be faster and more efficient to partner with relevant actors in the firm's ecosystem and build synergies beneficial to all partners. Finally, it is important to note that when a firm's board is a benchmark in both guiding AI operational capability and supervising AI governance capability, then it has the ability to shape the regulatory context, which may prove beneficial in continuing the development of its competitive advantage.

To help boards acquire a better understanding of where they are in terms of the development of their competence, we created a set of questions following the themes addressed in this study. Table 7.2 presents the questions relevant for guiding AI operational capability, following the sections: (3.1) guiding the gathering, harvesting and analysis of big data; (3.2) guiding AI-driven innovation; and (3.3) guiding the growth of a digital business ecosystem.

As mentioned previously, boards need to develop competence not only in regard to guiding AI operational capability but also to supervising AI governance. Table 7.3 presents a set of questions to support boards addressing sections: (4.1) supervising data management, ethics and black box decision-making, (4.2) supervising AI security and (4.3) supervising digital business ecosystem leadership.

6. Beyond competence to the future of board work

Not only will boards need to guide and supervise their firm's AI capabilities, but they will also need to rethink and redesign themselves and their tasks in the context of managing their business to meet the challenges brought on by digitalization (Bankewitz, Åberg and Teuchert, 2016). Through our research, we have identified several areas for board consideration.

One action for boards is to reflect on is how they themselves can become better resources for their organization. For example, should the board focus on personal development or should the board implement specific technical committees that will support the board's digitalization work? In the context of the implementation of an AI program by management, a subset of the board in a technology committee could have a role overseeing and supervising the implementation framework (Lauterbach and Bonime-Blanc, 2018).

Another action is to rethink how to best transition the workforce. One of the biggest issues with AI implementation in firms is job automation in society and the potential rise of unemployment and social unrest (Shewan, 2017). Boards will have an important role in guiding their organization through this important transition. Boards will need to monitor and oversee the decisions regarding the appropriate balance between the automation of processes and jobs versus the augmentation of job tasks, potentially reskilling workers and creating new jobs. Boards should think more broadly about automation and its displacement effect and propose how to create new tasks with AI, thereby engendering a *reinstatement effect* (Acemoglu and Restrepo, 2019).

Table 7.2 Guiding AI operational capability.

Guiding AI operational capability		
Capability level	*Area*	*Questions*
1. None: The board does not engage in guiding any AI operational capability development.	**Data**	How does the board evaluate the firm's internal data collection strategy? How does the board evaluate the firm's routines for exploring data sources?
	Innovation	How does the board assess the relationship between the firm's business strategy, innovation strategy and its data? How does the board challenge itself and management to learn about AI and its implications for the future of work?
	Ecosystem	How does the board rethink the firm's value chain? How does the board evaluate which partners may enable the firm to more quickly ramp up its AI operational capability for faster and more value creating operations and offerings?
2. Planned or piloting: The board has initial AI operational guiding capability (i.e., it has evaluated AI planning or early stage AI pilot[s]).	**Data**	How does the board monitor that the firm identifies and prioritizes both internal and external data to increase value for customers and the firm? How does the board ensure that the firm has the necessary competence to harvest, process and analyze data?
	Innovation	How does the board guide and challenge management to identify use cases to improve offerings or operations through AI in the firm? How does the board evaluate and guide which use cases to further explore?
	Ecosystem	How does the board monitor the exploration of possible partners with which the firm could collaborate within its value chain? How does the board monitor the exploration of possible partners with which the firm could collaborate beyond in the greater ecosystem?
3. In Operation: The board guides the active implementation of one or a few basic AI processes in the firm but still quite selectively.	**Data**	How does the board evaluate the use of third-party data sources by the firm, e.g., external data from an open third party, from the greater ecosystem? How does the board evaluate the firm's data storage, usage strategy and consider data as an asset?
	Innovation	How does the board continuously develop an understanding of the potential impact of AI on the firm's industry, offerings, business operations and business model? How does the board consider which aspects of the business could benefit from increased automation or machine-learning innovation?

	Ecosystem	How does the board ensure alignment with the firm's partners on shared values and communication standards? How does the board consider how AI might fit with other emerging technologies in the ecosystem?
4. Advanced: The board guides the development of many AI processes in the firm, some of which enable advanced tasks.	**Data**	How does the board ensure that the firm has a complete data management life cycle and comprehensive data strategy?
	Innovation	How does the board evaluate the firm's use of AI for incremental innovation in its operations?
	Ecosystem	How does the board ensure that the firm allows for all voices to be heard, including all stakeholders? How does the board monitor and guide which skills and talents to acquire/develop internally vs externally?
5. Differentiating: The board guides the development of AI practices in the firm and ecosystem that clearly differentiate the firm and its value-creation activities.	**Data**	How does the board work with guiding the development of the firm's data management strategy?
	Innovation	How does the board ensure that it continues to improve its guidance of the firm's data-driven innovation strategy and to evaluate how AI can help transform the firm's products or services? How does the board monitor and discourage AI silos and promote collaboration across the firm through policies, processes and systems?
	Ecosystem	How does the board guide the firm to ensure that the hierarchy (competence, titles, reporting lines, goals setting) and agility is compatible with the firm's ecosystem to ensure continued data-driven innovation? How does the board guide the firm to continuously explore partners in the ecosystem to move forward faster or better?
6. Leading: The board is seen as a world practice board and the firm as an AI leader with AI at the core of the firm and its offerings.	**Data**	How does the board guide and ensure that the data framework is applied to all aspects of the firm consistently and systematically? How does the board guide the work with data science talent management?
	Innovation	How does the board guide and ensure that the firm has a leading AI innovation portfolio? How does the board guide and ensure that the firm has the necessary innovation management capabilities to be harvesting both short- and long-term opportunities?
	Ecosystem	How does the board guide and ensure that the firm has the organizational muscle to manage complex relationships across an ecosystem (competing on the edge)?

Table 7.3 Supervising AI governance capability.

Supervising AI governance capability		
Capability level	*Area*	*Questions*
1. None: The board does not engage in supervising the development of any AI governance.	**Data governance**	How does the board learn about data and black box governance and introduce discussions in the board room and with management? How does the board supervise awareness of where AI and algorithms are used in the firm? How does the board supervise the alignment with GDPR?
	AI security	How does the board supervise AI security for the firm, and how is AI security reviewed and reported to the board?
	Ecosystem leadership	How does the board learn about the impact AI will have on different stakeholders and their expectations?
2. Planned or piloting: The board is supervising or planning to supervise early stage AI governance.	**Data governance**	How does the board keep pace of how the firm is registering where AI and algorithms are used in the organization? How does the board keep up-to-date on and supervise what the prioritized use cases and pilots are? How does the board define and agree on relevant data and AI governance to better prepare the firm for the future?
	AI security	How does the board ensure that the cyber risks and data privacy issues related to AI implementation are evaluated and monitored? How does the board review if the implications of a data breach have been fully understood and costed? How does the board encourage the firm to expedite learning related to AI security risks in the broader organization?
	Ecosystem leadership	How does the board continuously evaluate the impact AI will have on customers, employees and society? What are the expectations from shareholders on the board's insights and actions?
3. In Operation: The board actively supervises a shared AI governance policy implemented in one or a few businesses in the firm.	**Data governance**	How does the board discuss and agree on data governance priorities and policies for the firm? How does the board review which data and bias governance policy is relevant for the broader organization? How does the board ensure that the learnings are incorporated in the broader risk and governance system of the firm? How does the board ensure that the algorithms and models used are transparent and reviewed by an independent party?

	AI security	How does the board challenge management to respond strategically to risks associated with AI, including cyber risks? How does the board monitor that the agreed AI governance policy is implemented throughout the firm?
	Ecosystem leadership	How does the board ensure the firm continues to build stakeholder trust as the firm implements AI?
4. Advanced: The board actively supervises AI governance throughout the firm.	**Data governance**	How does the board remain updated on developments related to data governance, AI bias and ethics? How does the board ensure that the firm matures its data practices and runs them in the most effective way?
	AI security	How does the board monitor and supervise that the workforce is sufficiently and continuously equipped to exercise the emotional and ethical judgement required of them?
	Ecosystem leadership	How does the board ensure that the firm lives up to stakeholder expectations in its AI governance activities? How does the board supervise how stakeholders manage compliance with AI policies across the ecosystem?
5. Differentiating: The board actively pursues and develops AI governance, including active participation in e.g., AI committee/ council.	**Data governance**	How does the board ensure that the firm uses state-of-the-art AI, bias and ethics governance practices to create further differentiation for the firm?
	AI security	How does the board evaluate the firm's engagement in industry consortia to further learn and share its insights to increase society protection? How does the board monitor and supervise that the employees are engaged with continuous feedback?
	Ecosystem leadership	How does the board ensure collaboration and increased engagement with stakeholders on AI governance issues?
6. Leading: The board's supervision of AI governance is recognized as world leading, adds significant value to shareholders, and the board continues to actively develop new practices.	**Data governance**	How does the board ensure that the firm takes the lead and helps promote improved data, AI bias and ethical governance and research in the society?
	AI security	How does the board ensure the firm takes the lead and helps promote an intelligent, proactive and multilayered attitude toward research in society toward cyberattacks?
	Ecosystem leadership	How does the board ensure that the firm collaborates and leads the ecosystem in its understanding of AI governance? How does the board work with regulators regarding the shaping of AI governance policies and regulations?

A final point to consider when stewarding an organization is the importance for boards to maintain a focus on diversity. One growing challenge is that women might be at a disadvantage in the future due to higher barriers to transition in terms of time to reskill and due to their lower participation in the STEM fields (science, technology, engineering and mathematics) (Madgavkar et al., 2019). While entirely new occupations will be created, approximately 60% of the new US occupations created to date have been in male-dominated fields (Madgavkar et al., 2019). Boards need to bear this in mind and work even harder to ensure diversity in these emerging positions.

Moving forward, corporate boards will also need to develop the capability to work with AI at the board level. One of the biggest promises of AI is that it can be used to augment human intelligence, thereby changing how we work together, make decisions and manage organizations – from cognitive overload to intelligence augmentation (Rometty, 2016). Several large firms such as IBM and EY are working on digital boardroom solutions to improve board decision-making and time management. One interesting current development is the use of AI by recruiters to support nomination committees, to both assess the board's talent and to search for new board members (Biswas, 2019). Furthermore, at the board level, AI will be able to enable simpler tasks such as automatic speech transcription of board meetings. In the future, AI should be able to facilitate more complex strategic decision-making processes, such as track capital allocation patterns and highlight concerns, review and process press releases to identify potential new competitors, improve operational decision-making by analyzing internal communication to assess employee morale and predicting churn, and to identify subtle changes in customer preference or demographics impact on product development (Libert, Beck and Bonchek, 2017). Other areas include advice on board-relevant topics, such as acquisition candidates aligned with business strategy (Simonite, 2014).

Besides augmenting board members, AI may also augment the board itself by contributing in the role of a board member (Libert, Beck and Bonchek, 2017). As noted by Jeanne Ross, principal research scientist at the MIT Center for Information Systems Research, "companies are succeeding with AI by partnering smart machines with smart people who are learning to take advantage of what these machines can do" (Ross, 2018, p. 11). Already in 2014 an algorithm named Vital (validating investment tool for advancing life sciences) became the "world's first artificial intelligence company director" at Deep Knowledge Ventures, a Hong Kong-based venture capital firm (Zolfagharifard, 2014, para.1). Another example is the Finnish IT service and consulting company, Tieto, that appointed a bot called Alicia T. to be part of the leadership team and went so far to grant Alicia T. voting rights (Suni, 2016).

Boards will furthermore be under increased scrutiny from shareholders and the greater circle of stakeholders, using AI tools to monitor their performance. Investors are increasingly using AI to support their identification of investment objects, and both private and public investors are increasing their use of AI to analyze their portfolio companies in terms of both financial and sustainability performance. Examples such as digital AI analysts that leverage natural language processing and psycholinguistics to analyze nuanced speaking patterns of board members on earnings calls boards will be subject to increased transparency, and boards will need to learn how to act in such an AI world (Sansani, 2018). At this moment the efficiency of these examples could be debated and some even labeled as marketing and communication stunts, but they are still good examples of how AI could support in the creation of insights that will allow more efficient decision-making processes. Indeed, the World Economic Forum reported that 45% of the more than 800 global executives surveyed believed that the first AI machine would be part of a corporate board of directors. However, this would need a change in legal frameworks as the role as board member currently is reserved for natural persons (WEF, 2015).

In summarizing our findings, it becomes clear that boards will need to not only develop their competence to guide AI operational capabilities and supervise AI governance, but they will also need to challenge and adapt their traditional board processes to successfully steward their organizations into an AI future. Next we would like to highlight six additional areas that have emerged from our research to date:

1. Boards will need to better balance their time between development and control activities. Currently, the majority of board work is spent on control, but we foresee a need to move toward a more balanced commitment between development and control as well as the need to develop KPIs accordingly.
2. Boards will need to be more dynamic in their work. Traditional board work will need to be complemented with more flexible options that will allow faster pivoting and strategy adjustments.
3. Boards will need to ensure faster and more transparent insights based on indicators from the business and industry, allowing for better data-led decision-making.
4. Boards will need to expand their focus to include all stakeholders from a narrow focus on shareholders.
5. Boards will need to develop a clearer higher purpose for the firms, raising their ethical standards, versus the status quo of merely fulfilling the lowest legal threshold.
6. Boards will need to better balance the company's focus on the long term with the short term, combining scenario thinking with strategy development and implementation.

7. Conclusion

In this chapter, our purpose was to address one subset of labor – corporate boards – and discuss how one particular digital technology – AI – will influence this subset of labor in the future. More specifically, through extensive board work experience, a systematic review of academic and practitioner literature on corporate governance and AI implementation in firms and a series of interviews with board members of leading multinationals and global AI experts, we found that boards are aware of the importance of AI implementation as a key competitive advantage and that they do not see AI as replacing jobs in the boardroom. Furthermore, we found that boards need to develop two competence areas related to AI to best steward their companies within AI Leadership: (1) guiding AI operational capability – (a) guiding the gathering, harvesting and analysis of big data, (b) guiding AI innovation and (c) guiding the growth of a digital business ecosystem; and (2) supervising AI governance capability – (a) supervising data management, ethics and black box decision-making, (b) supervising AI security and (c) supervising business ecosystem leadership.

In order to facilitate a fruitful discussion among board directors to move toward developing these competence areas, we then proposed our *Boards 4 AI Leadership Matrix*. This tool supports the finding that if a board is only guiding a firm's AI operational capability, while not supervising AI governance, the firm will likely face high risk and strong regulatory headwinds in the future. We recognize that our approach is very general and does not address specific aspects of AI implementation, such as industry-specific questions or in-depth technology issues. Rather, we aim to contribute with a more general understanding of how boards can better develop their competence within guiding and stewarding AI implementation with the hope of further developing modern corporate governance.

Lastly, AI technology and implementation is an extremely dynamic field of research in which there are exciting developments nearly every day. For the next steps, the *Boards 4 AI Leadership Matrix* will be continuously tested and iterated under the project 4boards.ai. For example, it is likely that companies can learn from highly regulated industries, such as financial services or health care. Thus, testing the *Boards 4 AI Leadership Matrix* in these industries could be an interesting point of departure to establish an actionable strategy for AI implementation as supervising AI governance capability may be the preferable starting point. A further area for research is to test the *Boards 4 AI Leadership Matrix* across different governance models as these differ for ownership models, development stages and jurisdictions while keeping in mind specific national legislation and policies.

We conclude by inviting other scholars and practitioners to use the framework presented as well as to build insights and research on the propositions made in this chapter. We believe that the challenges put forward by AI are worthy of a societal discussion that should go beyond the boardroom.

Conflict of interest

The authors have received funding from Sweden's Innovation Agency, Vinnova, for their research project, "4boards.ai", under the coordination of Chalmers University of Technology. The 4boards.ai project aims to identify, codify and disseminate a set of best practices to enable corporate boards to more successfully leverage and govern AI and other exponential technologies in their innovation and sustainability efforts. The project's funding terms require a number of deliverables. This chapter, which summarizes the principal findings from the project's first phase, constitutes one of the deliverables of 4boards.ai.

Acknowledgements

We would like to thank all the interviewees who shared their valuable time and knowledge with us, and in particular we would like to thank FCG Group Board Member Johan Dahlfors and CEO Kristian Bentzer, Combient Chairman Tom Johnstone, CEO Mats Agervi and HR Director Jonas Wettergren, and Epiroc Chairman Ronnie Leten. Finally, we would like to sincerely thank Dr. Peter Crow – governance expert, Dr. Stephanie Woerner – research scientist at the MIT Sloan Center for Information Systems Research, Professor Stanislav Shekshnia – senior affiliate professor of entrepreneurship and family enterprise in INSEAD and Kathryn Myronuk – founding staff member at Singularity University, for their comments and feedback on this chapter as well as our research.

Note

1 A note on terminology. We use the term "competence" at the board level to be in line with previous research looking at board competence and incompetence, e.g., (Cebon, 2017). Competence is defined as "the quality or state of having sufficient knowledge, judgment, skill, or strength (as for a particular duty or in a particular respect)" (Merriam-Webster, 2019), and competence is generally used in the context of leadership. Capability, however, is generally used to describe the collaborative processes in a firm, e.g., "the ability of an organization to perform a coordinated set of tasks, utilizing organizational resources, for the purpose of achieving a particular end result" (Helfat and Peteraf, 2003, p. 999).

References

Accenture, 2017. *Technology Vision 2017 – Amplify You.* [online] Available at: <www.accenture.com/_acnmedia/Accenture/next-gen-4/tech-vision-2017/pdf/Accenture-TV17-Full.pdf> [Accessed 24 Jun. 2019].

Accenture, 2018. *Accenture Technology Vision 2018 – Data Veracity.* [online] Available at: <www.accenture.com/us-en/insight-data-integrity-veracity> [Accessed 22 Jun. 2019].

Acemoglu, D. and Restrepo, P., 2019. Artificial Intelligence, Automation, and work. In: A. Agrawal, J. Gans and A. Goldfarb, eds. *The Economics of Artificial Intelligence An Agenda.* Chicago, IL: National Bureau of Economic Research. pp. 197–237.

AI HLEG, 2019. *A Definition of AI: Main Capabilities and Disciplines – EU High-Level Expert Group on Artificial Intelligence.* [online] European Commission. Available at: <https://ec.europa.eu/digital-single-market/en/news/definition-artificial-intelligence-main-capabilities-and-scientific-disciplines> [Accessed 26 Jun. 2019].

Alphabet, 2018. *United States Securities and Exchange Commission – Alphabet Inc.* [online] United States Securities and Exchange Commission – FORM 10-K. Available at: <www.sec.gov/Archives/edgar/data/1652044/000165204419000004/goog10-kq42018.htm> [Accessed 14 Jun. 2019].

Bankewitz, M., Åberg, C. and Teuchert, C., 2016. Digitalization and Boards of Directors: A New Era of Corporate Governance? *Business and Management Research,* 5(2), pp. 58–69.

Berr, J., 2017. *WannaCry ransomware attack losses could reach $4 billion – CBS News.* [online] CBS News. Available at: <www.cbsnews.com/news/wannacry-ransomware-attacks-wannacry-virus-losses> [Accessed 16 Apr. 2019].

Biswas, S., 2019. *3 Use-Cases of AI in Recruitment.* [online] HR Technologist. Available at: <www.hrtechnologist.com/articles/digital-transformation/3-usecases-of-ai-in-recruitment> [Accessed 24 Jun. 2019].

Brown, S.L. and Eisenhardt, K.M., 1998. *Competing on the Edge – Strategy as Structured Chaos.* Boston, MA: Harvard Business School Press.

Bughin, J., Catlin, T., Hirt, M. and Willmott, P., 2018. *Why Digital Strategies Fail.* [online] McKinsey Quarterly. Available at: <www.mckinsey.com/business-functions/digital-mckinsey/our-insights/why-digital-strategies-fail> [Accessed 17 Apr. 2019].

Carlini, N., Mishra, P., Vaidya, T., Zhang, Y., Sherr, M., Shields, C., Wagner, D. and Zhou, W., 2016. Hidden Voice Commands. In: *SEC'16 Proceedings of the 25th USENIX Conference on Security Symposium.* pp. 513–530.

Cebon, P., 2017. *The 3 Company Crises Boards Should Watch For.* [online] Harvard Business Review. Available at: <https://hbr.org/2017/01/the-3-company-crises-boards-should-watch-for> [Accessed 21 Jun. 2019].

Chou, J., Murillo, O. and Ibars, R., 2017. *How to Recognize Exclusion in AI.* [online] Medium. Available at: <https://medium.com/microsoft-design/how-to-recognize-exclu sion-in-ai-ec2d6d89f850> [Accessed 14 Jun. 2019].

Cisco, 2018. *What Are the Most Common Cyberattacks?* [online] Cisco. Available at: <www.cisco.com/c/en/us/products/security/common-cyberattacks.html> [Accessed 10 May 2019].

Elsayed, G.F., Shankar, S., Cheung, B., Papernot, N., Kurakin, A., Goodfellow, I. and Sohl-Dickenstein, J., 2018. Adversarial examples that fool both computer vision and time-limited humans. In: *NIPS'18 Proceedings of the 32nd International Conference on Neural Information Processing Systems, December 2–8.* Montréal, Canada, pp. 3914–3924.

Else, S.R. and Pileggi, F.G.X., 2019. *Corporate Directors Must Consider Impact of Artificial Intelligence for Effective Corporate Governance.* [online] Business Law Today from ABA. Available at: <https://businesslawtoday.org/2019/02/corporate-directors-must-consider-impact-artificial-intelligence-effective-corporate-governance> [Accessed 16 Apr. 2019].

Engstam, L.H. and Caroan, K., 2016. *Digital Business & Board Leadership Progress Report.* Stockholm, Sweden: Digoshen.

EY, 2018. *When Boards Look to AI, What Should They See?* [online] Available at: <www.ey.com/Publication/vwLUAssets/ey-when-boards-look-to-ai-what-should-they-see/$FILE/ey-when-boards-look-to-ai-what-should-they-see.pdf> [Accessed 26 Jun. 2019].

Eykholt, K., Evtimov, I., Fernandes, E., Li, B., Rahmati, A., Xiao, C., Prakash, A., Kohno, T. and Song, D., 2018. Robust physical-world attacks on deep learning visual classification. In: *CVPR conference.* Salt Lake City, UT, June 18–22. pp. 1625–1634.

Featherstone, T., 2017. *Preparing Directors for Artificial Intelligence Whirlwind.* [online] Australian Institute of Company Directors. Available at: <https://aicd.companydirec tors.com.au/advocacy/governance-leadership-centre/external-environment/preparing-directors-for-artificial-intelligence-whirlwind> [Accessed 24 Jun. 2019].

Fenwick, M., McCahery, J.A. and Vermeulen, E.P.M., 2019. The end of 'Corporate' governance (Hello 'Platform' Governance). *European Business Organization Law Review,* 20(1), pp. 171–199.

Fleming, O., Fountaine, T., Henke, N. and Saleh, T., 2018. *Ten Red Flags Signaling Your Analytics Program Will Fail.* [online] McKinsey Analytics. Available at: <www.mckinsey.com/business-functions/mckinsey-analytics/our-insights/ten-red-flags-signal ing-your-analytics-program-will-fail> [Accessed 8 May 2019].

Fuller, J., Jacobides, M. and Reeves, M., 2019. The Myths and Realities of Business Ecosystems. *MIT Sloan Management Review Digital*, 60(3), pp. 1–9.

Galaz, V. and Moberg, F., 2015. *The Biosphere Code*. [online] The Biosphere Code. Available at: <http://thebiospherecode.com/> [Accessed 24 Jun. 2019].

Grasso, A., 2019. *Cybersecurity Challenges in a Digitally Transforming World*. [online] Ericsson Blog. Available at: <www.ericsson.com/en/blog/2019/2/cybersecurity-challenges-in-a-digitally-transforming-world> [Accessed 26 Jun. 2019].

Gregersen, H., 2018. *Digital Transformation Opens New Questions – and New Problems to Solve*. [online] MIT Sloan Management Review. Available at: <https://sloanreview.mit.edu/article/digital-transformation-opens-new-questions-and-new-problems-to-solve> [Accessed 26 Jun. 2019].

Helfat, C.E. and Peteraf, M.A., 2003. The dynamic resource-based view: Capability lifecycles. *Strategic Management Journal*, 24(10), pp. 997–1010.

Huse, M., 2007. *Boards, Governance and Value Creation: The Human Side of Corporate Governance*. Cambridge, UK: Cambridge University Press.

Kelnar, D., 2019. *The State of AI 2019: Divergence*. [online] Available at: <www.stateofai2019.com/introduction/> [Accessed 26 Jun. 2019].

Kenton, W., 2018. *Investiopedia – Bandwagon Effect*. [online] Available at: <www.investopedia.com/terms/b/bandwagon-effect.asp> [Accessed 9 May 2019].

Kolanovic, M. and Krishnamachari, R.T., 2017. *Big Data and AI Strategies Machine Learning and Alternative Data Approach to Investing*. [online] Available at: <https://faculty.sites.uci.edu/pjorion/files/2018/05/JPM-2017-MachineLearningInvestments.pdf> [Accessed 26 Jun. 2019].

Lauterbach, A., 2018. *Salzburg Questions for Corporate Governance – What Questions Should Boards Be Asking About AI?* [online] Salzburg Questions for Corporate Governance. Available at: <www.salzburgglobal.org/news/publications/article/anastassia-lauterbach-what-questions-should-boards-be-asking-about-ai.html> [Accessed 24 Jun. 2019].

Lauterbach, A. and Bonime-Blanc, A., 2018. *The Artificial Intelligence Imperative A Practical Roadmap for Business*. Santa Barbara, CA: ABC-CLIO, LLC.

Leblanc, R. and Gillies, J., 2005. *Inside the Boardroom: How Boards Really Work and the Coming Revolution in Corporate Governance*. New York, NY: John Wiley & Sons.

Libert, B., Beck, M. and Bonchek, M., 2017. AI in the Boardroom – The Next Realm of Corporate Governance. *MIT Sloan Management Review Blog*. [online] Available at: <https://sloanreview.mit.edu/article/ai-in-the-boardroom-the-next-realm-of-corporate-governance> [Accessed 26 Jun. 2019].

Madgavkar, A., Manyika, J., Krishnan, M., Ellingrud, K., Yee, L., Woetzel, J., Chui, M., Hunt, V. and Balakrishnan, S., 2019. *The Future of Women at Work – Transitions in the Age of Automation*. [online] Available at: <https://mckinsey.com/featured-insights/gender-equality/the-future-of-women-at-work-transitions-in-the-age-of-automation> [Accessed 22 Sep. 2019].

Main, A., McCormak, D. and Lamm, B., 2018. *On the Board's Agenda US – Not If, But How: Evaluating the Soundness of Your Digital Transformation Strategy*. [online] Available at: <https://www2.deloitte.com/us/en/pages/center-for-board-effectiveness/articles/on-the-boards-agenda.html> [Accessed 26 Jun. 2019].

McWaters, J., 2018. *The New Physics of Financial Services – Understanding How Artificial Intelligence Is Transforming the Financial Ecosystem*. [online] World Economic Forum. Available at: <www.weforum.org/reports/the-new-physics-of-financial-services-how-artificial-intelligence-is-transforming-the-financial-ecosystem> [Accessed 26 Jun. 2019].

Menn, J., 2019. *Microsoft Turned Down Facial-Recognition Sales on Human Rights Concerns.* [online] Reuters. Available at: <www.reuters.com/article/us-microsoft-ai-idUSKCN1RS2FV> [Accessed 10 May 2019].

Merriam-Webster, 2019. *Merriam Webster Dictionary.* [online] Available at: <www.merriam-webster.com/dictionary/competence> [Accessed 25 Jun. 2019].

Micheli, M., Blakemore, M., Ponti, M., Scholten, H. and Craglia, M., 2018. *The Governance of Data in a Digitally Transformed European Society. Second Workshop of the DigiTranScope Project.* [online] Available at: <https://ec.europa.eu/jrc/communities/en/community/digitranscope-digital-transformation-and-governance-human-society/document/digitranscope-0> [Accessed 26 Jun. 2019].

Microsoft, 2018. *United States Securities and Exchange Commission – Microsoft Corporation.* [online] United States Securities and Exchange Commission – FORM 10-K. Available at: <www.sec.gov/Archives/edgar/data/789019/000156459018019062/msft-10k_20180630.htm> [Accessed 10 May 2019].

Mitchell, M., 2019. *The Collapse of Artificial Intelligence.* [online] 2019 Science Board and Trustees Symposium – The End of Everything. Available at: <www.youtube.com/watch?v=4QBvSVYotVc> [Accessed 18 Jun. 2019].

Mueller, R.S., 2012. *Speech Robert S. Mueller.* [online] The FBI Federal Bureau of Investigation. Available at: <https://archives.fbi.gov/archives/news/speeches/combating-threats-in-the-cyber-world-outsmarting-terrorists-hackers-and-spies> [Accessed 10 May 2019].

Nanterme, P., 2016. *Digital Disruption has Only Just Begun.* [online] World Economic Forum. Available at: <www.weforum.org/agenda/2016/01/digital-disruption-has-only-just-begun> [Accessed 24 Jun. 2019].

NewVantage Partners, 2018. *Big Data Executive Survey 2018.* [online] Available at: <http://newvantage.com/wp-content/uploads/2018/01/Big-Data-Executive-Survey-2018-Findings.pdf> [Accessed 26 Jun. 2019].

Pagano, M., 2017. *Strategic Thinking in the Age of Digital Transformation.* [online] Board Agenda. Available at: <https://boardagenda.com/2017/09/11/strategic-thinking-digital-transformation> [Accessed 21 Jun. 2019].

Pauwels, E., 2018. *The Ethical Anatomy of Artificial Intelligence – Centre for Policy Research at United Nations University.* [online] Centre for Policy Research at United Nations University. Available at: <https://cpr.unu.edu/cpr-voices-the-ethical-anatomy-of-artificial-intelligence.html> [Accessed 18 Apr. 2019].

Perisic, I., 2018. *How Artificial Intelligence Is Shaking Up the Job Market.* [online] World Economic Forum. Available at: <www.weforum.org/agenda/2018/09/artificial-intelligence-shaking-up-job-market> [Accessed 15 Apr. 2019].

Proofpoint, 2019. *State of Phish.* [online] Proofpoint. Available at: <www.proofpoint.com/au/threat-reference/phishing> [Accessed 26 Jun. 2019].

Protiviti, 2019. *Competing in the Cognitive Age – How companies Will Transform Their Businesses and Drive Value Through Advanced AI.* [online] Available at: <www.protiviti.com/sites/default/files/united_states/insights/ai-ml-global-study-protiviti.pdf> [Accessed 21 May 2019].

PWC, 2019. *22nd Annual Global CEO Survey – CEO's Curbed Confidence Spells Caution.* [online] Available at: <www.pwc.com/gx/en/ceo-survey/2019/report/pwc-22nd-annual-global-ceo-survey.pdf> [Accessed 26 Jun. 2019].

Ringel, M., Grassl, F., Baeza, R., Kennedy, D., Spira, M. and Manly, J., 2019. *BCG The Most Innovative Companies 2019.* [online] Available at: <http://image-src.bcg.

com/Images/BCG-Most-Innovative-Companies-Mar-2019-R2_tcm9-215836.pdf> [Accessed 26 Jun. 2019].

Rolnick, D., Donti, P.L., Kaack, L.H., Kochanski, K., Lacoste, A., Sankaran, K., Ross, A.S., Milojevic-Dupont, N., Jaques, N., Waldman-Brown, A., Luccioni, A., Maharaj, T., Sherwin, E.D., Mukkavilli, S.K., Kording, K.P., Gomes, C., Ng, A.Y., Hassabis, D., Platt, J.C., Creutzig, F., Chayes, J. and Bengio, Y., 2019. *Tackling Climate Change With Machine Learning*. [online] arXiv.org. Available at: <http://arxiv.org/abs/1906.05433> [Accessed 26 Jun. 2019].

Rometty, G., 2016. *Digital Today, Cognitive Tomorrow*. [online] MIT Sloan Management Review. Available at: <https://sloanreview.mit.edu/article/digital-today-cognitive-tomorrow/> [Accessed 26 Jun. 2019].

Ross, J., 2018. The fundamental flaw in AI implementation. *MIT Sloan Management Review*, 59(2), pp. 10–11.

Sansani, T., 2018. *Brave New World: Equity Investing in the Age of Artificial Intelligence*. [online] American Century Investments. Available at: <https://institutional.americancentury.com/content/institutional/en/insights/asset/equity/meet-your-new-robo-analyst.html> [Accessed 24 Jun. 2019].

Sarrazin, H. and Willmott, P., 2016. *Adapting Your Board to the Digital Age*. [online] McKinsey Quarterly. Available at: <www.mckinsey.com/business-functions/digital-mckinsey/our-insights/adapting-your-board-to-the-digital-age> [Accessed 24 Jun. 2019].

Shewan, D., 2017. *Robots Will Destroy Our Jobs – And We're Not Ready for It*. [online] The Guardian. Available at: <www.theguardian.com/technology/2017/jan/11/robots-jobs-employees-artificial-intelligence> [Accessed 16 Apr. 2019].

Simonite, T., 2014. *The Room Where Executives Go to Get Help from IBM's Watson*. [online] MIT Technology Review. Available at: <www.technologyreview.com/s/529606/a-room-where-executives-go-to-get-help-from-ibms-watson> [Accessed 24 Jun. 2019].

Sinclair, A., Brashear, J. and Shacklady, J., 2018. *The Momentum Mindset*. [online] Available at: <www.accenture.com/_acnmedia/PDF-73/Accenture-Strategy-AI-Momentum-mindset-Exec-Summary-POV.pdf> [Accessed 26 Jun. 2019].

Snow, J., 2018. *"We're in a diversity crisis": Cofounder of Black in AI on What's Poisoning Algorithms in Our Lives – MIT Technology Review*. [online] MIT Technology Review. Available at: <www.technologyreview.com/s/610192/were-in-a-diversity-crisis-black-in-ais-founder-on-whats-poisoning-the-algorithms-in-our> [Accessed 18 Apr. 2019].

Spadafora, A., 2019. *Phishing Attacks on the Rise in 2018*. [online] TechRadar Pro. Available at: <www.techradar.com/news/phishing-attacks-on-the-rise-in-2018> [Accessed 16 Apr. 2019].

Statista, 2019. *Statista*. [online] Available at: <www.statista.com/statistics/471264/iot-number-of-connected-devices-worldwide> [Accessed 26 Jun. 2019].

Suni, A., 2016. *It-konsulten Tieto har valt en artificiell intelligens, Alicia T, i ledningsgruppen för enheten för datadrivet beslutsfattande [IT Consultant Company Tieto has Chosen an Artificial Intelligence, Alicia T, to Join the Management Team for the Unit for Data]*. [online] Dagens Industri. Available at: <www.di.se/nyheter/it-bolag-valjer-ai-i-ledningsgrupp> [Accessed 26 Jun. 2019].

Szegedy, C., Zaremba, W., Sutskever, I., Bruna, J., Erhan, D., Goodfellow, I. and Fergus, R., 2013. *Intriguing Properties of Neural Networks*. [online] arXiv.org. Available at: <https://arxiv.org/abs/1312.6199>.

Van Alstyne, M.W., Parker, G.G. and Choudary, S.P., 2016. Pipelines, Platforms, and the New Rules of Strategy. *Harvard Business Review*, 94(4), pp. 54–62.

Vinnova, 2018. *Artificial Intelligence in Swedish Business and Society – Analysis of Development and Potential.* [online] Available at: <www.vinnova.se/contentassets/29cd313d690e4be3a8d861ad05a4ee48/vr_18_09.pdf> [Accessed 26 Jun. 2019].

WEF, 2015. *Deep Shift Technology Tipping Points and Societal Impact – Survey Report.* [online] World Economic Forum. Available at: <http://www3.weforum.org/docs/WEF_GAC15_Technological_Tipping_Points_report_2015.pdf> [Accessed 26 Jun. 2019].

WEF, 2018. *Digital Transformation Initiative – Maximizing the Return on Digital Investments.* [online] World Economic Forum. Available at: <http://reports.weforum.org/digital-transformation/files/2018/05/201805-DTI-Maximizing-the-Return-on-Digital-Investments.pdf> [Accessed 26 Jun. 2019].

Weill, P., Apel, T., Woerner, S.L. and Banner, J.S., 2019. It pays to have a digitally savvy board. *MIT Sloan Management Review*, 60(3), pp. 41–45.

Zolfagharifard, E., 2014. *Would You Take Orders from a ROBOT? An Artificial Intelligence Becomes the World's First Company Director.* [online] Daily Mail. Available at: <www.dailymail.co.uk/sciencetech/article-2632920/Would-orders-ROBOT-Artificial-intelligence-world-s-company-director-Japan.html> [Accessed 26 Jun. 2019].

Part II

The role of the digital welfare state

8 Polarization, tax revenue and the welfare state

Digital disruption or still standing strong?[1]

Mårten Blix

1. Introduction

Some changes in society are significant enough to warrant a specific name. Digitalization is one of those and is sometimes described as the third industrial revolution. What can we learn from comparing the present situation to the state of society at the outset of the first Industrial Revolution, some two-and-half centuries ago?

From the late eighteenth century and onwards, industrialization led to an upheaval of work and livelihoods at a time when there were little in terms of social safety nets. The rapid transformation of economies and societies became an impetus to create new social and political institutions to manage and reduce the social costs of change. Universal education, social security and pension systems were introduced along with universal suffrage. Spurred by hazardous and difficult work conditions as well as strife over low pay, labor organized into trade unions to become a counterweight to employers and owners of firms. Societies developed methods to handle change and devised ways to resolve conflict mainly through rules and negotiations rather than through force. In Sweden, a general pension system was introduced in 1913, although less generous than today (Blix, 2017). Notably, today people live about twenty years beyond the retirement age compared to at the inception of the pension system, when at least half the population were not expected to enjoy any pension at all.

There is no need to reinvent the institutions and safety nets thus established. Indeed, the modern welfare state has shown remarkable resilience over the years. Especially in the 1980s, industrial action in Sweden was a big concern, with many days lost in strikes. In 1997, the system was reformed through an agreement with industry-wide bargaining, allowing local flexibility and yet retaining elements of centralized wage bargaining with informal coordination with the manufacturing sector in the lead (Driffill, 2006). After the agreement, industrial action declined markedly and the most recent round of collective wage bargaining has resulted in mainly three-year agreements, signaling trust in the institutions. The relative calm, however, may be challenged in the years ahead. Digitalization is now affecting some of the fundamental building blocks, and unless institutions are reformed, the social contract holding society together could crack.

For the welfare state, the balance of protection against a potentially destructive change and the promotion of innovations have from the outset been a central but fragile state of affairs. On the one hand, too onerous rules in the economy can dent productivity growth and undermine rising prosperity. On the other, strained social cohesion can erode the legitimacy of institutions.

The modern welfare state has managed change, but some countries have at times veered off course. Take the example of Sweden. Its welfare state expanded rapidly during the 1970s and 80s but high marginal tax rates dented incentives to work, and fiscal profligacy gradually created an untenable economic situation. Interest payments on public debt began to squeeze out social spending. Trust in the stability of the Swedish economy declined and reached an absolute low in the fall of 1992 when the *Riksbank* (the Swedish central bank) unsuccessfully defended the krona by raising the interest rate to 500%. The deep crisis spurred structural reforms and set the stage for reforming the welfare state during the 1990s.

The effects of digitalization are not dramatic in the short-run, compared to a fiscal or financial crisis when GDP can fall abruptly, and many jobs are lost. Indeed, so far, there is no compelling evidence that employment levels in OECD countries are declining. One reason for this is that the modern labor market has a high capacity for change and continuously creates new jobs, especially in services, as old ones are shed. In Sweden, for example, about 17% of all jobs were destroyed and created during the period 1990–2009 (Heyman, Norbäck and Persson, 2013). In OECD countries as a whole, employment levels have not fallen, though unemployment – and especially youth unemployment – is a considerable concern after the fallout of the financial crisis.

And yet, although the modern welfare state does not face an imminent crisis, over the medium-to-long term the changes due to digitalization will put a strain on existing institutions and labor market arrangements. In addition, the welfare state has to cope with unprecedented high levels of immigration. As I have argued elsewhere, the labor market is changing to such an extent that the social contract could begin to crack (Blix, 2017).

Most descriptions of the Swedish welfare state will at least include the following elements:

- Comprehensive social welfare spending (health care, education and care of the elderly) financed by taxes
- Social inclusion through universal education, progressive tax systems and transfer payments to reduce income inequality
- A balance of power between trade unions and employers through rules to manage and resolve conflicts and a trade union policy to decrease wage disparities by pushing up the lowest wages

Digitalization affects all of these pillars in both direct and indirect ways. Most will acknowledge that consumer behavior has changed due to digitalization, but the most prominent changes are those that affect the labor market.

The changes to the labor market tend to occur more gradually than in consumption, depending on the dynamics of young people entering the labor market, with older persons retiring and others switching jobs. The impact of technology and digitalization on the labor market comes from the accumulated changes of such dynamics. The main impact of technological change and digitalization has been an increase in polarization where middle-level workers have been the most affected (Goos, Manning and Salomons, 2014). Income has become more volatile, and uncertainty in the labor market has been rising (OECD, 2015).

With gradual changes, in principle, there should be ample time to adjust and reform. In practice, reforms necessary to accommodate changes may be too slow – or not made at all. First, the political system often has difficulties in managing reform when the political costs of action tend to be up front and the potential economic benefits come much later. Second, the reform of existing institutions often meets resistance from special interest groups, all from employer organizations to the professions and even regulatory bodies. Changes typically imply a shift in power, resulting in winners and losers.

The risk of not responding to rising labor-market uncertainty and income volatility is that disenfranchisement will continue to expand. Institutional legitimacy risks being damaged and, indeed, in some OECD countries the rise of populist parties may be seen as a sign of declining trust in the establishment and the institutions that represent it.

2. Rising inequality also in the welfare state

A standard measure of income inequality is the so-called Gini coefficient. As can be seen from Figure 8.1, the Gini coefficients have been trending upwards in many OECD countries since the 1980s. Although it is an established measure of income inequality, the Gini coefficient measure has some well-known drawbacks and can be measured in different ways (Blomquist, 1981; Yitzhaki, 1998). In the aftermath of the financial crisis, the relatively modest changes in relative incomes could mask more problematic absolute differences at low levels of income. In addition, the Gini coefficient does not account for publicly provided welfare services. For a country, such as Sweden with comprehensive benefits, this makes some – but not a huge – difference. Other measures such as the share of those earning below 60% of median incomes or measures of risk of absolute poverty can be better at capturing income inequality. However, notwithstanding the measure used, it is unequivocal that inequality has increased in most OECD countries.

Despite increases in income inequality, the Nordics and much of northern Europe (excluding the Anglo-Saxon countries) remain in the lower half in terms of Gini coefficients. But not all welfare states have fared the same. It is especially noteworthy that Sweden has experienced the most substantial increase in Gini coefficient since the 1980s. However, this is an increase from a suppressed low level that turned out to be unsustainable. Wages were compressed due to union priorities in wage-bargaining and due to strongly progressive taxation. Though income inequality was held low, economic incentives for entrepreneurship and

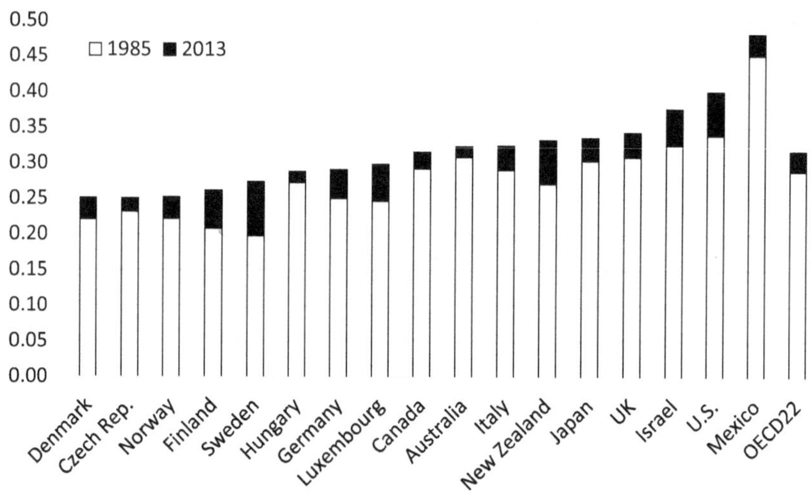

Figure 8.1 Gini coefficients in selected OECD countries. Levels in 1985 and in 2013. Note: The Gini coefficient is zero when everyone has an identical income. The Gini coefficient is one when a single individual has all of the income.

Source: OECD (2015).

work were eroded (Lindbeck et al., 2003). In particular, the 1970s and 80s was a period of economic stagnation in Sweden with a long-lasting decline in GDP per capita growth rates compared to other OECD countries.

Trade and globalization have likely led to lower income inequality in the world as a whole, but most arguments indicate that income inequality *within* countries will continue to rise. Rapidly aging populations will accelerate changes, and new technologies will compete with humans in many new areas, notably also in advanced services and result in damped wage growth for those without special skills: the polarization of labor markets noted in the literature (Goos, Manning and Salomons, 2014). One interpretation is that digitalization results in a common shock that drives up income inequality in some countries. At the same time, other countries with high inequality (such as Chile and Mexico) have seen some reduction but this development is likely linked to other factors. The overall effect may appear as a form of convergence (OECD, 2015) but it is a bit early to make such an assessment. More urgently, however, countries with increasing inequality need to find ways to address these changes or risk see further deterioration in their institutional legitimacy and further populism.

The economist Andre Sapir presents a straightforward way to summarize different models of growth and social inclusion (Sapir, 2006). In Table 8.1, some countries and regions are divided into combinations of low-high equity and efficiency. A useful way to think about the different country models is to interpret the labels rather broadly. Efficiency can be thought of as productivity growth, per

Table 8.1 Combinations of efficiency and equity.

		Efficiency	
		Low	*High*
Equity	Low	Southern Europe	US, UK
	High	Northern Europe	Scandinavia

Source: Sapir (2006).

capita growth or capacity for innovation; Equity can be considered as measuring income inequality or, better yet, equality of opportunity.

The characterization is not meant to imply that there is a growth-equity trade-off. An IMF study finds no such pattern is supported by data (Ostry, Berg and Tsangarides, 2014). Also, the OECD (2017b) emphasizes that there are several policy levers that support both equity and growth, such as promotion of product market competition. Instead, a country may find it hard for political economy reasons to pursue the reforms that would lead to improvements in either long-term productivity growth or equity, not least when the social costs are often up-front.

Most of Table 8.1 capturing the state of affairs in 2005 stands the test of time, but not all. Several countries have been experiencing declining productivity growth. For the UK, the decline actually began before the financial crisis. Even with rising inequality, Sweden remains a country with one of the most favorable combinations of equity and growth. Will the Swedish welfare state be better at coping with technological change than other systems?

3. The social contract in the welfare state is threatened

The welfare state can be seen as a particular type of social contract between different groups: The young and the old; workers and owners of capital; cities and regions. Those in work and good health pay large shares of their income in tax to get social support when they are old or fall sick. Those living in the regions are often subsidized by more prosperous regions.

The challenge for all countries is that substantial relative changes in fortune for some groups or areas can lead to discontent and undermine the willingness to take part in intergenerational transfers or geographical redistribution. Arguably, political events during 2016–18 could be a sign of such developments. The list is becoming long: The election of President Donald Trump in the US, the Brexit-referendum in the UK, Catalonia's unilateral declaration of independence from Spain, Germany's procrastinated negotiations of forming a coalition government and Italy's continued drift toward yet more political fragmentation. Welfare states in the north of Europe are by no means immune, as evidenced by the recent upsurge of populism even in prosperous countries with medium-to-low inequality. This is evidenced by the contemporary developments in Sweden.

Most notably the case of the political fringe party, the *Sweden Democrats* (Swe: *Sverigedemokraterna*), which went from having failed to reach past the election threshold prior to 2010 to becoming the third largest party following the Swedish general election of 2014. Some pre-election opinion polls also anticipated that the Sweden Democrats would increase their mandate following the 2018 general election and become the second largest – or even the largest – political party in Sweden.

Resentment against the elites that are perceived to benefit from changes can, in turn, lead to undermining the social contract that holds the welfare state together. This is especially the case in countries with aging populations and significant immigration levels. Stagnant wages thus risks fanning the flames of disenfranchisement even further.

3.1. The labor market and stagnant wages

The labor market is essential to the welfare state. Without a well-functioning labor market prosperity cannot increase and support for the social contract may wane.

Productivity growth and slack in labor markets are traditional explanations for understanding how wages develop. One reason for concern in recent years is that wage growth has been stagnant in much of the advanced economies. According to the International Monetary Fund (2017a), these can account for a significant share of the recent stagnant wages. As can be seen in Figure 8.2, wages in advanced

Figure 8.2 Nominal wage growth in advanced economies compared to the level of wage growth in 2007, percentage points. Note: Wage growth is normalized by subtracting the change in 2007.

Source: International Monetary Fund (2017a, p. 78).

economies have been in gradual decline; a process that started well before the financial crisis.

Though low productivity growth and the ample availability of workers can explain some of the stagnant wages, they cannot explain the full slowdown. Other explanations include advances in technology and automation that result in stronger competition between humans and machines (OECD, 2017a). Even if past technological advances have had far-reaching influence on work, advances in digitalization are being implemented faster than before (Comin and Ferrer, 2013, p. 14).

An overall effect of digitalization on the labor market is to reduce the bargaining power of workers. In many professions, the "middle man" is a function that is under pressure from robots. Such pressures are in evidence in banking, insurance and retail just to name a few. In banking, for example, the continued fallout from the financial crisis in combination with technology is leading many banks to reduce staff and automate a range of services. In Sweden, the Financial Supervisory Authority has granted licenses to financial institutions that provide automated advice. Back-office operations are especially prone to automation, as they are routine and occur on a regular basis. Such automation can also incorporate better risk-management as well as regulatory compliance. Some banks are testing so-called "Robo-branches" which are in effect local bank branches largely without professional staff. There are examples of insurance companies introducing completely automated claims-processes.

At the aggregate level, jobs are not disappearing. Rather, technology is creating additional downward pressure on wage growth. Other parts of the economy are also set to be affected. The increase in e-commerce is affecting many retail stores and boutiques. Semi-autonomous checkouts where customers scan their own goods have been available for many years and are growing more common. The next step is completely automated checkouts. Amazon has been experimenting with such technology for some time and opened its first such grocery store in Seattle, Washington in the beginning of 2018 (Wingfield, 2018). Though the technology is thus far in its infancy, it may ultimately obliterate the need for cashiers altogether.

Shopping for goods and clothes online has become large commerce. As the e-commerce companies become better at knowing their customers and can deliver goods quickly, the pressures on physical shops will grow. The company Zalando has plans to let their customers order tailor-made clothes from measures deduced body scanning (Bränström, 2018), which could help reduce costly returns and make ordering online even more attractive. In other words, technology is set to further increase the push toward e-commerce.

Advances in technology have reignited the angst that automation will destroy jobs. For example, in an oft-quoted paper, Frey and Osborne (Frey and Osborne, 2017) argue that about half of US jobs can be automated within the next two decades. Others have used different methodology and found substantially lower estimates (Arntz, Gregory and Zierahn, 2016; Nedelkoska and Quintini, 2018). More generally, evidence for EU countries continues to point to the labor market's ability to adapt (Gregory, Salomons and Zierahn, 2016): Job losses in one area are compensated by demand spillovers in other areas so that the net effect is mostly stable employment levels. Overall, there is so far no support for the notion that human work is disappearing.

However, there is ample evidence for the notion that the content of work is changing (Acemoglu and Autor, 2011). Improvement in technology has led to a process favoring those with high-skills regarding cognitive or social abilities, so-called skilled-biased technological change. For such workers, wage developments have been positive, and the share of such work has increased in the economy (see Figure 8.3). By contrast, routine work has been in decline. The overall result has been an increased polarization of the labor market that has been occurring over an extended period (Goos, Manning and Salomons, 2014).

The polarization of work has occurred in most OECD countries. We can expect that automation of work will put further pressures on wages for those with middle-level skills. The tools and technology that are now available could accelerate polarization compared to previous periods. There is a risk that those who are slow to upgrade their skills will experience further wage stagnation. Admittedly, there are historical examples where new technology did not cause downgrading of skills. For example, when automated teller machines (ATMs) were introduced, bank cashiers often moved up the skill ladder by instead providing financial advice to customers (Bessen, 2015). But this is not an inevitable development. For instance, jobs that disappear in stores might instead become software programming jobs elsewhere and thus much less likely to occur.

At the overall level, a combination of developments could lead to a decline in the wage-bargaining power of labor. Apart from technology, both demography,

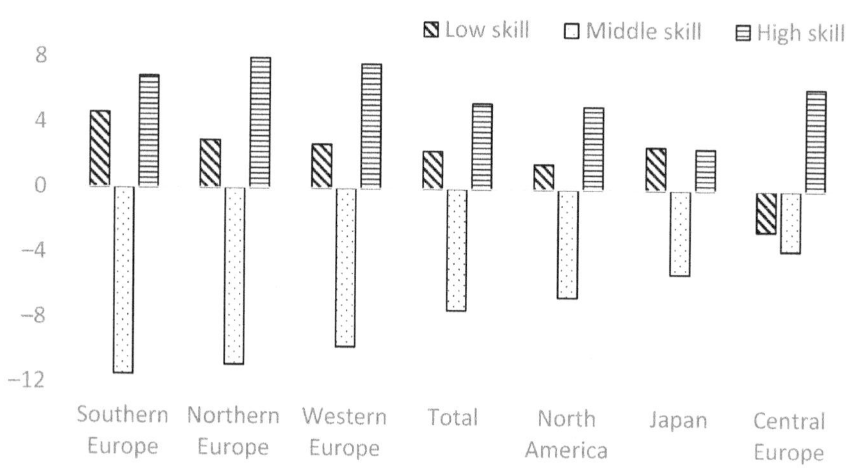

Figure 8.3 Percentage point change in share of total employment 1995–2015.

Source: OECD (2017c, p. 86).

and more flexible employment legislation protection serve to accelerate changes in the labor market. Aging populations imply fewer young compared to the old, and so in principle, the young could fill the jobs of those retiring. With large cohorts leaving the labor market, some areas will even experience scarcity of workers. In practice, young workers can only seldom directly replace older workers, especially not in positions where on-the-job experience is essential. What this means is that the incentive to automate work will be stronger due to aging populations, as firms find it hard to find workers with the right skills.

Technology is of course not the only thing that affects the bargaining power of labor (OECD, 2017a). In many OECD countries, protection for temporary or fixed-term contracts has been in decline since the 1990s. By contrast, permanent positions have remained mostly unchanged. As a result, the *duality* of labor markets has increased, and especially so in Sweden, for example (Cahuc, 2010, pp. 150–53). Young people are overrepresented among temporary workers, and their share has increased. OECD calculates that in 2015 about 40 million youth or 15% of those in the ages 15–24 are neither in education nor employment, so-called NEET (OECD, 2016).

Technology is not only changing the landscape of work through automation and robots. With so-called platform-based labor market, non-standard work is on the rise. Platform-based work has been given many names, such as the sharing economy or gig work. In what follows, I will use the term gig work to denote a situation where a worker performs tasks organized through the conduit of a digital platform and where the platform owner does not take employer responsibilities, such as paying payroll taxes and value-added tax (VAT).

Gig work has always existed, notably in entertainment, such as in music, art or television. Non-standard work without employment protection is also prevalent in journalism. Non-standard work contributes to rising inequality (OECD, 2015). For example, the self-employed enjoy fewer benefits in social security. Besides, the self-employed are also excluded from additional benefits in collective wage bargaining agreements, such as topped-up pensions, parental leave and sick leave.

Gig work is increasing on broad fronts (Sundararajan, 2017; Katz and Krueger, 2016). A common misconception is that gig work is only about simple tasks, such as driving taxis (for example Uber) or household services (such as TaskRabbit). The services are much broader, all from medical to legal professions. While it has increased sharply over the last few years, in terms of overall share of employment it remains small in Sweden. Despite its limited size, it could be set to affect the labor market in fundamental ways. By creating a situation where work is on permanent standby, 24 hours a day, seven days a week, it lessens the need for permanent workers. One of the largest platforms is Upwork. It has more than 12 million workers worldwide – doing tasks ranging from web design to data analysis (Sundararajan, 2017).

Consider the thought experiment that today's digital gig platforms had existed for as long as there have been firms. In such a world, would firms have hired workers to the extent reflected by today's medium and large size enterprises? Probably not. Ronald Coase, recipient of the Nobel Memorial Prize in Economic Sciences in 1991, argued that the existence of the firm supersedes the price mechanism of hiring individual workers on an atomistic market (Coase and Coase, 1937). When the cost of individual contracts is higher than organizing work into employment, the existence of the firm can be explained. With gig platforms, the cost of hiring temporary staff on a needs-only basis is much smaller than in the past. Hence, it is likely that permanent works would be much fewer in numbers.

What are the possible implications? The main channel of change is through the regular churn of the labor market: retirement of older workers, hiring of new workers as well as voluntary or involuntary employment changes. These changes occur slowly and mostly without drama. In countries with collective wage agreements, bargaining over wages and benefits may occur over various yearly intervals. In Sweden, for example, some wage agreements cover two-to-three years.

Gig markets pose a direct threat to the Swedish labor market model where the trade unions and the employer organizations are responsible for setting wages (Blix, 2017). Gig contracts bypass entirely collective wage bargaining agreements and the transaction occurs in the cloud. Moreover, the buyer and seller of services can even be in different countries. As a consequence, the traditional trade union threat of a boycott is more difficult to use compared to a shop or a factory. Also, non-payment of taxes is an issue for the government. A tilted playing field in taxation can lead to unfair competition, where tax and regulatory differences have an outsized role in success compared to the efficiency of services.

So far, the changes are occurring gradually, but most of the incentives point to a clear direction of change toward work and jobs becoming more loosely tied to a single employer and with a shrinking share of permanent employment. Exactly how far this process will continue is hard to say. It will, among other things, depend on the policy responses of governments, employers and trade unions.

For the welfare state, it means more flexible labor markets and also that security through work will be lower than in the past. In Sweden, the collective wage bargaining agreements cover about 90% of the labor market today. A system of collective wage bargaining can likely survive a small share of gig work in the economy but begins to lose its legitimacy if gig work becomes large.

3.2. Financing the social welfare state: tax base on labor becoming more mobile

The mobility of capital has been a feature of world economies for a long time. Of course, workers have a long tradition of moving to jobs, even if not as readily as capital. But as outlined in the previous section, technology is now increasing the mobility of labor in ways that were not possible before. Technology makes it easy to outsource work with the simple press of a button to global gig markets. Moreover, the expanding possibilities of automating all from simple to advanced services will make it easier for firms to substitute away from labor to machines. This substitution has consequences for government revenue, as the tax on labor is one of the largest tax bases. On average, about 50% of government revenue (in 2013) stems from tax on labor in OECD countries (Blix, 2017). The implications may be even more significant in countries with high tax rates on human work; most notably, of course, welfare states. It is not that governments will not be able to collect revenue. Instead, the challenge is that the distortions of a high tax on labor may increase further, which poses risks to productivity growth.

The threat to government revenue and the advent of rising distortions are not immediate. Instead, labor markets are likely to change over many years, but there are already some indications that the relation between machines and humans have shifted. As illustrated in Figure 8.4, the wage share of national income has fallen in most industrialized countries during the last three decades (Karabarbounis and Neiman, 2014; International Monetary Fund, 2017b). This result implies that as the GDP is expanding, humans are no longer keeping the same share of the pie.

The IMF calculates that about half the decline in the wage share of labor can be explained by technology (International Monetary Fund, 2017b). Notably, this development has been observed years before smartphones became ubiquitous and before the so-called "Frightful Five" of big tech, i.e., Amazon, Apple, Facebook, Google and Microsoft, gained dominance in global markets (Manjoo, 2016). Since the capacity of software has significantly expanded, it stands to reason that the wage share of labor is set to fall further. The result could be an even more significant shift away from human labor to machines. Evidence from other areas shows that high tax rates can give rise to significant shifts. High tax rates can lead to a sizeable substitution between the legal and the shadow economy as well as between unpaid household production and market production (Davis and Henrekson, 2005). The effects of automation could be even more substantial.

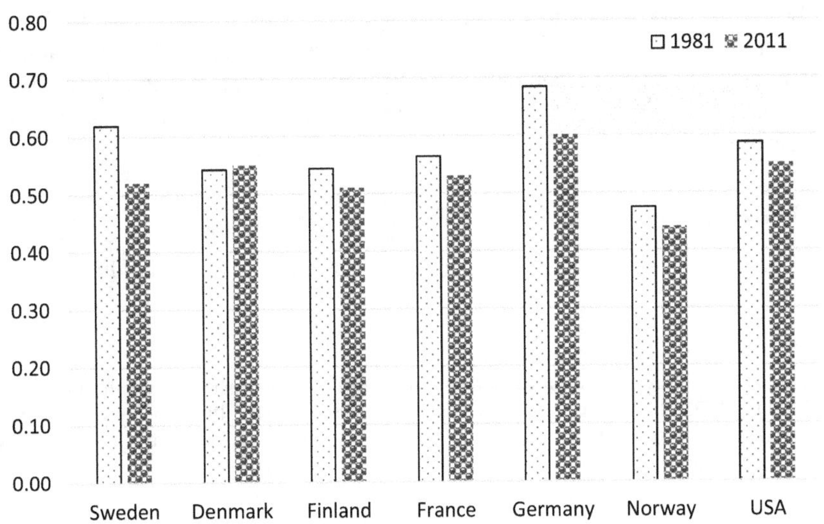

Figure 8.4 Wage share of national income. Percent.

Source: Karabarbounis and Neiman (2014).

4. Conclusions

As labor markets are becoming more polarized, inequality increases, and income uncertainty becomes more pronounced. What happens to the legitimacy of institutions when a large number of persons get fewer of the benefits of growth and when the share of labor market outsiders grows?

Welfare states may be more resilient to these changes than other countries. Notably, they have more well-developed and comprehensive social safety nets. They are geared toward providing social security and support workers to find new jobs through retraining and education.

But the welfare state also carries some weaknesses: The high level of taxes supporting the welfare spending creates even stronger incentives for firms to automate work or to buy services on global *gig* markets. This results in the bypassing of the high taxes and collective wage agreements that are vital pillars of the Nordic labor markets.

The outcome of the welfare state depends on policy responses of governments, trade unions and employer organizations. Trade unions that adapt and provide new forms of support and safety to its members could remain relevant to workers and serve as a counterweight to some of the increases in income uncertainty. Governments may also try to broaden tax bases to support welfare ambitions, especially for the self-employed.

It is hard to say how likely institutions are to step up to the challenge. One political difficulty is that the changes tend to be gradual and it may be tempting to postpone reforms rather than address the hard choices early on. Reform of institutions may also be hampered by special interest groups and lobbyists that act to protect the status quo.

Low inequality is core to the welfare state, yet it is set to rise even further in the years ahead. Without judicious reforms, the welfare state will not be immune from cracks in the social contract. One way or another, the outcome for the welfare states hangs in the balance in the years ahead. Will the welfare state be able to reinvent itself once again?

Acknowledgments

I am grateful to Marianna Blix Grimaldi for comments on an earlier draft.

Note

1 This chapter draws upon an earlier iteration originally published under the title "The Effects of Digitalisation on Labour Market Polarisation and Tax Revenue" in *CESifo Forum 2017–4*, vol. 18 no. 4, pp. 9–14. Permission for reprint has been granted by the copyright holder.

References

Acemoglu, D. and Autor, D., 2011. Skills, tasks and technologies: Implications for employment and earnings. In: O. Ashenfelter and D. Card, eds., *Handbook of Labor Economics*. Amsterdam, Netherlands: Elsevier, pp. 1043–1171.

Arntz, M., Gregory, T. and Zierahn, U., 2016. *The Risk of Automation for Jobs in OECD Countries: A Comparative Analysis*. OECD Social, Employment and Migration Working Papers, No. 189. Paris, France.

Bessen, J., 2015. Toil and technology: Innovative technology is displacing workers to new jobs rather than replacing them entirely. *Finance and Development, International Monetary Fund*, 52(1), pp. 16–19.

Blix, M., 2017. *Digitalization, Immigration and the Welfare State*. Cheltenham, UK: Edward Elgar Publishing.

Blomquist, N., 1981. A comparison of distributions of annual and lifetime income: Sweden around 1970. *Review of Income and Wealth*, 27(3), pp. 243–264.

Bränström, S.L., 2018. *Zalandochefen: 'T-shirt för 50 kr inte hållbart' [Head of Zalando: "T-Shirt for SEK 50 Not Sustainable]*. [online] Svenska Dagbladet. Available at: <www.svd.se/zalandochefen-t-shirt-for-50-kr-inte-hallbart-for-nagon> [Accessed 18 Sep. 2019].

Cahuc, P., 2010. *Det svenska anställningsstödet [The Swedish Job Protection]*. [online] Suppl. 6 to LU2011, SOU 2010:93, Government Offices of Sweden. Available at: <www.regeringen.se/rattsliga-dokument/statens-offentliga-utredningar/2010/12/sou-201093> [Accessed 18 Sep. 2019].

Coase, R.H. and Coase, R.H., 1937. The Nature of the Firm. *Economica*, 4(16), pp. 386–405.

Comin, D.A. and Ferrer, M.M., 2013. *If Technology Has Arrived Everywhere, Why has Income Diverged?* NBER Working Paper, No. 19010. Cambridge, MA.

Davis, S.J. and Henrekson, M., 2005. Tax effects on work activity, industry mix and shadow economy size: Evidence from rich-country comparisons. In: R. Gómez-Salvador, A. Lamo, B. Petrongolo, M. Ward and E. Wasmer, eds., *Labour Supply and Incentives to Work in Europe*. Cheltenham, UK: Edward Elgar Publishing, pp. 44–104.

Driffill, J., 2006. The centralization of wage bargaining revisited: What have we learnt? *JCMS*, 44(4), pp. 731–756.

Frey, C.B. and Osborne, M.A., 2017. The future of employment: How susceptible are jobs to computerisation? *Technological Forecasting and Social Change*, 114, pp. 254–280.

Goos, M., Manning, A. and Salomons, A., 2014. Explaining job polarization: Routine-biased technological change and offshoring. *American Economic Review*, 104(8), pp. 2509–2526.

Gregory, T., Salomons, A. and Zierahn, U., 2016. *Racing With or Against the Machine? Evidence from Europe*. ZEW Discussion Papers. Mannheim, Germany.

Heyman, F., Norbäck, P.J. and Persson, L., 2013. *Var skapas jobben? En ESO-rapport om dynamiken i svenskt näringsliv 1990 till 2009 [Where are the Jobs Created? An ESO-Report About the Dynamics in the Swedish Enterprise 1990 to 2009]*. Stockholm, Sweden.

International Monetary Fund, 2017a. Recent wage dynamics in advanced economies: Drivers and implications. In: *World Economic Outlook: October 2017*. Washington, DC: International Monetary Fund (IMF), pp. 73–116.

International Monetary Fund, 2017b. Understanding the downward trend in labor income shares. In: International Monetary Fund, ed., *World Economic Outlook, April 2017: Gaining Momentum?* Washington, DC: International Monetary Fund (IMF), pp. 121–172.

Karabarbounis, L. and Neiman, B., 2014. The global decline of the labor share. *The Quarterly Journal of Economics*, 129(1), pp. 61–103.

Katz, L.F. and Krueger, A.B., 2016. *The Rise and Nature of Alternative Work Arrangements in the United States, 1995–2015*. NBER Working Paper, No. 22667. Cambridge, MA.

Lindbeck, A., Molander, P., Persson, T., Petersson, O., Swedenborg, B. and Thygesen, N., 2003. *Turning Sweden Around*. Cambridge, MA: MIT Press.

Manjoo, F., 2016. *Tech's 'Frightful 5' Will Dominate Digital Life for Foreseeable Future*. [online] *The New York Times*. Available at: <www.nytimes.com/2016/01/21/technology/techs-frightful-5-will-dominate-digital-life-for-foreseeable-future.html> [Accessed 18 Sep. 2019].

Nedelkoska, L. and Quintini, G., 2018. *Automation, Skills Use and Training*. OECD Social, Employment and Migration Working Papers, No. 202. Paris, France.

OECD, 2015. Overview of inequality trends, key findings and policy directions. In: *In It Together: Why Less Inequality Benefits All*. Paris, France: OECD Publishing, pp. 19–58.

OECD, 2016. *Society At a Glance 2016 – A spotlight on Youth*. Paris, France: OECD Publishing.

OECD, 2017a. Collective bargaining in a changing world of work. In: *OECD Employment Outlook 2017*. Paris, France: OECD Publishing, pp. 125–186.

OECD, 2017b. *Economic Policy Reforms 2017: Going for Growth*. Paris, France: OECD Publishing.

OECD, 2017c. *OECD Employment Outlook*. Paris, France: OECD Publishing.

Ostry, J., Berg, A. and Tsangarides, C., 2014. *IMF Staff Discussion Note: Redistribution, Inequality, and Growth*. Washington, DC: International Monetary Fund (IMF).

Sapir, A., 2006. Globalization and the reform of European social models. *Journal of Common Market Studies*, 44(2), pp. 369–390.

Sundararajan, A., 2017. The Future of Work. *Finance and Development*, 54(2), Jun., pp. 6–11.

Wingfield, N., 2018. *Inside Amazon Go, a Store of the Future*. [online] *The New York Times*. Available at: <www.nytimes.com/2018/01/21/technology/inside-amazon-go-a-store-of-the-future.html> [Accessed 18 Sep. 2019].

Yitzhaki, S., 1998. More than a Dozen alternative ways of spelling Gini. *Research on Economic Inequality*, 8, pp. 13–30.

9 Welfare states and digitalization

Bent Greve

1. Introduction

The aim of this chapter is to discuss whether welfare states will be prepared and able to cope with the possible strong transformations on the labor market and how this will interact with the ability to finance the welfare states in the years to come. This will be done within a framework where the focus is on using the existing studies related to possible changes on the labor market when new technologies are integrated in production, including the digitalization of work in several sectors of the economies.

Based on the discussion of change on the labor market, focus will be on how this might, in a variety of ways, influence the ability to finance the welfare states using the classical welfare regime approach (for a recent overview see Von Kersbergen [2019] and Vis [2019]). This as the possible impact can be expected to vary dependent on the welfare regime a country belongs to (Greve, 2018), and for an overview of individual countries see Kuhlman, Schubert and de Villota (2016). The expectation being that universal welfare states, as the Nordic, to a larger degree will be influenced by the development and their ability to finance, relatively more generous welfare states, than liberal welfare states with less state influence on the societal development. The reason for this expectation will be explained more in Section 3, which presents a few data on overall spending and ways of finance welfare states in Europe.

Section 2 will present possible changes as a consequence of the fourth industrial revolution. This scenario is, in turn, based on several studies presented in recent years connected with the theoretical understanding of "insiders" and "outsiders" on the labor market (Schwab, 2016). This will include how there might be different viewpoints on the impact the use of new technology might have. Section 3 will briefly depict spending and financing across welfare states as part of the risk for welfare states in the wake of strong technological changes.

Section 4 will thereafter connect these debates as a way to depict the possible connection between technological change and welfare states development. This is mainly done in an explorative way, given that the expected changes are more extensive than the previous actual changes.

There are as always delimitations. This includes what kind of competences is expected in the future (Kaplan, 2015). The consequence of trade globalization, albeit having many similarities with the technological development, will only narrowly be included. This, despite that the impact, such as that "displacement destroys industry-specific human capital, leaving affected workers in positions for which they are poorly suited relative to non-displaced workers" (Autor, Dorn and Hanson, 2016, p. 232), is also a possible impact on workers of technological changes on the labor markets. How ideas influence the choice of labor market and social policy is also outside the scope of the chapter (Greve, 2018).

2. Impact of technological change

Labor markets have been, at all times, under change and reconstruction. Technological impact on job has been discussed since the Luddites, and albeit jobs have been wiped away, new jobs have so far been created so that high unemployment has mainly been related to the overall economic fluctuations (Graetz and Michaels, 2015). Anxieties of technology's impact on jobs are a long history – from the Industrial Revolution to the Great Depression and onwards (Mokyr, Vickers and Ziebarth, 2015). The recent anxiety relates to computing power, artificial intelligence and robotics (Autor, 2015), also including, as will be later explained, the stronger polarization on the labor markets.

The consequences of the fourth industrial revolution will presumably be large all over the world. A recent study indicated that until 2030 between 75 and 375 million people should change job, and between 4–800 million people should find complete new types of jobs as a consequence of the implementation of new technology (Bughin et al., 2017). Countries that have thus far only achieved limited progress in automatizing their industrial production due to already low labor cost would be the countries where one could expect the largest degree of change. Another estimation for developed countries shows a variation from around 44% (Slovakia) and 42% (Slovenia) to 22% (Finland and South Korea) (Hawksworth, Berriman and Goel, 2018). It is naturally difficult to predict the future development. In 2003, Autor, Levy and Murnane (2003, p. 1283), three experts, wrote that "navigating a car through city traffic or deciphering the scrawled handwriting on a personal cheque – minor undertakings for most adults – are not routine task by our definition" and that truck driving were areas with "limited opportunities for substitution or complementarity". The development since then indicates that changes in these jobs are now within reach – and a personal check is already almost completely archaic. Thereby, also, those more optimistic about development might perhaps be too optimistic about future job development. Still, their distinction between routine and not routine related to types of task is central to the understanding of the possible impact on the labor market of the changes is important (see Table 9.1).

Routine work is, not surprisingly, mainly at risk of being automated, whereas non-routines are less likely of being at risk. Since the article by Autor et al. was written in 2003, it has been possible to split work-function into even smaller units and thereby increasing the number of functions that can be considered routine tasks. The possible consequence for the labor market has thereby been stronger since, still this distinction has been behind many of the studies trying to depict the

Table 9.1 Skills, risk and consequences of automation split between routine and non-routine work.

	Routine work	*Non-routine work*
Analytical and interactive tasks	Substantial substitution	Strong complementarities
Manual tasks	Substantial substitution	Limited opportunities for substitution or complementarity

Source: Based on Autor et al. (2003, p. 1286).

possible change in number of jobs on the labor market, with Frey and Osborne's (2013) study being central. They expected that within 10–15 years half of the job-functions we know today to be gone. Several subsequent studies have been done on this subject (Arntz, Gregory and Zierahn, 2016; Manyika et al., 2017; Hawksworth, Berriman and Goel, 2018). These studies are, with variations, often close to the results in Frey and Osborne (2013). Refinement since using more detailed knowledge on educational level has helped in understanding types of jobs at risk of being automated.

Figure 9.1 shows the jobs at risk of automation split into high risk of being automated and jobs of high risk of substantial change (OECD, 2017a).

Overall the Figure 9.1 points to than on average 9% of all jobs are in high risk of being automated, whereas jobs at high risk of substantial change is 25%, so that more than one-third of all jobs are in danger. The figure varies across countries, although still so that dramatic changes will be on the way for the labor market in many countries. There are natural methodological challenges with these calculations and how to measure and split jobs into tasks, however, this is not the point here, as despite disagreement about the size, there is a consensus about that dramatic changes will take place. There is a stronger disagreement about whether sufficiently new jobs will be created (Greve, 2017). Still, there will be changes, and presumably fewer jobs, and, at the same time a possible continued development with a split between insiders and outsiders, and stronger polarization on the labor market.[1] Polarization will be a continuation of the trend that has already been toward stronger diversities on the labor markets over the last 20–25 years (see Figure 9.2) and is expected to continue (OECD, 2017b; Goos, Manning and Salomons, 2014).

Figure 9.2 points to the polarization so that especially middle-skill jobs are in danger, and despite this is mainly estimated based on change in wage level indicates profound changes on the labor market, and, there is no indication that this will not continue, despite that one of the possible competence in the future and types of jobs can include different types of care (children and elderly) and also primary education. This is where the level of qualifications might be more in the middle, however, often with a relatively low level of wages. Still overall, it indicates in most professions growing insecurity on the labor markets, but presumably also in the ability to finance welfare states (see more in Section 3).

Polarization will also imply a negative impact on the distribution implying a continuation of the rise in inequality (Brynjolfsson and McAfee, 2014; Reich, 2015).

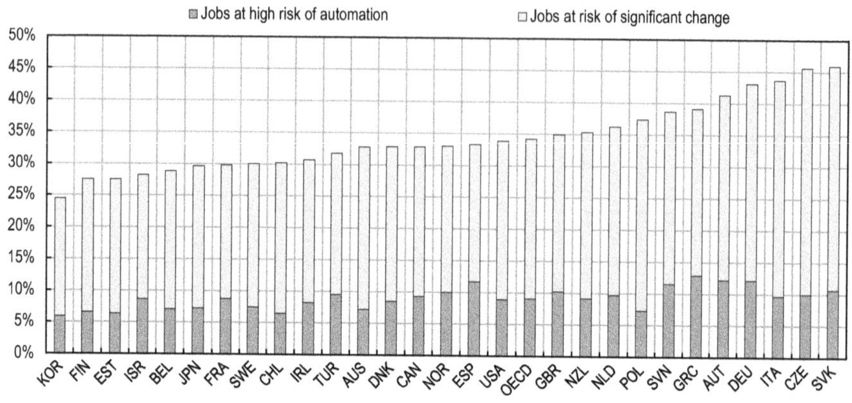

Figure 9.1 The risk of automation in OECD countries.

Source: OECD (2017a), Figure 3.10.

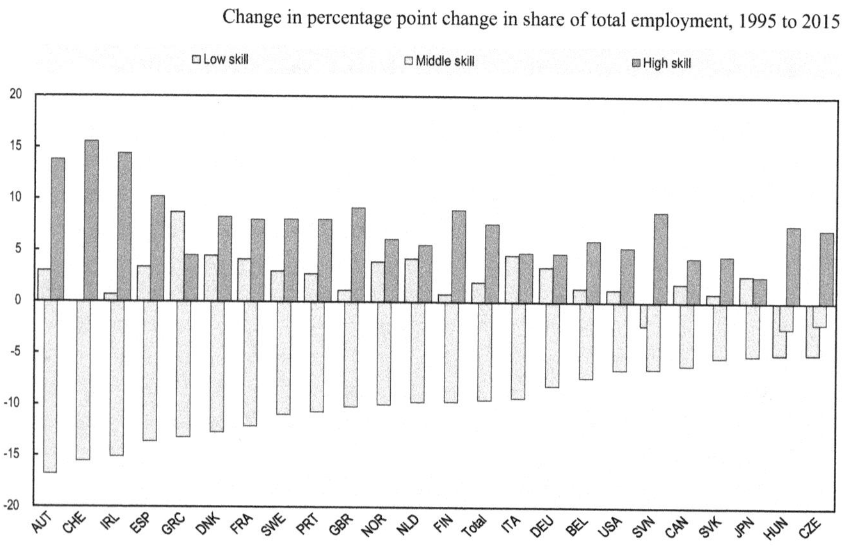

Figure 9.2 Polarization in OECD labor markets 1995–2015.

Source: OECD (2017a, p. 121).

Furthermore, even if jobs will be created, there will be times when some will be left outside the labor market lacking the necessary skills to participate and get jobs. Therefore, this also challenges the ability to finance the welfare states.

Overall, the studies strongly indicate that the level of routinization and education influences the possibility of automation, but also increasingly that middle and high-skilled jobs are at risk of automation. The change in jobs seems to be stronger in countries where industrial production still is central, and, thus Eastern and Southern Europe are under more pressure than the Nordic and Western parts of Europe. This should, all else being equal, in general imply that the pressure on welfare states will be higher in east and south of Europe. There seems further, even if disagreement on the overall impact on the number of jobs in danger, agreement on that it might increase inequalities and that there as least can be a long transition time where the welfare states have a strong role in ensuring the necessary qualifications for workers in need of moving from one sector or type of job to another. The possible impact on different welfare state types is a topic that is in focus in the next section.

3. Financing and spending in welfare states – impact on preparedness

Welfare states are in the comparative welfare state literature clustered into different "welfare regimes". For a more in-depth discussion of "regimes", see Greve (2019). The presentation here will follow the classical line of *Three Worlds of Welfare Capitalism*, adding Southern Europe and Eastern Europe as clusters. Welfare regimes have been a way to systematize our knowledge on how different countries have some similarities with regard to their welfare state (Von Kersbergen, 2019). The division into these regimes reflects, in relation to financing and spending strong variations, also in the actors involved, and a different mix of state, market and civil society. Just to give a few indicators, Table 9.2 shows the

Table 9.2 Spending on social protection, overall level of taxation and inequality (GINI).

Country	*Social protection % GDP, 2015*	*Taxes and duties as % of GDP, 2016*	*Gini-coefficient, 2016*
Nordic welfare states			
Denmark	32.3	47.3	27.7
Sweden	29.2	44.6	27.6
Continental welfare states			
Germany	29.1	40.4	29.5
France	33.9	47.6	29.3
Liberal welfare states			
UK	28.6	35.1	31.5
Ireland	16.3	23.8	29.5
Southern Europe			
Italy	29.9	42.9	33.1
Spain	24.6	34.1	34.5
Eastern Europe			
Czech Republic	19.0	34.8	25.1
Poland	19.1 (2014)	34.4	33.9

Source: Eurostat (2018).

spending on social protection as percentages of GDP, overall level of taxes and duties and the GINI coefficient for a selection of countries within the EU (Eurostat, 2018).[2]

Table 9.2 shows the simple fact that the Nordic and Continental welfare states often spend more on welfare than in the liberal countries, with Southern (with Italy as the exception) and Eastern European countries spending even less (Adema, Fron and Ladaique, 2011).[3] Whether this reflects the very precise level of social spending is not the central issue in this article. Still, the high level of spending indicates a stronger need for financing as witnessed in the table by higher levels of taxes and duties, and, thus at the outset put more pressure on the Nordic and Continental welfare states with the higher level of spending and taxation than in the other types of welfare state (Bussemeer, Krell and Meyer, 2016). This has, historically, not been a strong issue in the sense that there has been a willingness to pay. Moreover this has been possible with merely a limited negative impact on and possible distortion of the choice between, for example, work and leisure. The legitimacy of the welfare state has also influenced the ability to finance the welfare states, but given the relatively high legitimacy, in especially the Nordic and Central European welfare states, it has been possible even for these types of welfare states to cope herewith (Morel and Palme, 2019).

There are differences in use of state, market and civil society among the welfare states, and, again this points also toward that liberal with more focus on the market and Southern and Eastern Europe with strong familiaristic approaches will be less influenced by the development.

The degree of inequality is also different. And, as shown in Table 9.2, and given that it typically involves a progressive tax system and/or public transfers to ensure a high degree of equality, this implies that if this aim shall be continued to be fulfilled in the future, a stronger pressure on welfare states with aims of redistributing will be needed.

The overall pressure on the income taxation as a consequence of globalization has been discussed for some time, for an early example (Ganghof, 2006). However, it seems still to have been having an only more limited impact, whereas there has been a tendency to lower taxes on companies, albeit at the same time with a broadening of the tax base (Brys et al., 2016). Broadening the tax base is also argued to be part of the solution of the pressure on the tax system related especially to taxation of companies. Here the labor market change might have more profound impact, especially in welfare states with high level of personal income taxation, as people who to a lesser degree are on the labor market also will have less stable income. This, in combination with more income from the digital economy possibly being generated offshore as royalty and capital income (Trepelkov, Tonino and Halka, 2015), increases the pressure. If this is the case, the ability to fund social protections will be weaker than previously.[4] This, however, is not a problem that is exclusive to only the developed welfare states. This development follows the traditional discussion on whether or not (and indeed how) to tax immobile production factors higher than mobile production factors as a way of circumventing the external pressure of funding the welfare state.

A possible way to cope herewith is to introduce a robot tax, as proposed by – among others – Bill Gates, Elon Musk and Stephen Hawking, who also informs us that South Korea has imposed a robot tax (Pieterson, 2018). This is not the place to enter into these issues in detail, but indicates that new ways to ensure financing might be needed. A possible argument being that of Guerreiro, Rebelo and Teles (2017, p. 32): "[A] tax on robots decreases the wage rate of non-routine workers and increase the wage rate of routine workers". However, they also argue that it might influence the distribution negatively. Naturally, those gaining from use of new technologies to have better paid jobs might pay more in tax and duties, but still this will not be able to off-set the income from those losing out if the continued polarization reduces the numbers of jobs, as presented back in Section 2.

4. Which welfare states are most exposed?

This question – whether countries in certain welfare regimes are most exposed – is based upon changes in the labor market and the ability to finance welfare states. Naturally, if looking into individual countries within the EU, there can indeed be variations. Nevertheless, given the size of the welfare states, the need to collect taxes and duties and the degree of inequality, a possible implication is that the countries most exposed are those in the Nordic and Continental welfare regimes.

However, there is a contradiction in the sense that, on the one hand, one could expect the more generous and expensive welfare states to be more exposed due to the possible difficulties in financing the welfare states. This as when fewer are on the labor market, and also that those who are on the labor market often will have lower income due to increased income polarization, which will reduce the ability to finance expenditures. Finally, more people might be in need of economic support from the welfare state. On the other hand, the more developed welfare states seem to be witnessing dwindling changes as a consequence of them having already either offshored or having used technology to reduced employment in the more traditional industrial sector, as shown in Section 2. Thereby the changes and polarization in the years to come might be less strong than in those countries where both the continuing restructuring of the industrial production as well as technological change might imply a stronger pressure on the ability to finance welfare states.

On the financing side, it is not just the risk of lower revenue due to the possibility of fewer employed citizens, but also that more people will be performing activities in the platform economy. While the income there may be lower, there is also a greater risk of it being derived from work done in the "hidden economy", where the digital platforms' owners do not necessarily pay tax in the countries where they offer their services (Greve, 2017). Therefore, even if there is a high number of jobs available, there is a risk that many of those jobs are temporary, working (e.g., even if having a job then the income will be below the poverty line) poor or a type of underemployment.

Overall, the discussion and data as presented in Section 2 and 3 can be systematized as done in the following Table 9.3.

Table 9.3 Impact on countries in different welfare regimes.

	Change on labor market	*Pressure on taxes and duties*	*Pressure on spending*	*Degree of equality*
Nordic	Low	High	High	High
Continental	Low	Middle	High	Middle
Southern	High	Low	Low	Low
Liberal	High	Low	Low	Low
Eastern	Middle	Low	Low	Low

Source: Author's own depiction.

The difference between low, middle and high is the author evaluation based upon the presentations and tables in sections 2 and 3. Where change on labor market is related to Figures 9.1 and 9.2, pressure on taxes and duties and spending follows in Table 9.2. The degree of equality, see Table 9.2, indicates the variation and emphasis on the goals toward equality in different welfare regimes/levels, as well as the size of the public sector and taxes and duties, and thereby the willingness to use the welfare state to counteract the impact of the market.

The implication is that the pressure from digitalization can vary, presumably due to the financing and size of the welfare state as it is today, and the loss of jobs. However, it will also be influenced based upon how the flexibility on the labor market and the educational system (especially lifelong learning) functions as a way of mediating the consequence of technological changes on the labor market. As upgrading, upskilling and development can be important, this is also recommended by those who do not necessarily agree that there will be a lack of jobs (Kaplan, 2015), or even that investment in education reduces the risk of the changes (Oesch, 2013), although so far there is seemingly no indication that this takes place within the European countries (Bengtsson, de la Porte and Jacobsson, 2017).

A core reason for the strong importance of employability of the workforce is that, if more generous welfare states simultaneously receive a lower level of income through the tax system and a higher level of expenditure to income transfers, the possible growing public sector deficit will incur difficulties in keeping the spending consistent, unless other types of financing can be found. The risk of a vicious circle will also be prevalent, as the income transfers in many ways has functioned as automatic stabilizers. If this is reduced, the economy might become more volatile and could also influence the degree of inequality. By the same token, if countries would like to continue having low levels of inequality, then the ability to use the tax system is important, given that welfare states with a high level of taxation have the largest degree of redistribution (Avram, Levy and Sutherland, 2014), and this is due to the more generous benefits and welfare states.

Overall, this indicates that universal and generous welfare states (especially the Nordic and Continental welfare states) are the ones under the most pressure from the changes. At the same time, those countries have already been witnessing strong changes in the industrial sector, which they have managed to cope with.

So, it might be that the consequences are stronger in countries where the recent development toward digitalization comes at a time where there is still a strong need for changes in the industrial sector, and perhaps also a need for change in the future voter expectation for additional welfare.

5. Some concluding remarks

There will be strong and persistent changes on the labor markets in the years to come and there will be challenges in all welfare states, albeit for different reasons.

Southern and Eastern Europe will see challenges stemming from the transformation of the industrial production that has been slower in these countries, which in turn has been a consequence of a lower wage level than in the more mature welfare states.

The Nordic and Continental welfare states face challenges due to the pressure on spending and the ability to finance their welfare states.

The liberal has a position in between as they have less welfare than others, but at the same time have had several of the change in the production structure.

Across welfare states a challenge will be the transformation process, so that even if the loss of jobs will be limited, the welfare states will have to undergo a transition period. If a large number of citizens should be able to keep their jobs or get a new one, they will need to be trained. Thus, lifelong learning seems to be an important aspect in the welfare states' future development.

Notes

1 This article will not go into and describe these concepts; see instead Greve (2018) for an overview.
2 Choice of countries so as they represent the five regimes (in total ten countries) also in order to ensure an overview.
3 Naturally, there are caveats with this presentation as it do not included occupational and fiscal welfare.
4 The EU-commission has in March, 2018 proposed a three-percent-point tax on turnover in companies, presumably as a first step in acknowledging that the existing tax system is under pressure from platforms generating revenues, but paying less tax than traditional companies.

References

Adema, W., Fron, P. and Ladaique, M., 2011. *Is the European Welfare State Really More Expensive?: Indicators on Social Spending, 1980–2012; and a Manual to the OECD Social Expenditure Database (SOCX)*. [online] OECD Social, Employment and Migration Working Papers, No. 124. Available at: <www.oecd-ilibrary.org/social-issues-migration-health/is-the-european-welfare-state-really-more-expensive_5kg2d2d4pbf0-en> [Accessed 17 Sep. 2019].

Arntz, M., Gregory, T. and Zierahn, U., 2016. *The Risk of Automation for Jobs in OECD Countries: A Comparative Analysis*. OECD Social, Employment and Migration Working Papers, No. 189. Paris, France.

Autor, D.H., 2015. Why are there still so many jobs? The History and future of workplace automation. *Journal of Economic Perspectives*, 29(3), pp. 3–30.

Autor, D.H., Dorn, D. and Hanson, G.H., 2016. The China shock: Learning from labor-market adjustment to large changes in trade. *Annual Review of Economics*, 8, pp. 205–240.

Autor, D.H., Levy, F. and Murnane, R.J., 2003. The skill content of recent technological change: An empirical exploration. *The Quarterly Journal of Economics*, 118(4), pp. 1279–1333.

Avram, S., Levy, H. and Sutherland, H., 2014. Income redistribution in the European Union. *IZA Journal of European Labor Studies*, 3(1), p. 22, pp. 1–29.

Bengtsson, M., de la Porte, C. and Jacobsson, K., 2017. Labour market policy under conditions of permanent austerity: Any sign of social investment? *Social Policy & Administration*, 51(2), pp. 367–388.

Brynjolfsson, E. and McAfee, A., 2014. *The Second Machine Age. Work, Progress, and Prosperity in a Time of Brilliant Technologies*. London, UK: W.W. Norton & Company.

Brys, B., Perret, S., Thomas, A. and O'Reilly, P., 2016. *Tax Design for Inclusive Economic Growth*. OECD Taxation Working Papers, No. 26. Paris, France.

Bughin, J., Staun, J., Andersen, J.R., Schultz-Nielsen, M., Aagaard, P. and Enggaard, T., 2017. *Digitally-Enabled Automation and Artificial Intelligence: Shaping the Future of Work in Europe's Digital Front-runners*. [online] McKinsey & Company. Available at: <https://mckinsey.com/~/media/McKinsey/Featured Insights/Europe/Shaping the future of work in Europes nine digital front runner countries/Shaping-the-future-of-work-in-Europes-digital-front-runners.ashx> [Accessed 17 Sep. 2019].

Bussemeer, T., Krell, C. and Meyer, H., 2016. *Social Democratic Values in the Digital Society: Challenges of the Fourth Industrial Revolution*. [online] Available at: <www.socialeurope.eu/wp-content/uploads/2016/01/OccPap10.pdf> [Accessed 17 Sep. 2019].

Eurostat, 2018. *Gini Coefficient of Equivalised Disposable Income*. [online] Available at: <http://appsso.eurostat.ec.europa.eu/nui/show.do?dataset=ilc_di12> [Accessed 17 Sep. 2019].

Frey, C. and Osborne, M., 2013. *The Future of Employment: How Susceptible are Jobs to Computerisation?* [online] Available at: <www.oxfordmartin.ox.ac.uk/downloads/academic/The_Future_of_Employment.pdf> [Accessed 17 Sep. 2019].

Ganghof, S., 2006. *The Politics of Income Taxation: A comparative Analysis*. Colchester, UK: European Consortium for Political Research.

Goos, M., Manning, A. and Salomons, A., 2014. Explaining job polarization: Routine-biased technological change and offshoring. *American Economic Review*, 104(8), pp. 2509–2526.

Graetz, G. and Michaels, G., 2015. *Robots at Work*. CEP Discussion Paper. London, UK.

Greve, B., 2017. *Technology and the Future of Work: The Impact on Labour Markets and Welfare States*. Cheltenham, UK: Edward Elgar Publishing.

Greve, B., 2018. *Social and Labour Market Policy: The Basics*. London, UK: Routledge.

Greve, B., 2019. Routledge Handbook of the Welfare State, 2nd ed. Oxon, UK: Routledge.

Guerreiro, J., Rebelo, S. and Teles, P., 2017. *Should Robots be Taxed?* NBER Working Paper Series, No. 23806. Cambridge, MA.

Hawksworth, J., Berriman, R. and Goel, S., 2018. *Will Robots Really Steal Our Jobs?: An International Analysis of the Potential Long Term Impact of Automation*. [online] PwC. Available at: <www.pwc.com/hu/hu/kiadvanyok/assets/pdf/impact_of_automation_on_jobs.pdf> [Accessed 17 Sep. 2019].

Kaplan, J., 2015. *Humans Need Not Apply: A Guide to Wealth and Work in the Age of Artificial Intelligence*. London, UK: Yale University Press.

Kuhlman, J., Schubert, K. and de Villota, P., 2016. Recent developments of European welfare systems: Multiple challenges and diverse reactions. In: K. Schubert, P. de Villota and J. Kuhlman, eds., *Challenges to European Welfare System*. Heidelberg, Germany: Springer, pp. 1–8.

Manyika, J., Lund, S., Chui, M., Bughin, J., Woetzel, J., Batra, P., Ko, R. and Sanghvi, S., 2017. *Jobs Lost, Jobs Gained: Workforce Transitions in a Time of Automation*. [online] McKinsey Global Institute. Available at: <www.mckinsey.com/featured-insights/future-of-organizations-and-work/jobs-lost-jobs-gained-what-the-future-of-work-will-mean-for-jobs-skills-and-wages> [Accessed 17 Sep. 2019].

Mokyr, J., Vickers, C. and Ziebarth, N.L., 2015. The history of technological anxiety and the future of economic growth: Is this time different? *Journal of Economic Perspectives*, 29(3), pp. 31–50.

Morel, N. and Palme, J., 2019. Financing the welfare state and the politics of taxation. In: B. Greve, ed., *The Routledge Handbook of the Welfare State*, 2nd ed. London, UK: Routledge, pp. 401–409.

OECD, 2017a. Annex 3.A1 – Additional Evidence on Polarisation. In: OECD, ed., *OECD Employment Outlook 2017*. Paris, France: OECD Publishing, pp. 120–122.

OECD, 2017b. How technology and globalisation are transforming the labour market. In: *OECD Employment Outlook 2017*. Paris, France: OECD Publishing, pp. 81–124.

Oesch, D., 2013. *Occupational Change in Europe: How Technology and Education Transform the Job Structure*. Oxford, UK: Oxford University Press.

Pieterson, W., 2018. *Digitization and Work: How Governments are Responding to Changing Labour Markets?* DG Employment, Social Affairs and Inclusion. Brussels, Belgium.

Reich, R.B., 2015. *Saving Capitalism: For the Many, Not the Few*. New York, NY: Alfred E. Knopfs.

Schwab, K., 2016. *The Fourth Industrial Revolution*. New York, NY: Crown Publishing Group.

Trepelkov, A. Tonino, H. and Halka, D., 2015. *United Nations Handbook on Selected Issues in Protecting the Tax Base of Developing Countries*. New York, NY: United Nations.

Vis, B., 2019. How to analyse welfare states and their development. In: B. Greve, ed., *The Routledge Handbook of the Welfare State*, 2nd ed. Oxon, UK: Routledge, pp. 267–277.

Von Kersbergen, K., 2019. What are the welfare state typologies and how are they useful, if at all? In: B. Greve, ed., *The Routledge Handbook of the Welfare State*, 2nd ed. London, UK: Routledge, pp. 115–123.

10 "Gig patients"

Health and dental care in the gig economy

Anthony Larsson and Dominika Sabolová

1. Introduction

The "gig economy" (or the "shared economy", as it is also known) suggests a new style of employment where workers sustain themselves by performing a number of "gigs" on a freelance basis (often for several different contractors) rather than having a fixed-term employment or a permanent contract at a particular employer (Petriglieri, Ashford and Wrzesniewski, 2018; Shibata, 2019). In a majority of cases (63% according to Petriglieri, Ashford and Wrzesniewski [2018]), the "gig workers" have chosen this type of employment on their own volition. The main reason for doing so is often that it seemingly provides the workers with the possibility to choose their own projects and set their own schedules.

In principle, the "gig economy" allow workers to pick up temporary jobs anywhere in the world, although the percentage of the population engaging in "gig work" varies depending on geographical region. For instance, in the United States alone, between 34% to 36% of the workforce currently freelance, or "gig"; a figure that is expected to have grown into a majority by the year 2027 (Taylor, 2018; Spinner, 2019). In Australia, on the other hand, the numbers are sizably lower (although the exact figures vary) with between 7% to 25% of the workforce expected to find jobs through the "gig economy" (Offer, 2019; Bailey, 2018). In Europe, the numbers are for more fragmented, ranging from an estimated 9% in Germany, to 22% in Italy (New Europe, 2017).

Nevertheless, one major and often overlooked downside in the contemporary debate, is that a consequence of the "gig economy" is that in the future there will be a sizable number of adults who will be without regulated health care and/or dentistry benefits and legislative protection. Moreover, workers will invariably need some kind of insurance coverage as the welfare system (which exists to varying degrees in different countries) may not always provide the necessary healthcare benefits for the workers (DePillis, 2018; Taylor, 2018). However, freelancers are twice as likely as permanent employees to report that they lack traditional health insurance (Spinner, 2019).

Various measures have sometimes been taken in different countries to deal with this situation, but limitations in various forms of regulation has often entailed that this has had limited success. For instance, in the United States, the Affordable Care

Act (ACA), often nicknamed "Obamacare", was first approved in 2010 and fully implemented in 2014 (Courtemanche et al., 2018). Among many other things, the ACA made it illegal for insurance companies to deny US-citizens coverage, refuse to cover treatment or charge higher premiums for pre-existing health conditions (Thompson, 2015; Blumenthal, Abrams and Nuzum, 2015; Rovner, 2018). However, the crux of the matter is that the ACA only stipulates that no higher premiums are to be charged for pre-existing conditions. There are, however, other routes the insurance companies may take. For instance, they may opt for applying high-deductible health plans (HDHPs)[1] and/or increased "patient copays" models, etc. This would entail that the patient's financial responsibility would escalate and, in turn, undoubtedly lead to a new phenomenon of "gig patients" that will severely impact the future health-care and dentistry industries.

Proponents of HDHPs argue that by holding patients financially responsible through high copayments and deductibles, it helps decrease "moral hazard" by stymieing detrimental behavior by patients that would otherwise waste health-care resources and drive up the overall societal costs (Wilensky, 2006). Notwithstanding, such a development would have detrimental effects on some gig workers, as many of them will not be earning much more than minimum wage-level pay (or even less). Even in some countries in Europe, where there is no legally mandated minimum wage and where wages are negotiated between the employees and the employers through so-called "social partners", the "gig economy" impacts welfare by pressing back the wages of the employees and sometimes depriving workers of the full welfare benefits that come with an employment status (Worstall, 2017; Coyle, 2017; Kessler, 2018).

Fact of the matter is that the issue of minimum wage for "gig workers" has been a fiercely debated topic in recent years (Healy, Nicholson and Pekarek, 2017). In 2019, the labor advocacy group Working Washington instituted a campaign for a USD15-an-hour minimum wage for "gig workers" (Eisenberg, 2019). However, a problem is that these propositions only relate to hourly paid work when in many cases "giggers" are paid on the basis of a fixed price per project. Calculated in terms of time worked versus pay, the effective wage can actually be far below the suggested minimum wage level. While it is true that some minimum wage regulations can include so-called "piece work", it is important to remember that the rate is generally estimated on the time taken to complete various tasks undertaken by an "average" worker (Rubin and Perloff, 1993; Gittleman and Pierce, 2015). However, in many cases these tasks may vary too much in order to estimate them to the extent that it can be said that they truly and accurately represent a standard "average". For instance, a telemarketer who gets paid by number of calls made, rather than by commission on successful sales, can have a productive or a slow day depending on how many people pick up the phone. Also, software programming varies immensely depending on task and software, and likewise, the time taken to conduct identical programming procedures may vary from time to time, depending on the hardware used, internet connection, software glitches, etc. This makes average productivity for "piece rates" very difficult, if not impossible, to measure fairly (Shearer, 2004; Lazear, 2000).

More than this, many countries lack proper legislation to fully cover the "giggers". For instance, in the United Kingdom, being classified as employees would mean automatic enrollment in workplace pension and employer contributions, but the legislation defines "giggers" as "self-employed", meaning that more than a million "gig economy" workers in the UK risk missing out on £22,000 (≈ USD27,000) of pension (Partington, 2017). Also, "giggers" will oftentimes find that they will need to sign multiple insurance plans depending on how and where they work. "Giggers" will sometimes also have to acquire a "liability insurance" at their own expense in case something at work goes awry. However, this is a slippery slope, because many insurance policies are not valid for commercial use, and oftentimes local regulations and requirements vary depending on in what city or country the work is being performed. The multiplicity of actors involved may sometimes make it difficult to identify all the actors present and to delineate their areas of responsibilities (which is often needed whenever making a claim to the insurance companies) (Prassl, 2018).

Drawing upon available literature in the field, this chapter serves as an analytical commentary on the phenomenon of "gig patients". Specifically, the aim of this chapter is to investigate the wider ramifications this group of people could have to the welfare society and the future of labor, and the possible courses of actions that can be taken to deal with the emergent situation of "gig patients".

2. Discussion

2.1. The cause of "gig patients"

For many workers, the "gig economy" offers an irresistible allure where they are free to set their own schedule. Likewise, whenever entrepreneurs need to enlist help, or as their entrepreneurial company grows, the "gig economy" offers a way out of the expense of having payroll insurance, employee benefits, sick leave and vacations. Thus, entrepreneurs only need to pay for the work they need, whenever it is needed. Since the advent of digital technology along with information and communications technology (ICT) devices, an increasing number of workers and entrepreneurs alike have begun to identify themselves as "digital nomads" (Sisson, 2017). This means that they no longer depend on work in traditional workplaces, but are free to work anywhere and whenever, as long as there is access to a laptop computer or a tablet computer and an adequate internet connection (Müller, 2016). Conversely, it is also possible to gain many different short-term workplaces via the "gig economy". That is to say, the "gig economy" offers a wide array of freelance jobs available on the online market for various professions and skillsets.

Unshackled from the constraints placed by managers and corporate norms, workers can choose the assignments they feel put their talents to the greatest use while choosing only the assignments that they find appealing. The workers thus have a sense of agency in what they produce and how to manage their own life situation. However, while the personal freedom is far-reaching, the stakes are also exceptionally high and the cost may be far more than a financial one.

The workers are invariably expected to always produce and deliver and to always be on top of their game (Petriglieri, Ashford and Wrzesniewski, 2018). Keeping one's level of productivity at top capacity at all times is a constant struggle, and invariably means working long hours (Sinicki, 2018). Distress and distractions can erode one's level of productivity, as can various forms of obstacles that may appear along the way. Of course, various forms of health-related issue also act as impediments, which risks creating a vicious circle where the worker is less productive due to stress and/or illness, but yet has to compensate by putting in more hours to make up for lost productivity, which consequently serves to worsen the stress and/or illness (O'Connor, 2018).

As previously mentioned, one of the greatest boons for those contracting labor is not having to pay for various forms of welfare benefits. On the flipside, this is also one of the greatest banes to the workers. That is to say, the "gig economy" rarely or never offer workers any benefits of any kind that is commonly found among the traditional social safety net, such as sick pay, parental leave, paid vacations, paid course training, etc. Thus, "giggers" lack the possibility to take paid leave and receive such low wages that they are unable to take time off and fund their own health care. The aforementioned tendency of "giggers" lacking traditional health insurance serves to further deepen the problem. The ramifications of this situation is that it risks creating a vicious circle for "giggers" who are sick becoming even sicker without the means of doing much about it. For example, La Duke (2016, para.5) writes that "the increasing loss of dental benefits means that many in the "gig economy" decide to stop going to the dentist or, at the very least, reduce dental visits". Of course, this holds true for virtually any other health-care procedure as well, such as optometry, physiotherapy, and in extreme cases, possibly even procedures such as oncological treatments. Some reports even suggest that the stress levels "giggers" are subjected to can lead to an increase of cardiovascular diseases and even cancer (Ahuja, 2017; Tran and Sokas, 2017; University of Oxford, 2018). A 2019 study by the American Heart Association indicates that those working long hours (more than ten hours for at least 50 days a year) may run a 29–45% greater risk of stroke (Fadel et al., 2019).

These groups of people are unlikely to seek out the medical attention they need in time due to the constraints of their work-life situation in the "gig economy". These people will be known as "gig patients", and can be expected to increase in numbers unless preemptive action is taken before the problem becomes widespread. Already now, a staggering 54% of adult US-citizens claim that they have delayed health-care treatment because they cannot afford it, and this number can be expected to grow steadily as the pool of "giggers" grows larger (Carter, 2018).

2.2. The "gig patients'" effects on the welfare services

Beyond the personal tragedies of the "gig patients", there is also the aspect of them adding further strain on the welfare resources due to the more advanced care needed in order to treat them when they finally do seek treatment as opposed to if they had sought treatment sooner. Another factor in this is that in many of the more popular "gig companies", the workers can make less than USD500 per month, while the typical US-consumer in 2017 was responsible for USD1,820 in deductible payments and another USD4,400 in out-of-pocket costs (Dyrda, 2017; Bloom, 2017; New, 2017). Naturally, these numbers do not add up, which means that the "gig patients" are falling deeper in debt. Sixty-nine percent of adults in the US claim are paying for significant health-care expenses using different means such as savings, credit cards, disposable income, loans from family and friends, etc. (Lagasse, 2018). Moreover, this also means that the health-care providers have difficulties in getting paid for their services. In the US, 73% of health-care providers have reported that it takes them at least one month (or longer) to collect payment from their patients and that 68%, or more than two-thirds, of all patients in 2016 could not pay their bill balances in full (TransUnion, 2017). For dentistry, the figures are also grim. Recent studies have shown that several European countries lack sufficient insurance coverage for patients and that fewer Europeans with low incomes tend to visit the dentist today, especially following the 2008 European financial crisis (Elstad, 2017).

This raises the question as to whether or not the health-care service and dentistry, in its current format, can survive the surge of "gig patients" in the long run. For the labor market, "gig patients" are a very negative occurrence, as they signal that the workplace in question is unsafe. As La Duke (2016, para.7) states:

> Gigs are often the result of larger companies outsourcing the most dangerous jobs to individuals. Most individuals lack the resources to acquire proper regulatory training, and what's more, many small companies are actually exempt from regulatory protections for workers.

Given this context, it would be in society's interest to curb the growth of the "gig patients", but the question is how to best deal with the situation?

2.3. Dealing with the "gig patient" situation

In some cases, a consequence of trying to deal with the current situation is that some "giggers" who operate under the "digital nomad" way of life, are able to take up jobs for different companies in the world while they themselves move to a low-cost country (Dal Fiore et al., 2014; Reichenberger, 2018; Gaid, 2019). The minimum wage-level payment they receive may not last longer in one of these countries than it would in their home countries, the "digital nomads" may even be able to find more affordable health and dental-care in these low-cost countries (Backe, 2018). While "digital nomads" do come in all different shapes, forms and sizes, many of them are young, able-bodied, single and without a family to support or any other strings tying them down to any particular place (Tynan, 2015). This, however, is not the case for everyone in the "gig economy", and even for many of those who do who fit the archetypical description of a "digital nomad", that particular lifestyle may not present itself as an option. Moreover, many "digital nomads" still run the risk of being laid off and losing their "gig", irrespective of where they are (Backe, 2018). Likewise, they may also fall on hard times and end up becoming "gig patients" in the countries in which they are currently residing.

In terms of dealing with the "gig patients" that already exist, there are a number of measures the health-care and dentistry industry can take. On a more immediate level, it is important to catch the "gig patients'" affliction as soon as possible (possibly in connection when seeking treatment for a different ailment), and make sure a treatment plan is administered before the malady escalates into a worse (and more difficulty treated) condition. Given the fact that "gig patients" are often already deep in debt, payment options will need to be transparent and clearly defined up front so that there is no shock for the patient when the bill arrives. Ideally, there should also be some flexibility in arranging a payment plan, where smaller installments are paid over a longer period of time, with the possibility of adjusting the payments alternatively shortening or extending the time period as necessary, as long as there is a steady installment being continuously paid (Spinner, 2019). Moreover, health-care services will need to be open for securing ways of further simplifying the payment process. One practical way is to offer access to a secure and easy online payment system, which would eliminate credit checks before delivering the services. Another way of providing simplified payments is to explore ways of consolidating bills from multiple visits by the same "gig patient" so that the procedure in setting up a payment plan becomes more seamless and automatized (Spinner, 2019).

A lingering issue has been whether or not "giggers" are workers or self-employed, which in itself is a critical distinction on which a number of key entitlements hinge, including the aforementioned issues of holiday pay, sick leave, possible minimum wage and so forth. To this end, there needs to be clear legislation from the national government delineating between the two concepts, and where it will be made clear into which category of the two "gig work" falls under. There also needs to be clear-cut regulations in regards to liability insurances, and whether the contractor or the worker should be responsible for arranging them. To this end, there also needs to be an overview of the available insurances available to the "giggers" to ensure that the terms are fair to this category of workers, and that potential legislative loopholes that do exist in this space must be remedied.

Another issue is that societies in many countries today rely too much on "giggers" to enforce their own workplace rights. This becomes a problem inasmuch that the people that are most likely to have reason for grievance are also the ones who are the least able to assert themselves. As an example, in December 2018, the UK Employment Tribunal registered a mere 0.2% of claims made by agency workers, even though they comprise close to three percent of the workforce (Judge, 2018a). To this end, there will be a need for labor unions to stay more attuned to the developments in the digital space and the "gig economy", as this remains a weak spot for many unions (which is to a large part tied to the unclear legal regulations on "gig work") (Matthias, 2019).

In addition, it is essential for governments to allocate enough resources to their respective labor market enforcement agencies, so that they can seek out firms that make cynical use of dubious labor contracts or keeping substandard records of their workers and their respective purview, etc. Ideally, these labor market enforcement agencies can even take preemptive action in rooting out the companies that do not comply with the rules and labor laws of the country. Moreover, it is essential that the labor market enforcement agencies seek to be as strategic as possible in their activities. Recent findings suggest that "giggers" in areas with weak labor markets are most vulnerable to non-compliant behavior (Judge, 2018b). Specifically, this means that people who lack other local options for work are unable to "vote with their feet" when they are handed a raw deal or are subjected to unlawful practices. To this end, a recommendation would be to employ a place-based enforcement strategy, targeting various hotspots in areas that have little competition within a particular market. In that way, the labor market enforcement agencies could prioritize the most affected places while also maximizing their impact, and hopefully causing ripple effects to surrounding areas until these agencies have mustered enough resources to gain a wider geographic coverage.

It is true that the "gig economy" is spreading fast, perhaps leaving people with few options in the future, as traditional employment may well decrease. However, the "gig economy" can only thrive as a format if enough people seek themselves to it to begin with. While added taxation on actors seeking to enlist "gig labor" might have a stymieing effect on the wider spread of the "gig economy", it is important that citizens are informed, ideally at an early stage, what it means to be a "gig worker" and all the ramifications that follows, so that they can make an informed decision of whether or not to pursue this work format.

3. Conclusion

The aim of this chapter was to investigate the wider ramifications that "gig patients" could have to the welfare societies and the future of labor. The results of this study indicate that while the "gig patients" are not yet a widespread problem, they can be expected to increase substantially in the future unless preemptive action is taken soon. The "gig patients" to date only affect the workers partaking in the "gig economy", but may have more far-reaching impact on society at large, and in particular regard to the welfare sector.

While the problem in part stems from the desire of employers to rid themselves of any added cost of labor short of wages, another factor is mankind's need to feel in control of their own time and being free to choose their own assignments. In this sense, it may in some cases risk becoming an "illusion of choice", where the worker is not actually free, but rather caught in a vicious circle where they cannot afford to tend to their health, thereby eventually becoming "gig patients" (Sinicki, 2018). The problem is complex inasmuch that it involves several different parties. On the one hand, it involves both the concerned parties themselves, i.e., the contractors and the workers. On the other hand, it also involves several external actors, such as the insurance companies, the national governments, national labor laws, labor unions, the health-care services, etc. As such, there needs to be a clearer legal definition identifying the role of the "giggers", if they are to have the legal status of "self-employees" (with the tax deductions and benefits that entails) or if they should be considered employees (with the social net and welfare benefits that entails). Once in place, the other practical labor-law issues will need to be tethered out, and social partners among the "gig economy" would need to be established in order to for there to be a community of practice established, from which people could make their own informed decisions about the "rules of the game" and as to whether or not the "gig economy" is a type of work that caters to their interests and needs. That is not to say that the current "gig economy" does not provide for optimal solutions for some people. For instance, for people looking for a part-time job, or for an extra job on the side of their regular employment during the odd hours, the "gig economy" can provide a solid opportunity (Ravenelle, 2019; Sinicki, 2018). Nevertheless, workers intending to find a means to support themselves on a fulltime basis need to operate under clearly defined premises in order to avoid the risk of becoming "gig patients". By instituting a community of practice, a clear legal framework and fair standards, the risk of facilitating "gig patients" should be mitigated.

Note

1 A high-deductible health plan (HDHP) entails a health-insurance plan that has lower premiums and higher deductibles than a traditional health plan would have (Buntin et al., 2011; Kullgren et al., 2010).

References

Ahuja, A., 2017. *Why 'Gig Health' Matters*. [online] Financial Times. Available at: <www.ft.com/content/bdc90c22-408f-11e7-82b6-896b95f30f58> [Accessed 17 Sep. 2019].

Backe, C., 2018. *Becoming a Digital Nomad: Your Step By Step Guide To the Digital Nomad Lifestyle – 2019 Edition*. Chicago, IL: Worthy Go.

Bailey, M., 2018. *The Gig Economy is Growing Whether You Like It or Not*. [online] The Australian Financial Review. Available at: <www.afr.com/boss/gig-economy-explainer-the-phenomenon-rocking-the-way-we-work-20180514-h101t9> [Accessed 17 Sep. 2019].

Bloom, E., 2017. *Here's How Much Money Americans are Making from the Gig Economy*. [online] CNBC. Available at: <www.cnbc.com/2017/06/19/heres-how-much-money-americans-are-making-from-the-gig-economy.html> [Accessed 17 Sep. 2019].

Blumenthal, D., Abrams, M. and Nuzum, R., 2015. The Affordable Care Act at 5 years. *The New England Journal of Medicine*, 372(25), pp. 2451–2458.

Buntin, M.B., Haviland, A.M., McDevitt, R. and Sood, N., 2011. Healthcare spending and preventive care in high-deductible and consumer-directed health plans. *The American Journal of Managed Care*, 17(3), pp. 222–230.

Carter, S.M., 2018. *Over Half of Americans Delay or Don't Get Health Care Because They Can't Afford It – These 3 Treatments Get Put Off Most*. [online] CNBC. Available at: <www.cnbc.com/2018/11/29/over-half-of-americans-delay-health-care-becasue-they-cant-afford-it.html> [Accessed 17 Sep. 2019].

Courtemanche, C., Marton, J., Ukert, B., Yelowitz, A. and Zapata, D., 2018. Effects of the Affordable Care Act on health care access and self-assessed health after 3 years. *INQUIRY: The Journal of Health Care Organization, Provision, and Financing*, 55, p. 0046958018796361, pp. 1–10.

Coyle, D., 2017. *The Gig Economy and the Welfare State*. [online] National Institute of Economic and Social Research. Available at: <www.niesr.ac.uk/blog/gig-economy-and-welfare-state> [Accessed 17 Sep. 2019].

Dal Fiore, F., Mokhtarian, P.L., Salomon, I. and Singer, M.E., 2014. "Nomads at last"? A set of perspectives on how mobile technology may affect travel. *Journal of Transport Geography*, 41, pp. 97–106.

DePillis, L., 2018. *Gig Economy Workers Need Benefits. These Companies Are Popping Up to Help*. [online] CNN Business. Available at: <https://money.cnn.com/2018/08/23/technology/gig-economy-worker-benefits/index.html> [Accessed 17 Sep. 2019].

Dyrda, L., 2017. *12 Trends in Patient Responsibility Payments, Up 29.4% Since 2015*. [online] Becker's ASC Review. Available at: <www.beckersasc.com/asc-coding-billing-and-collections/12-trends-in-patient-responsibility-payments-up-29-4-since-2015.html> [Accessed 17 Sep. 2019].

Eisenberg, R., 2019. *How Well is the Gig Economy Working for Gig Workers?* [online] Forbes. Available at: <www.forbes.com/sites/nextavenue/2019/02/18/how-well-is-the-gig-economy-working-for-gig-workers> [Accessed 17 Sep. 2019].

Elstad, J.I., 2017. Dental care coverage and income-related inequalities in foregone dental care in Europe during the great recession. *Community Dentistry and Oral Epidemiology*, 45(4), pp. 296–302.

Fadel, M., Sembajwe, G., Gagliardi, D., Pico, F., Li, J., Ozguler, A., Siegrist, J., Evanoff, B.A., Baer, M., Tsutsumi, A., Iavicoli, S., Leclerc, A., Roquelaure, Y. and Descatha, A., 2019. Association between reported long working hours and history of stroke in the CONSTANCES cohort. *Stroke*, 50(7), pp. 1879–1882.

Gaid, A., 2019. *Digital Nomad Cities: How to Choose Your First Destination*. [online] Oberlo. Available at: <www.oberlo.com/blog/cheapest-places-to-live-in-the-world> [Accessed 17 Sep. 2019].

Gittleman, M. and Pierce, B., 2015. Pay for performance and compensation inequality: Evidence from the ECEC. *ILR Review*, 68(1), pp. 28–52.

Healy, J., Nicholson, D. and Pekarek, A., 2017. Should we take the gig economy seriously? *Labour & Industry: A Journal of the Social and Economic Relations of Work*, 27(3), pp. 232–248.

Judge, L., 2018a. *Life for Gig Economy Workers Will Only Improve If the Government's Rules Can be Enforced*. [online] New Statesman America. Available at: <www.newstatesman.com/politics/economy/2018/12/life-gig-economy-workers-will-only-improve-if-government-s-rules-can-be> [Accessed 17 Sep. 2019].

Judge, L., 2018b. *The Good, the Bad and the Ugly: The Experience of Agency Workers and the Policy Response*. [online] Resolution Foundation. Available at: <www.resolutionfoundation.org/publications/the-good-the-bad-and-the-ugly-the-experience-of-agency-workers-and-the-policy-response> [Accessed 17 Sep. 2019].

Kessler, S., 2018. *The Gig Economy: Lower Wages, More Injuries, Horrible Benefits*. [online] Literary Hub. Available at: <https://lithub.com/the-gig-economy-lower-wages-more-injuries-horrible-benefits> [Accessed 17 Sep. 2019].

Kullgren, J.T., Galbraith, A.A., Hinrichsen, V.L., Miroshnik, I., Penfold, R.B., Rosenthal, M.B., Landon, B.E. and Lieu, T.A., 2010. Health care use and decision making among lower-income families in high-deductible health plans. *Archives Internal Medicine*., 170(21), pp. 1918–1925.

La Duke, P., 2016. *Is the Gig Economy Sustainable?* [online] Entrepreneur Magazine. Available at: <www.entrepreneur.com/article/278826> [Accessed 17 Sep. 2019].

Lagasse, J., 2018. *Americans Stressing Out Over Soaring Healthcare Costs This Enrollment Season.* [online] Healthcare Finance. Available at: <www.healthcarefinancenews.com/news/americans-stressing-out-over-soaring-healthcare-costs-enrollment-season> [Accessed 17 Sep. 2019].

Lazear, E.P., 2000. Performance pay and productivity. *The American Economic Review*, 90(5), pp. 1346–1361.

Matthias, H., 2019. *Problems Within the Gig Economy.* [online] Iona Journal: The Exchange. Available at: <www.ionajournal.ca/exchange/2019/2/25/problems-within-the-gig-economy> [Accessed 17 Sep. 2019].

Müller, A., 2016. The digital nomad: Buzzword or research category? *Transnational Social Review*, 6(3), pp. 344–348.

New, C., 2017. *How Much Are People Making From the Sharing Economy?* [online] Earnest. Available at: <www.earnest.com/blog/sharing-economy-income-data> [Accessed 17 Sep. 2019].

New Europe, 2017. *European Gig Economy Report: Crowd Workers Are Not Self-employed.* [online] Available at: <www.neweurope.eu/article/european-gig-economy-report-crowd-workers-not-self-employed> [Accessed 17 Sep. 2019].

O'Connor, S., 2018. *Workplace Exhaustion is a Vicious Cycle in the UK.* [online] Financial Times. Available at: <www.ft.com/content/f959a19c-d095-11e8-a9f2-7574db66bcd5> [Accessed 17 Sep. 2019].

Offer, K., 2019. *Australians Flock to Gig Economy for Work.* [online] Daily Liberal. Available at: <www.dailyliberal.com.au/story/6222601/australians-flock-to-gig-economy-for-work> [Accessed 17 Sep. 2019].

Partington, R., 2017. *Gig Economy Workers in UK Risk Missing Out on £22,000 of Pension.* [online] The Guardian. Available at: <www.theguardian.com/business/2017/dec/07/gig-economy-workers-uk-missing-out-22200-status-workplace-pension> [Accessed 17 Sep. 2019].

Petriglieri, G., Ashford, S.J. and Wrzesniewski, A., 2018. *Thriving in the Gig Economy.* [online] Harvard Business Review. Available at: <https://hbr.org/2018/03/thriving-in-the-gig-economy> [Accessed 17 Sep. 2019].

Prassl, J., 2018. *Humans as a Service: The Promise and Perils of Work in the Gig Economy.* Oxford, UK: Oxford University Press.

Ravenelle, A.J., 2019. *Hustle and Gig: Struggling and Surviving in the Sharing Economy.* Oakland, CA: University of California Press.

Reichenberger, I., 2018. Digital nomads – a quest for holistic freedom in work and leisure. *Annals of Leisure Research*, 21(3), pp. 364–380.

Rovner, J., 2018. *Fact Check: Who's Right About Protections for Pre-Existing Conditions?* [online] NPR. Available at: <www.npr.org/sections/health-shots/2018/10/11/656503264/fact-check-whos-right-about-protections-for-pre-existing-conditions?> [Accessed 17 Sep. 2019].

Rubin, D.K. and Perloff, J.M., 1993. Who works for piece rates and why. *American Journal of Agricultural Economics*, 75(4), pp. 1036–1043.

Shearer, B., 2004. Piece rates, fixed wages and incentives: Evidence from a field experiment. *The Review of Economic Studies*, 71(2), pp. 513–534.

Shibata, S., 2019. Gig work and the discourse of autonomy: Fictitious freedom in Japan's digital economy. *New Political Economy*, (Ahead of Print), pp. 1–17.

Sinicki, A., 2018. *Thriving in the Gig Economy: Freelancing Online for Tech Professionals and Entrepreneurs*. Berkeley, CA: Apress.

Sisson, N., 2017. *The Suitcase Entrepreneur: Create Freedom in Business and Adventure in Life*. 3rd ed. New York, NY: North Star Way.

Spinner, M., 2019. *How the Gig Economy Stands to Change Healthcare*. [online] MedCity News. Available at: <https://medcitynews.com/2019/04/how-the-gig-economy-stands-to-change-healthcare> [Accessed 17 Sep. 2019].

Taylor, C., 2018. *Healthcare, Freelanced: Where Will Gig Economy Workers Get Coverage?* [online] Reuters. Available at: <www.reuters.com/article/us-world-work-health care/healthcare-freelanced-where-will-gig-economy-workers-get-coverage-idUSKCN 1IG1C8> [Accessed 17 Sep. 2019].

Thompson, T., 2015. *The Affordable Care Act*. Farmington Hills, MI: Greenhaven Press.

Tran, M. and Sokas, R.K., 2017. The gig economy and contingent work: An occupational health assessment. *JOEM*, 59(4), pp. e63–e66.

TransUnion, 2017. *Patients May Be the New Payers, But Two in Three Do Not Pay Their Hospital Bills in Full*. [online] Available at: <https://newsroom.transunion.com/patients-may-be-the-new-payers-but-two-in-three-do-not-pay-their-hospital-bills-in-full> [Accessed 17 Sep. 2019].

Tynan, K., 2015. *Free Agent: The Independent Professional's Roadmap to Self-Employment Success*. Boca Raton, FL: CRC Press.

University of Oxford, 2018. *Digital Gig Economy is Bad for Your Wellbeing, New Research Suggests*. [online] Phys.org. Available at: <https://phys.org/news/2018-08-digital-gig-economy-bad-wellbeing.html> [Accessed 17 Sep. 2019].

Wilensky, G.R., 2006. Consumer-driven health plans: Early evidence and potential impact on hospitals. *Health Affairs*, 25(1), pp. 174–185.

Worstall, T., 2017. *Gig-economy Employers Aren't Free Riding on the Welfare State – Employees Are*. [online] Forbes. Available at: <www.forbes.com/sites/timworstall/2017/05/01/gig-economy-employers-arent-free-riding-on-the-welfare-state-employees-are> [Accessed 17 Sep. 2019].

Part III

Digital disruption of status quo

11 GDPR

What are the risks and who benefits?

Anthony Larsson and Pernilla Lilja

1. Introduction

The General Data Protection Regulation (GDPR) was implemented across the European Union (EU) on May 25, 2018. In its most basic form, it is a regulation on data protection and privacy for all individuals within the (EU) and the European Economic Area (EEA) territories. In addition, it also restricts the export of personal data outside these geographical areas (European Parliament and Council of the European Union, 2016). The underlying intention of implementing the GDPR was chiefly to provide the European residents a level of control over their personal data, while also harmonizing the regulatory environment for international business within the EU. Even though the United Kingdom voted to leave the European Union via a referendum on June 23, 2016, the Westminster Government has confirmed that the GDPR will be brought into UK law (NLA, 2018). Although an EU directive, GDPR actually transcends national borders, making its relevance an international concern. That is, for any company (European or otherwise) whose business touches the EU and those failing to comply with the GDPR, regardless of national origin, faces a fine potentially equal to four percent of their company's global revenue, or €20 million (≈ USD22.3 million), whichever is greater (Ashton, 2018; Hart, 2017; Hon, 2016).

However, the implementation of this new directive has caused some concerns and will continue doing so for the foreseeable future. While it is possible to regard GDPR as a punitive construct, it may also serve as a catalyst in prompting companies to transform the way they handle data and manage risk and compliance that will enable them to become more competitive in the digital economy (Ashton, 2018). Given its recent introduction into the European legal framework, there is currently a dearth of available scientific research in regards to the ramifications of GDPR. Thus, this chapter will serve as a phenomenological/speculative study based on the available literature and best-practices in order to anticipate the future ramifications of GDPR in a labor market that is becoming increasingly digitalized (Kim, Sefcik and Bradway, 2017; Cooper and Endacott, 2007; Elliott and Timulak, 2005; Murphy and Dingwall, 1998). The overarching research question is: *In what way may GDPR influence the labor market of tomorrow, and what businesses are at risk?*

2. Background

The origin of GDPR is rooted in an ambition of creating a framework that safeguards personal data. According to the European Commission and Article 8 of the Charter of Fundamental Right of the European Union, everyone has the right to the protection of personal data (European Commission, 2018; European Communities, 2000; Fuster and Gellert, 2012). Within the EU, personal data may only be collected and used for legitimate purposes and must be protected from misuse (Ashton, 2018).

As several incidents have shown, concerns about the misuse of personal data have been far from unwarranted. A case in point was the *Cambridge Analytica*-scandal that was unraveled in 2018 (some two month before the implementation of GDPR) (Schneble, Elger and Shaw, 2018; Townsend, 2018). To this point, a study by the Ponemon Institute (2017) found that the total average cost of a data breach for businesses is USD3.62 million (≈ €3.08 million). Adding to this is that the likelihood of being breached is continuously increasingly rising, all the while companies are dealing with an "information explosion", where they keep collecting increasingly more data about a growing number of people (Ashton, 2018).

3. The aim of GDPR

In light of the developments mentioned in the introduction, GDPR seemingly presents itself as the perfect antidote as its jurisdiction can reach far beyond the geographical confines of the EU. Still, there are several problems associated to GDPR, as it carries a slow and unwieldy implementation process. A study by McKinsey has shown that it can take some companies years before completing all the necessary implementations, and depending on the starting position of the company, the cost can be significant (in many cases more than €10 million/≈ USD11.2 million) (Mikkelsen, Soller and Strandell-Jansson, 2017).

GDPR affects any and all companies operating with European customers, although those affected hardest are those holding and processing vast amounts of consumer data, such as technology firms, marketers and the data brokers who connect them (Hern, 2018). Although an increasing number of companies are becoming increasingly aware of the existence of the GDPR mandate, many are not sure about how to proceed with the implementation itself. While some first-movers have raced ahead, they have acted in manners that have, and continue to, incur unnecessary costs. McKinsey surveyed 60 major European companies and found that only ten percent had mature cybersecurity risk-management practices, while 45% of respondents replied that they would need to make significant investments in basic tools to comply with GDPR requirements (Mikkelsen, Soller and Strandell-Jansson, 2017).

Given the fact that roughly 90% of all surveyed companies to varying degrees lack the readiness for GDPR, it is important to consider that its most essential end-goal is privacy (even though the law as such does not make use of that word explicitly) and how to deal with the protection of sensitive data when processing client and employee data (Håkansson, 2017). However, that is not all. Beyond the "privacy" aspect, there is also the matter of "trust" and "risk". Essentially, these three concepts can be considered the cornerstones of GDPR (Cameron, 2018; Garber, 2018).

3.1. Privacy

The concept of "privacy" has had different meanings depending on context. At the heart of the matter, privacy involves the control, use and disclosure of personal information (Solove, 2008). Privacy can be understood as existing on a continuum, meaning that a person's level of privacy can either increase or decrease, and by varying amounts at that (Cofone and Robertson, 2018). Privacy may also be something that the individual chooses to surrender, and by different amounts, depending on the situation and the individual's preferences, often in exchange for perceived benefits. Often personal privacy, such as divulging one's name to someone, is sacrificed by means of gaining or improving trust with another actor, or to get them to reciprocate the same action (Gaudeul and Giannetti, 2017). Moreover, individuals tend to be more willing to freely sacrifice privacy if the reason for doing so appears transparent and it is clear for what and how the information sacrificed will be used (Oulasvirta et al., 2014).

At bedrock, GDPR seeks to aims to permeate privacy, while at the same time allowing for different sectors to contribute to new norms and best practices that may apply to the new specific, and often digitalized, circumstances (Buttarelli, 2016). Privacy is essential, because in a digitalized society, breaches of privacy, or "privacy harms" becomes far more salient. To this end, there are numerous ways in which personal privacy may become violated. According to Salie (2017), there are six inherent risks of personal privacy violation:

1 **Discrimination**: *Use predictive analytics for determination on individuals.*

The use of predictive analytics by the public and private sector can be used by actors to make determinations about the people's propensity to fly, find jobs, obtain clearances or get a credit-card approval. The use of associations in predictive analytics may carry negative impacts on certain individuals, which can lead to discrimination of these people from various services.

2 **Embarrassment of breaches**: *Create public awareness by exposing personal information – identity theft.*

This includes data breaches at various businesses and institutions that may serve to expose personal information of thousands of customers, employees, patients, clients, etc. Adding to this is the all-time high occurrence of credit-card fraud and identity theft.

3 **Abolishment of anonymity**: *Removing only a few data sets can lead to re-identification.*

Barring rules for anonymized data files, it is possible to combine data sets. Given the circumstances, this might make it possible to re-identify certain people by combining various subsets of data.

4 **Government exemptions**: *Collecting and adding more and more personal information to government databases.*

For instance, various government databases will collect personally identifiable information (PII). This includes name, potential aliases, ethnicity, gender, date and place of birth, social security number, passport and driver's license numbers, home address, telephone numbers, photographs, fingerprints, various financial information such as bank accounts, employment and business information, etc.

5 **Data brokerage**: *Selling of unprotected and incorrect data profiles.*

Some companies will collect and sell consumer profiles that are not explicitly protected under the legal frameworks. The data files used for various forms of big data analysis may contain invalid data about individuals. They may also use data models that are faulty as they relate to individuals, or simply be flawed algorithms.

6 **Data misinterpretation**: *Having more data is no substitute for having high-quality data.*

While it is possible to find any number of political expressions on various social media platforms, these statements do not constitute a reliable representation of voters. To this effect, it has been know that a substantial share of tweets and Facebook posts about politics around the world have in fact been computer-generated.

As illustrated above, the key success factor for organizations will be the role and importance of information management and governance in data privacy. However, in the strictly legal sense, "privacy harm" will often take the form of a "visceral and vested approach", which means that it involves some dimension of palpable physical injury or financial loss and that the harm must have *de facto* occurred (i.e., is real rather than perceived) (Solove, 2014). However, that is not to say that privacy issues cannot cause various degrees and forms of emotional distress. Since these cases are difficult to prove and/or measure, the individuals are often subjected to the concept of "trust".

3.2. Trust

The basis of "trust" implies that the individual has trust in various data controllers (the person responsible for all personal data contained by the organization) to treat personal information justly and professionally (Buttarelli, 2016; European Commission, 2019). The amount of data stored throughout the various cloud services are for 2019 expected to be in the ranges of ten zettabytes (ZB), corresponding to approximately ten trillion gigabytes (Hofman et al., 2017). Much of this data includes crucial records that make it possible for individuals, businesses and even governments to continue functioning. This includes, but is not limited to, identity and vital statistics records, financial records, legal documents, contracts, ownership and land records. In addition, records related to the Internet of Things (IoT) will also be stored on cloud services (Hofman et al., 2017). This stresses the need of secure servers and trust in the ability of the data holders to guarantee due and proper management of data, so that information does not fall into the wrong hands. As expressed by Duranti and Rogers (2014, p. 203) "[e]ven as we have ever greater access to untold stores of information, our right to know comes at a rising cost to our privacy and anonymity, due to a complex web of data collection and surveillance, benign and not. These stores of information, furthermore, are accumulated and extracted from sources we often cannot know or evaluate".

However, there is reason to question the security of cloud-based record-keeping, even in the absence of malice. This includes issues such as managing trans-jurisdictional data flows, establishing accountability for data breaches, and establishing due process for when a cloud service provider ceases to operation/function (Duranti and Rogers, 2012). Given the potential risks, establishing trust then becomes essential for those service providers wishing to attract customers. Likewise, establishing a bond of trust with the service provider becomes essential for individuals seeking to utilize that particular service above a different one, as individuals will seek to maximize their benefits while at the same time reducing the perceived risks.

3.3. ***Risk***

Privacy risk generally refers to a situation that involves the potential loss of control over one's personal information. That is to say, when such information about you is used without your knowledge or permission (Lee, 2010). To this end, privacy risks exist wherever an actor collects, uses, shares and/or manage personal information relating to their staff, customers, clients, patrons, students, etc. (New Zealand Government, 2016). Risks commonly exist in terms of two separate values, "likelihood" and "severity". In the context of privacy, the "likelihood" is characterized by the probability that harm may be caused by the processing system. In turn, "severity" characterizes the magnitude of the impact on the victims (De and Le Métayer, 2016).

According to the OECD (2016), it is impossible to entirely eradicate digital security risk when carrying out action that rely on the digital environment, although the risk may be mitigated through the implementation of digital risk management. On this account, it is incumbent on Europe's independent data protection authorities to foster risk management through the implementation of GDPR, so that there is transparency and accountability to all individuals and businesses enterprises (Buttarelli, 2016). Nevertheless, the price of conformity demanded by GDPR may be a steep one for certain types of businesses. As such, it is important to tether out what future business will find themselves challenged by the implementation of the GDPR-framework, and what businesses that can thrive from it.

4. Business challenged by GDPR

One of the greater challenges can also be an opportunity. While true that business may be aided by GDPR as they are forced to structure and harmonize their customer data, the process in doing so may indeed be a costly and time-consuming endeavor. It is estimated that roughly 90% of the companies lack readiness for GDPR (Håkansson, 2017). Thus, many organizations will need to commit to undertaking large investments in revamping their information structure. On the outset, it may appear that such investments would only be done out of necessity rather than a means of generating business value. However, it is important to remember that business value of such investments are not simply squandered, as the pertinent data is polished, validated and organized in a new structure that allows for easier adherence to GDPR standards while also ensuring that the data is collected, stored and used more wisely for data analytics and possibly even for generating knowledge that may be usable for artificial intelligence (AI) solutions. As stated, the main challenge going forward is that so many organizations lack the readiness for GDPR, and possibly even the ability to afford to commit to the necessary investment needed to do so. To make matters worse, GDPR came into effect at the last week in the month of May, a time of year when large parts of the industry (especially the bank and finance sector) are reluctant to take on large-scale transformations as they tend to start scaling down their activities in preparation for the summer holidays. Adding to the burden was that the launch of GDPR came in the backwater of the recent implementation of MiFID 2, another large regulations project that has incurred large ramifications on the bank and finance sector (Finansinspektionen, 2019; Prorokowski, 2015). Many organizations in the bank and finance sector were already stretched thin following this reform that the implementation of GDPR came across as an anticlimax that did little to boost the motivation among the employees.

Organizations processing large volumes of data and sensitive data material may also face challenges in implementing GDPR (Ashton, 2018). Obvious examples of actors affected by GDPR are the "big tech" companies, such as Google and Facebook, but it is likely that smaller companies will be hit more severely, such as studios for online game developers (Kottasová, 2018). GDPR is expected to have a huge impact on both landlords and tenants, as both categories will gain greater control over their personal data (NLA, 2018). Other examples of affected industries are the health-care sector, as well as the aforementioned bank and finance sector. These sectors are all examples of industries that routinely collect and store personal data about their customers/clients; data that will need to be anonymized. To that extent, these organizations (and any other organization processing sensitive personal information) will need to take meticulous care and establish a rigid set of routines in order to ensure that the data is kept safe.

Another aspect to consider is that due to GDPR, all companies are subject to data cleansing, regardless of whether or not the data per se is considered "sensitive". Data cleansing (or "data scrubbing") refers to the process in which an actor identifies and removes/amends data within a database that is either incorrect, incomplete and/or duplicated. Under GDPR, there is an added stipulation that data that is either irrelevant or unnecessary should be removed as well (GDPR Report, 2017). This is generally done in a bulk process, and with the GDPR implementation, this needs to be performed on a regular basis.

The process of data cleansing will force organizations to preemptively consider what types of data they wish to store about the customers, as too much stored data may lead to added bureaucracy down the line. GDPR will also prompt organizations to consider if any of the data they have stored is actually relevant to the organization's enterprise or if it borders on any potential gray areas, as the threat of sanctions may make the organizations more selective about what types of information they seek to collect and store (Ashton, 2018).

Another challenge to the organization's employees is that the new GDPR rules and its associated procedures may have a perceived steep learning curve. This problem should, however, be transient and will likely dissipate once the new routines become more commonplace. Furthermore, this challenge is inherent to organizations that have predated GDPR. New organizations founded after the implementation of GDPR will become "natives" of the new regulatory system and should be more equipped to handle the routines that follow in order to ensure GDPR compliance.

5. Businesses aided by GDPR

The onset of GDPR has oversaturated the market with services purporting to sell GDPR-compliant solutions and services (Ashton, 2018). These solutions tend to be comprehensive and seek to structure the documentation on the database systems the organization aims to use as well as help the organization keep track of what type of information that is kept in each database section, what classification the data has, and the reason for storing each particular dataset, along with the processes for data cleansing.

Another type of business that is bolstered by the implementation of GDPR, are the companies providing cloud storage. "Cloud storage" in this context refers to a model of computer data storage that stores digital data in logical pools. The physical storage spans multiple servers (possibly even in multiple locations), and the physical environment is generally owned and managed by a hosting company (Mohamed, 2018). The advantages for opting for a cloud-based solution is that it allows for a better optimization of IT resources as cloud solutions contain virtually unlimited scalability and have great flexibility, while generally also being cost-effective (Tolsma, 2018). Cloud services are also used to a large extent to store various GDPR-compliant agreements between, for instance, the data controller and the data processor (the person who processes data on behalf of the data controller) (Voigt and von dem Bussche, 2017; European Commission, 2019). Hence, we can expect to see an increasing number of cloud-service providers cropping up in the future, marketing themselves toward keeping GDPR data safe.

Needless to say, legal experts and lawyers are also expected to see a boost to their businesses following the implementation of GDPR. While software and automated digital systems may guide organizations through the process easily and make data processing manageable, legal advisors/lawyers will help companies understand the implications of GDPR for their specific enterprise, while ensuring full GDPR-compliance (Rishikof and Sullivan, 2017).

The emergence of GDPR has given rise to a new profession, namely that of "data protection officer" (DPO). The function of a person carrying this title is to ensure that the organization handles data in a manner that complies with GDPR standards. That is not to say that a DPO is a required function in all organizations. Rather, the need of a DPO is contingent on a number of factors such as organizational size, if the organization is a governmental agency, or if the organization's core business is centered on large-scale and/or systematic monitoring of personal data, etc.

GDPR may also result in new assignments for consultants in digital strategy, as it opens up areas in which there may be need for various forms of advisory services, especially so given the urgency that many companies are faced with in ensuring complete GDPR compliance throughout their organizational processes.

Other professions benefitted by GDPR tend to be those that are veered toward structuring and/or building up foundations for data management. This may include professions such as software architects, solution architects, software developers and data-analytics professionals. In the future, it is not unlikely that AI-solutions will gain a greater foothold in the management of data, and thus, professionals who possess skills related to AI-programming are likely to benefit from GDPR, at least in the not-too-distant future.

By the same token, following the digitalization process, customers will indubitably expect "smarter" (i.e., more customized) products and services, which in turn is enabled by the possibility of companies to combine different pools of data that they may have on their customers. To that extent, the companies will have to walk a tightrope between delivering perceived value to the customers, while at the same time not acting in a manner that makes the customer feel uneasy and supervised. That is to say, upholding integrity and trust will be paramount for companies seeking to secure customer loyalty. In this respect, GDPR may help to strengthen the bond of trust between the companies and their customers in the sense that the customers feel if not empowered, then at least informed, by the ability to choose what information to share with whom, while also possessing the ability to withdraw their consent at their own behest.

6. Legal ramifications

The legal ramification of GDPR is that consent is much harder to obtain and prove due to the fact that the directive permits use of personal data only in limited and clearly delineated circumstances (Villers, Vonner and Nédélec, 2016). This, in turn, will most certainly prompt organizations to examine and re-examine how they collect and use personal data. The will results in there being much larger quantities of bureaucracy and paperwork documenting what personal data is used and in what way the organization uses it. The fines imposed by GDPR for non-compliance are administered by individual member state supervisory authorities. If an organization incurs several infractions on the GDPR, it is fined according to the gravest infraction, rather than being penalized separately for each individual infraction. Nevertheless, this does not mean that recalcitrant companies are let off lightly, given the possible sums involved. Specifically, the following ten criteria are used when determining the sum of the fine issued due for infractions against GDPR (European Parliament and Council of the European Union, 2016).

- **Nature of infringement**: This denotes the number of people affected, the damage incurred, the duration of the infringement, and the purpose of processing.
- **Intention**: This determines whether the infringement is caused intentionally or by negligence.
- **Mitigation**: This concerns the actions taken by the organization to mitigate the damage to the data subjects.
- **Preventative measures**: This looks at how much technical and organizational preparation the organization had previously implemented in order to prevent its non-compliance (if any at all).
- **History**: This considers past relevant infringements, and past administrative corrective actions taken under the GDPR, ranging from warnings to bans on processing and fines.
- **Cooperation**: This considers how cooperative the organization has been with the authorities in remedying the infraction.
- **Data type**: This concerns what types of data the infraction impacts.
- **Notification**: This is contingent on whether the infringement was proactively reported to the supervisory authority by the organization itself or by a third party.
- **Certification**: This is determined by whether or not the firm had previously been qualified under approved certifications and/or if it had adhered to approved codes of conduct.
- **Other**: Aggravating or mitigating factors may include financial impact on the organization from the infraction.

As previously mentioned, at the upper level, organizations risk facing fines potentially equal to four percent of their company's global revenue, or €20 million (≈ USD22.3 million), whichever is greater (Ashton, 2018; Hart, 2017; Hon, 2016). This fine entails infractions against: (1) the basic principles for processing, including conditions for consent; (2) the subjects' rights (as outlined in the directive); (3) the transfer of personal data to a recipient in a third country or an international organization; (4) any obligations pursuant to EU Member State law; (5) any non-compliance with an order by a supervisory authority (European Parliament and Council of the European Union, 2016).

However, while the aforementioned denotes the higher level, there is also a lower level of infractions that carries a fine. This fine constitutes €10 million

(≈ USD11.2 million), or two percent of the worldwide annual revenue of the prior financial year, whichever is higher. This includes infractions of: (1) the controllers and processors as outlined in the directive; (2) the certification body as outlined in the directive; (3) the monitoring body as outlined in the directive (European Parliament and Council of the European Union, 2016).

7. Concluding discussion

This chapter sought to answer the research question: *In what way may GDPR influence the labor market of tomorrow, and what businesses are at risk?*

In answering this question, it should be stated that while GDPR had been in the pipeline for years prior to its implementation in May 2018, a 2017 survey showed that approximately 90% of all companies were poorly prepared for GDPR (Håkansson, 2017). Even still, many organizations struggle with securing full compliance throughout their data systems. According to a study conducted by Deloitte, one fifth of organizations only aimed for bare minimum compliance at the outset of GDPR (Gooch et al., 2018). However, the recruitment of DPOs appears to have increased internationally in recent times, with the UK in the lead with 92% of respondent companies having assigned a DPO. Since then, organizations have by and large been taking measures in continuing their GDPR implementation, with 92% of the respondents feeling confident in their long-term ability to comply with GDPR standards (Hawker, 2018). Notwithstanding, in the immediate term, many companies will still need to address today's challenges of responding to data requests, especially in terms of online tools, as these enable consumers to make mass data requests, which may in turn help the companies secure long-term customer loyalty (Gooch et al., 2018; Hawker, 2018).

Moving forward, GDPR aims to ensure that the personal data stored by companies is structured, which means it is machine-readable and stored in an interoperable format (Voigt and von dem Bussche, 2017). This can in turn aid the future development toward AI solutions. Still, ensuring that the data is structured can be a laborious and resource-consuming process that can have detrimental effects on some actors on the market. Specifically, this affects companies that collect and process large volumes of personal data. While the larger companies (i.e., the "big tech" companies) will likely manage to weather the storm, smaller companies processing large volumes of personal data (such as studios for online game developers) are likely to be hit harder by GDPR. By the same token, GDPR may also provide new business opportunities in other categories of ventures, especially for consultants in digital strategy and professionals with skill sets in analytics and software architecture.

References

Ashton, C., 2018. *GDPR: What You Need to Know and What it Means for the Digital Economy*. [online] Digitalist Magazine. Available at: <www.digitalistmag.com/cio-knowledge/2018/01/02/gdpr-what-you-need-to-know-and-what-it-means-for-the-digital-economy-05653998> [Accessed 17 Sep. 2019].

Buttarelli, G., 2016. The EU GDPR as a clarion call for a new global digital gold standard. *International Data Privacy Law*, 6(2), pp. 77–78.

Cameron, S., 2018. GDPR: Valuing data, assessing risk and consent services. *Journal of Data Protection & Privacy*, 2(1), pp. 41–52.

Cofone, I.N. and Robertson, A.Z., 2018. Privacy harms. *Hastings Law Journal*, 69(4), pp. 1039–1098.

Cooper, S. and Endacott, R., 2007. Generic qualitative research: A design for qualitative research in emergency care? *Emergency Medicine Journal*, 24(12), pp. 816–819.

De, S.J. and Le Métayer, D., 2016. *Privacy Risk Analysis*. San Rafael, CA: Morgan & Claypool Publishers.

Duranti, L. and Rogers, C., 2012. Trust in digital records: An increasingly cloudy legal area. *Computer Law & Security Review*, 28(5), pp. 522–531.

Duranti, L. and Rogers, C., 2014. Trust in online records and data. In: J. Lowry and J. Wamukoya, eds., *Integrity in Government Through Records Management: Essays in Honour of Anne Thurston*. Farnham, UK: Ashgate, pp. 203–216.

Elliott, R. and Timulak, L., 2005. Descriptive and interpretive approaches to qualitative research. In: J. Miles and P. Gilbert, eds., *A Handbook of Research Methods in Clinical and Health Psychology*. Oxford, UK: Oxford University Press, pp. 147–160.

European Commission, 2018. *Protection of Personal Data.* [online] Policies, Information and Services. Available at: <https://ec.europa.eu/info/aid-development-cooperation-fundamental-rights/your-rights-eu/know-your-rights/freedoms/protection-personal-data_en> [Accessed 17 Sep. 2019].

European Commission, 2019. *What Is a Data Controller or a Data Processor?* [online] Policies, Information and Services. Available at: <https://ec.europa.eu/info/law/law-topic/data-protection/reform/rules-business-and-organisations/obligations/controller-processor/what-data-controller-or-data-processor_en> [Accessed 17 Sep. 2019].

European Communities, 2000. *Charter of the Fundamental Rights of the European Union (2000/C 364/01)*. [online] Official Journal of the European Communities. Available at: <https://eur-lex.europa.eu/legal-content/en/ALL/?uri=OJ%3AC%3A2000%3A364%3ATOC> [Accessed 17 Sep. 2019].

European Parliament and Council of the European Union, 2016. *Regulation (EU) 2016/679 on the Protection of Natural Persons with Regard to the Processing of Personal Data and On the Free Movement of Such Data, and Repealing Directive 95/46/EC [General Data Protection Regulation]*. [online] Official Journal of the European Union. Available at: <https://eur-lex.europa.eu/legal-content/EN/TXT/HTML/?uri=CELEX:32016R0679> [Accessed 17 Sep. 2019].

Finansinspektionen, 2019. *Om Mifid/Mifir [About Mifid/Mifir]*. [online] Available at: <https://fi.se/sv/marknad/vardepappersmarknad-mifidmifir/om-mifidmifir> [Accessed 17 Sep. 2019].

Fuster, G.G. and Gellert, R., 2012. The fundamental right of data protection in the European Union: In search of an uncharted right. *International Review of Law, Computers & Technology*, 26(1), pp. 73–82.

Garber, J., 2018. GDPR – compliance nightmare or business opportunity? *Computer Fraud & Security*, 2018(6), pp. 14–15.

Gaudeul, A. and Giannetti, C., 2017. The effect of privacy concerns on social network formation. *Journal of Economic Behavior & Organization*, 141, pp. 233–253.

GDPR Report, 2017. *The Time is Now: Cleansing Data for Improved Customer Engagement and GDPR Compliance*. [online] Available at: <https://gdpr.report/news/2017/05/26/time-now-cleansing-data-improved-customer-engagement-gdpr-compliance> [Accessed 17 Sep. 2019].

Gooch, P., Luysterbourg, E., Sponselee, A., Frank, D.P., Dewitt, B., Sehgal, M. and Batch, D., 2018. *A New Era for Privacy: General Data Protection Regulation ("GDPR") Six Months On*. [online] Deloitte. Available at: <https://www2.deloitte.com/uk/en/pages/risk/articles/gdpr-six-months-on.html> [Accessed 17 Sep. 2019].

Håkansson, M., 2017. *GDPR: A Driver for Digitalization*. [online] Axiomatics. Available at: <www.axiomatics.com/blog/gdpr-driver-digitalization> [Accessed 17 Sep. 2019].

Hart, J., 2017. *Avoiding the Compliance Complications of GDPR*. [online] Compliance Week. <Available at: https://www.complianceweek.com/avoiding-the-compliance-complications-of-gdpr/2463.article> [Accessed 17 Sep. 2019].

Hawker, E., 2018. *Businesses Struggling with GDPR Compliance*. [online] Accountancy Age. Available at: <www.accountancyage.com/businesses-struggling-with-gdpr-compliance> [Accessed 17 Sep. 2019].

Hern, A., 2018. *What is GDPR and How Will it Affect You?* [online] The Guardian. Available at: <www.theguardian.com/technology/2018/may/21/what-is-gdpr-and-how-will-it-affect-you> [Accessed 17 Sep. 2019].

Hofman, D., Duranti, L., How, E., Hofman, D., Duranti, L. and How, E., 2017. Trust in the balance: Data protection laws as tools for privacy and security in the cloud. *Algorithms*, 10(2), p. 47, pp. 1–11.

Hon, K., 2016. *GDPR: Potential Fines for Data Security Breaches More Severe for Data Controllers than Processors, Says Expert*. [online] Out-Law.com. Available at: <www.out-law.com/en/articles/2016/may/gdpr-potential-fines-for-data-security-breaches-more-severe-for-data-controllers-than-processors-says-expert> [Accessed 17 Sep. 2019].

Kim, H., Sefcik, J.S. and Bradway, C., 2017. Characteristics of qualitative descriptive studies: A systematic review. *Research in Nursing & Health*, 40(1), pp. 23–42.

Kottasová, I., 2018. *These Companies are Getting Killed by GDPR*. [online] CNN Business. Available at: <https://money.cnn.com/2018/05/11/technology/gdpr-tech-companies-losers/index.html> [Accessed 17 Sep. 2019].

Lee, I., 2010. *Encyclopedia of E-Business Development and Management in the Global Economy*. Hershey, PA: IGI Global.

Mikkelsen, D., Soller, H. and Strandell-Jansson, M., 2017. *The EU Data-protection Regulation – Compliance Burden or Foundation for Digitization?* [online] McKinsey & Company: Risk. Available at: <www.mckinsey.com/business-functions/risk/our-insights/the-eu-data-protection-regulation-compliance-burden-or-foundation-for-digitization> [Accessed 17 Sep. 2019].

Mohamed, A., 2018. *A History of Cloud Computing*. [online] Computer Weekly. Available at: <www.computerweekly.com/feature/A-history-of-cloud-computing> [Accessed 17 Sep. 2019].

Murphy, E. and Dingwall, R., 1998. Qualitative methods in health services research. In: N. Black, B. Reeves, J. Brazier and R. Fitzpatrick, eds., *Health Services Research Methods: A Guide to Best Practice*. London, UK: BMJ Books, pp. 129–140.

New Zealand Government, 2016. *Privacy Risk and Opportunity Identification*. [online] Guidance on Privacy Management. Available at: <www.ict.govt.nz/guidance-and-resources/privacy/guidance-on-privacy-management-issued-by-the-government-chief-privacy-officer> [Accessed 17 Sep. 2019].

NLA, 2018. *GDPR – What You Need to Know as a Landlord*. [online] Available at: <https://landlords.org.uk/news-campaigns/news/gdpr-what-you-need-know-landlord> [Accessed 17 Sep. 2019].

OECD, 2016. *Managing Digital Security and Privacy Risk*. OECD Digital Economy Papers. Paris, France.

Oulasvirta, A., Suomalainen, T., Hamari, J., Lampinen, A. and Karvonen, K., 2014. Transparency of intentions decreases privacy concerns in ubiquitous surveillance. *Cyberpsychology, Behavior, and Social Networking*, 17(10), pp. 633–638.

Ponemon Institute, 2017. *Ponemon Institute's 2017 Cost of Data Breach Study: Global Overview*. [online] Ponemon Institute. Available at: <https://www-01.ibm.com/common/ssi/cgi-bin/ssialias?htmlfid=SEL03130WWEN> [Accessed 17 Sep. 2019].

Prorokowski, L., 2015. MiFID II compliance – are we ready? *Journal of Financial Regulation and Compliance*, 23(2), pp. 196–206.

Rishikof, H. and Sullivan, C., 2017. Legal and compliance. In: D. Antonucci, ed., *The Cyber Risk Handbook: Creating and Measuring Effective Cybersecurity Capabilities*. Hoboken, NJ: Wiley, pp. 255–270.

Salie, E., 2017. *Big Data Privacy Risks and the Role of the GDPR: Part 1*. [online] Digitalist Magazine. Available at: <www.digitalistmag.com/finance/2017/04/19/big-data-privacy-risks-role-of-gdpr-05028203> [Accessed 17 Sep. 2019].

Schneble, C.O., Elger, B.S. and Shaw, D., 2018. The Cambridge Analytica affair and Internet-mediated research. *EMBO Reports*, 19(8), p. e46579, pp. 1–2.

Solove, D., 2014. *Privacy and Data Security Violations: What's the Harm?* [online] Teach Privacy. Available at: <https://teachprivacy.com/privacy-data-security-violations-whats-harm> [Accessed 17 Sep. 2019].

Solove, D.J., 2008. *Understanding Privacy*. Cambridge, MA: Harvard University Press.

Tolsma, A., 2018. *GDPR and the Impact on Cloud Computing: The Effect on Agreements Between Enterprises and Cloud Service Providers*. [online] Deloitte. Available at: <https://www2.deloitte.com/nl/nl/pages/risk/articles/cyber-security-privacy-gdpr-update-the-impact-on-cloud-computing.html> [Accessed 17 Sep. 2019].

Townsend, K., 2018. *Would Facebook and Cambridge Analytica be in Breach of GDPR?* [online] SecurityWeek.Com. Available at: <www.securityweek.com/would-facebook-and-cambridge-analytica-be-breach-gdpr> [Accessed 17 Sep. 2019].

Villers, V., Vonner, F. and Nédélec, C., 2016. *GDPR: The Legal Issues in Jargon-free English*. [online] PwC Luxembourg. Available at: <www.pwc.lu/en/general-data-protection/docs/pwc-gdpr-legal-issues-in-jargon-free-english.pdf> [Accessed 17 Sep. 2019].

Voigt, P. and von dem Bussche, A., 2017. *The EU General Data Protection Regulation (GDPR)*. Cham, Switzerland: Springer.

12 Players for hire

Games and the future of low-skill work

Edward Castronova

1. The coming wave of inequality

The technological system within which we live has irresistible imperatives to promote the development of more efficient means of production. Intelligent systems are increasingly deployed to perform work that low-skill humans used to do. At the moment I write, people are no longer employed to handle money, and those who drive vehicles for pay are in the cross-hairs of automated driving systems. It is apparent that most jobs that require little skill will eventually be done by a digital entity.

In the past, the invention of new machines for doing low-skill work caused disruption, but also led to an explosion of new employment opportunities for low-skill workers. In the Industrial Revolution, automated looms put weavers out of work, but they also created a factory system whose explosive growth provided employment and wages for millions of low-skill workers. It is often overlooked that the Industrial Revolution caused a population explosion of unprecedented magnitude. That would not have been possible if low-skill people were unable to find work that would feed many new mouths.

When a technological change puts millions of people out of work, a social change must necessarily result, one in which some work becomes available. With the factory system, the work that did become available was dirty, oppressive, harsh and crushing to the spirit. But it was work. People survived. They survived because, having lost their incomes from farming or weaving or whatever old craft they used to do, they sought about for some way to make money, some way to eat. In the nineteenth century, they found their way to factories. And so, populations migrated from countryside to towns, cities exploded in size, the urban proletariat was born, and so on and so forth. Automation put people out of work, but those people went off and did *something* else.

In our current moment we must ask, where will all the drivers go? Forrester predicts that automation will cause a net loss of seven percent of current jobs over the next decade (Forrester, 2016). That is just the beginning. If we think about the low-skill people who are about to be displaced, we can wonder, what sort of work will they be able to find? Or, be driven into? Their children will come of age knowing full well that they cannot live by the usual jobs of low-skill people. By what means will these new generations of low-skill workers gather up enough food and housing to stay alive? For if there are any means at all, they will surely seek them and secure them. They will find something to do, something they can do that machines cannot, and that others who have means will pay them for.

Who are those others, the ones with means? They are the people with high skills. As the digital age progresses, there will continue to be high demand for the services of bright people, the creative ones, the emotionally intelligent ones and the best of the best human servants. Entertainers, technicians, engineers and performers of all stripes, will do better and better. Anyone whose special gifts are impossible for a machine to recreate will find themselves doing very well indeed. And then the other group who will do well, of course, are those who own the machines themselves. Those who own a share of the profits produced by machines will gain income at the same rate that the machines displace low-skill workers. Every time a self-driving car puts a taxi driver out of work, the owner of the car earns more money. Thus there will be three types of wealthy people: Those who own machines, those who design/build/operate machines, and those who provide services that only people can provide.

The system as a whole will get richer and richer. Every time a machine replaces a person, it does so because it is better at the job. It produces more at the same cost, or the same amount at a lower cost. Otherwise, there would be no reason to make the replacement. But this means that each time a machine does a person's job, the economy as a whole gets richer. More stuff is made at lower cost. Automation is a force for efficiency. It induces economic growth. It makes money. And that money will go to the people at the top, the people with ownership or irreplaceable skills.

Automation thus necessarily induces inequality, at least in the first moment. One man is out of a job, another man makes more money on his assets. But these unequal slices are part of a growing pie. The question becomes, what happens in the second instance – after the driver loses his job, after the car owner gains more wealth. What happens then? What new social arrangement might happen? Is there some new way that the rich person can hire the poor person? A new system whereby the money of the wealthy moves into the pockets of those who are looking for anything to do in return for money. If the car driver can no longer provide car-driving, can he provide something else?

In the very near future, we will be facing the reality of hordes of low-skill workers with no meaningful work. Indeed, this is already happening. In the late 1970s, 7% of men in their 20s with less than a bachelor's degree did no work at all in the preceding year. In 2015, the number was three times higher, 22% (Swanson, 2016). Unpublished research by economists suggests that large numbers of low education young men have abandoned the world of work simply because of the joys of video-game play. Facing a choice between seeking work or staying at home playing games, many young, low-skill men seem to be choosing the games. It is a sign that the real world of work is becoming less rewarding for those with low skills.

2. Options for the low-skilled

In the nineteenth century, low-skill workers went to work in factories. What will they do in the twenty-first century?

One common idea is that education can change these dynamics. It cannot. It is patently unfair to try to address this problem by pressuring people to go to engineering school. For surely all this will do is start an arms race among high-skilled workers. If the technological economic system requires only ten percent of the workforce to operate at full capacity, then only ten percent of the workforce will be employed as operators. Training up millions more people to have the necessary operating skills will only unleash a vicious competition among them, driving down the wages of those who do get the jobs. Consider: If you have 100 people whose skill is ranked from zero points to 100 points of ability, and you will hire only the top ten, it does not help anybody to teach them all five more points of skill. You will still only hire the top ten, the people with skills from 95 points to 105 points.

Skilling up the workforce only makes sense if there will be an increased need for high-skill people. But the dynamic of technological explosion in which we live is such that for every new high-tech job that is created, two lower-tech jobs will be destroyed. This is simply because people are more expensive than machines. The system will do everything in its power to get its work done with the minimum number of people. It is going to try to shed people whenever and wherever it can. Teaching low-skill people to be somewhat better technicians than they currently are is a losing strategy for them.

Instead, we must think of other ways that a low-skill person can do paid work for others. What kinds of things can low-skill people do that machines simply cannot? This question forces us to unpack the concept of "low skill". What we really mean, by now, is that low-skill people are the ones that machines can replace. They are currently working in a way that a machine can do. Whatever is special and unique about human thinking, whatever we have that machines cannot have, may well be present in the minds of the low-skill workers; but right now, that special sauce is not a part of their work. Low-skill workers may have high skills, at something; but the work they do today is work that a machine will be doing soon enough.

If low-skill workers are becoming defined as "those whose jobs will be automated", it follows that if those people are to survive, are to eat, they must bring up within themselves, and express to the world, some set of skills that cannot be automated. It also follows that the term "low skill" applies to a lot more of us than would have been expected under traditional definitions of the term.

Earlier I argued that the only people who will be compensated in the automated future are (a) those who own machines, (b) those who operate machines, and (c) those who perform services that only people can perform. If the low-skill displaced masses do not own machines and are unable to become operators, only one avenue of compensation is open to them: Service.

The question is, what kind of human-only services can low-skilled people provide, in massive numbers? For there will surely be huge masses of displaced low-skill people. The answer here is not going to do tennis instruction or lawn care. In the future we are facing, if every rich person had 10,000 butlers and maids, it would still not be enough employment. The technological disruption will therefore drive some kind of huge social disruption that creates new ways for a few privileged people to move income toward massive numbers of the technologically unemployable population.

3. "Low-skill" people can play games

All of us can play video games, regardless of skill. The game industry excels at creating systems in which both angry lawyers and stoned teenagers can make their way. Games offer a range of skill challenges, by design. Games are designed so that all the players have fun, including those whom the outside world would call "low skill".

Skill effects within games are tightly managed by the designers. Designers make acutely conscious choices about when, where and how the real-life cognitive and physical skills of the players will have an influence on game outcomes. Games are now designed to adjust dynamically and automatically to the perceived abilities of the players. Is this player dying quite a lot? Remove some of the zombies. Not dying enough? More zombies! It could be argued that games are designed to reward skills that the market finds less valuable, precisely in order to capitalize on the way people feel underappreciated. The point is that whatever the outside world may think of skills, game design seeks ways to make every player appreciated, regardless of skills.

Designers also carefully manage when, where and how the real-life monetary resources of the players can affect the game. In many games, it is possible to spend extra money to get a better sword, or more life potions or unlock a faster horse or a new level. Whereas in other games, no amount of outside money can affect what you can achieve in the game.

Designers manage skill, money and time as well: Some players have quite a lot of time to spend in a game, others do not. It is up to the designers how long it takes to achieve things in the game.

As a result of these designer decisions, the game industry finds itself unwittingly serving as a vast global agora of skills, time and money. Designers can increase or decrease the impact of player skill, money and time input, and in so doing they give their games a certain profile. Different games appeal to different players. Players can move from game to game, according to their resources and tastes.

Some games are so big in scope, time length and space that they can accommodate widely different types of players. Different parts of a single game may appeal to the skill-rich, the money-rich and the time-rich. This is seen most clearly in the free-to-play revenue model. A game run on the F2P model opens its doors to anyone. Anyone can play the game, free of charge. At some point, however, some aspect of the game either requires or encourages payment. Perhaps the game has 100 levels, and the first 50 are free. You have to pay to unlock the other 50. Or the game has 100 levels, all completely free, but it takes two years to get to level 100, unless the player buys some sort of special equipment. F2P games make money by charging for extra game features.

The free-to-play model has exploded across the industry in the last decade. It has turned out to be fabulously lucrative. A free game has the lowest possible barriers to entry, and can quickly gain a large population of players. Those players create buzz and excitement; they confirm to the world that it is a good game. Production costs are such that even if only a fraction of players ever pay for something, the revenues gained far exceed the costs of providing the game free to all the others.

It is understood colloquially among game designers that the revenues of F2P games follow the same patterns as casinos, in that a few big spenders are enough to make the casino turn a profit – even if you give free drinks to hundreds of low-spending people. These big spenders are known as "whales". The game industry has whales as well. There are people who spend thousands of dollars every month on the most trivial of game items. A very large portion of revenues are provided by a small percentage of the player base. Gamasutra reports that less than one-fourth of one percent of F2P players generate almost half the revenue (Rose, 2014). "Conversion rates" – the percentage of players who spend any money at all – can be as low as three percent or one percent, yet the game still turns a profit. The free spending of a few is sufficient to support the game.

What then is the role of all the other players? What purpose is served by all those people in the casino who drink their free drinks and gamble away comparatively tiny amounts of money? Those people are a critical part of the revenue model, for they form the social environment within which the whale can make friends, and compared to whom the whale looks like the awesome person he wishes to be. Without the free-drink gamblers, the casino whale has nobody to talk to and nobody to impress with his casino privileges and private rooms. The same holds for F2P games: Without the free gamers, the spending gamer has nobody to talk to and nobody to impress.

Thus there are two elements that explain why big spenders will provide revenues to companies that allow free access: A communion aspect and a comparison aspect. The communion aspect is easily enough understood: Rich people get lonely too, and a thriving game community gives them a social world for making friends. The comparison aspect is a little more complicated, so let us unpack the social dynamics of a F2P game. Consider three ways that a person might spend real money:

1 You can spend money on cosmetic items – better looking hats, a neat new car, a sword that glows or a cute pet. You look better in the game.
2 You can spend money to overcome or get by annoying parts of game play. Perhaps in order to play the next puzzle, you have to wait 20 seconds. That is annoying! So you can spend real money to reduce the wait time. Or, there may be a certain item that you need to get to the next bit of fun. Perhaps getting the item requires a lengthy, boring quest. So you can go through the lengthy, boring quest, or, you can just buy the item at the store.

3 You can spend money to enhance your power. This is usually known as "pay-to-win". In a PTW system, real money has to be spent to get a car with competitive speed, or armor decent enough to withstand serious combat. You can still play without these things, but unless you have great skills as a player, you will consistently lose.

Note that each of these ways of spending enhances a comparative difference between the spending player and everybody else. Cosmetic spend (1) enhances the social prestige of the spender, just as flashy diamonds do in the real world. Convenience spend (2) allows people who are short on time, or who are bored, to move quickly to more entertaining parts of the game. They get ahead quicker and look more competent with respect to the game systems. Competitive spend (3) allows people to dominate other people, to beat them and to get ahead.[1] In all of these cases, the spender looks better than everyone else. Compared to people who do not spend, he has better-looking items, he skips the annoying parts and he directly puts a beatdown on them. The spender is playing the game in first class.

Psychologically and socially, first class only exists if there is a second class. Thus the comparison role of all the free players is simply to be the second class. Their job is to sit in second class precisely so that the first-class passengers can feel good about having been able to board first, get better food and have more space. In an airplane, there are physical reasons why not everyone can have the privileges of the first-class passengers. But in a game, there is no reason at all. There is no reason whatsoever why a game company could not give all players the same privileges they give to spenders. Thus the lines that are drawn between the spenders and the non-spenders are wholly artificial. They exist only because drawing them induces the spenders to spend. And therefore, one of the main purposes of the free players is simply to be present in the environment as those-who-did-not-spend. Their job is to form a comparison group, against which the spender looks impressive indeed. And then, alongside that comparative context, to be available for friendship, camaraderie or romance. The free players make it a real social world, one worth paying for.

4. Playing games for money

Free-to-play games illustrate that there is a role in games for people who do nothing but play. They do not have to be smart, they do not have to be good, they do not have to be pretty, they do not have to be dedicated, honest or loving. They just have to hang around, and – this is important – be PEOPLE. The humanness of the members of a game society is essential to the formation of meaning in that society. Only human beings can serve. It is a job only people can do.

True, a savvy game operator could try to skunk his whales by populating his game with lifelike AI. But word would get around, and one whiff of people-faking would drive all the whales out overnight. The game Barbie Online had troubles because no real boys were signing up, so the company added artificial boys. It did not work; players knew the boys were not real and the game fizzled. The same thing happened with Canadian dating site Ashley Madison: Many of the "women" on the site were bots, but it only took a few email leaks to bring the scam to light. It is one thing for one computer to trick one person in one interaction – the "Turing test." It will be a long, long time before a crowd of AIs can fool a crowd of people for years. It will be strange times, though, as game companies will guarantee sentient players the way grocery stores guarantee fresh vegetables. On the other hand, nothing kills a game like empty servers. Locations are designed to keep the real people bumping up against one another. Critical mass is critically important.

In games, we already see how valuable pure humanity is. Whatever the markets may say, pure simple humanity is a good thing. People with means are willing to pay money to be in community with other folks who get in for free. But how much will they pay? Is the willingness of big spenders to spend enough to send the street cost of game-playing into negative territory? Consider: Twenty years ago, all players paid the company for the right to play the game. These payments fell and fell. Now they are at zero across wide swathes of the industry. Why should not the price trajectory continue into negative territory? Can the spenders support not only the game-playing of poorer people, but their incomes as well?

Consider again the F2P revenue model. The iron equation of F2P is eLTV > eCPA + Ops, where eLTV is the expected long-run revenue value of a player, eCPA is the cost of acquiring that player, and Ops is the operations cost of the game. Generally, what seems to happen is that large numbers of free players are attracted into the game initially, but then most of them fall away. Over time, the paying players make up an increasing share of the player base. As this happens, companies spend less on acquisition.

What does "acquisition" mean? Typically, this is a marketing cost. Advertising – on mainstream TV, for example – can drive millions of players into a game. So long as the marketing costs (plus operations) are exceeded by the revenues that these players bring in, the marketing is worth doing.

We are considering here a second possibility, one that has not yet been tried but is surely around the corner: Paying players to play. If a large player base is important for a game, and if the main source of revenue comes from a comparatively small segment of the player base, it might well make sense for companies to pay players to come in. Paying players directly is just another acquisition cost. Instead of marketing to people, you pay them.

Initially, paying players will look like the lame maneuver by developers who lack confidence in their product. But the same was once said of developers who gave their games away for free. The stigma against F2P fell when the industry realized how much money it made. One can expect a similar acceptance of play-for-hire.

Some games are doing this already, though indirectly. In *EVE Online*, it has become possible for a player to earn enough in-game currency to purchase a pass that allows them to play for free for one month. The pass card replaces what would have been a real-money cash payment. Since the company sets the rate at which in-game currency can be earned, it is effectively paying the players for their in-game labor, which it then allows them to translate into a good – the pass – that has real-world cash value. Another game, *World of Warcraft*, has a similar system. Indirectly, these games are hiring people to play.

And why not? Economics teaches that transactions should occur whenever they are beneficial to both parties. Companies will almost surely gain from paying players; it is probably more efficient than marketing at getting people into the game. As for the player, it is surely the case that for some people, earning money in a game is just better (almost as lucrative and much more enjoyable) than earning it out of game. The in-game compensation is better than the out-of-game compensation. It is a better job, all things considered: For for-hire players, one suspects, the money would be just so-so, but the work environment would be outstanding.

This applies especially to the situation of a low-skill worker who has just lost a job to automation. His choices are to take a worse job (that he will surely lose in the near future), try to skill up or go begging. What person facing this situation would not leap at the chance to earn money while playing a video game? Thus it stands to reason that the number of people willing to work in games, even for small amounts, will increase as automation does. Paying players to pay generally makes sense for the game companies, and will make sense for more and more workers as time goes on. It is a practice destined to grow.

The typical dynamics of player bases – with the free players coming early but then leaving, while whales stick around – suggest that the practice of paying players to play games will be concentrated at the launch of a game. The company will allocate budget to the hiring of millions of players, anticipating that they will be in the game for a relatively short period. Over time, these player wage costs will fall. Yet while the for-hire players are there, the whales will become dedicated to the game for the long haul. The spending of the whales will then sustain the game for a long period. As for the for-hire players, their lifestyle will be one of migrating from game to game, picking up rewards during the launch phase of new games and then moving on when the money starts to dry up.

5. How play for hire will emerge

Right now, the environment for compensated play is clouded by overlapping institutions and regulations. Games often have multiple currency systems in place. The real world does too, of course. Laws and regulations constrain this space in ways that nobody yet understands. Yet some things are clear.

The practice of working for money online has become widespread if not yet common. New research indicates that large numbers of people around the world participate in a global market for online jobs (Lehdonvirta et al., 2015). Oxford University has recently launched the world's first Online Labour Index (University of Oxford, 2016). The online labor market exists and seems to be growing.

Precedent also exists for the idea that wealthy people can pay extra money to make a game more enjoyable. Games have their shops, and game companies accept credit cards, debit cards, PayPal and various kinds of points and credits and time cards and what have you. There should be no problem sending purchasing power from the vaults of the rich into the vaults of the game developers.

It is the ability of poorer people to get money *out* of games that faces legal questions. In principle, a game company could simply hire players as employees. This is highly unlikely to be a first step (or a second or a third or a seventeenth), however, since the regulatory overhead on hiring is incredibly burdensome. What is more likely is a system that evolves from payment practices that already exist.

Under current US law, anyone who makes money in a game has to report the earnings when they are liquidated in the form of real-world currency. It is not clear what would happen if someone were to earn gold inside a game and then trade that gold for something in-kind, like rent. But this could happen. It could happen today, in fact.[2] A player can already earn in-game money and give it to his landlord in lieu of rent. Players can already sell their game assets for real-world cash. Thus there is already some precedent for players to be able to transfer some of their in-game earnings into outside purchasing power.

Also it is important to realize that the world of finance is facing a wave of disintermediation. Technology is beginning to allow an almost unlimited variety of ways to transfer purchasing power among people. Technology also allows game companies almost unlimited ways to craft incentives without falling afoul of the law. For example, today, one would not want to allow whales to directly hire underlings within a game for anything that could be liquidated, because this would enable money laundering. But there is no opportunity for money laundering if whales pay USD1 million to the game company and the game company provides one million piles of random, sellable loot worth USD1 on the eBay market.

There might evolve a market that translates game time into game assets, such as time cards, that can then be quickly and easily traded for real money on the outside. Or a game company could make a deal with a credit-card company so that game time yields bonus points on the card. Since the card can be used to pay rent and buy groceries, the game time translates quickly into real purchasing power. Current law covering these points systems does not handle them as income, so there would not apparently be any tax or reporting burden. Another possibility would be to pay players with an undercover currency, such as Bitcoin. These types of changes will represent a minimal disruption to current practices, but they will be the first signs of an emerging general change in employment patterns.

Once the basic idea is established of paying players to play, there will be some sort of legal and regulatory push to normalize the practice and make it explicit. There will be controversy. Initial objections will come from different parties. Developers, or at least the purists among them, will say that paying players to play undermines the whole idea of a game. Perhaps it does. But since it makes economic sense, it will happen anyway. Pundits and lawmakers will say that players are being exploited, that games are becoming sweatshops. Perhaps. But as inequality reaches soul-crushing levels, sweatshop employment within a game will look like a fairly humane alternative for people who cannot find work. Academics will lament the transfer of inequality and oppression from the real world into the game world. Yes, that will be sad. But since it will allow some people to eat who otherwise would starve, it will have to be accepted. The transition will be no different from that which drove farmhands into the factories. Sad; much was lost; but it was the best alternative. Social critique, policy and aesthetics will have to give way. That is why it is called *disruption* and not merely *change*.

After the play-for-hire model has been normalized, low-skill work will shift into games on a large scale. It will be found that for most people who do not have a particular skill or an inherited lump of wealth, one of the best ways to spend time is to play games and, with the meager earnings, make the best life one can. Consider the minimal lifestyle necessary to sustain a full-time job playing games:

- 2000 calories a day
- 100 square feet of living space
- Internet access
- Game gear
- A few pieces of furniture

Given that the work can be performed anywhere with internet, the costs of space and food can be very low. The needs of a working gamer will cost no more than a few dollars a day. Although, to us, this looks like a terrible lifestyle indeed, remember that this person is spending all of his time inside a game, a place designed to make him feel quite good. True, there will be other people in that space who are doing even better. The people with money to spend in there will do so, and they will have the most powerful weapons, the fastest cars and the best-looking outfits. Yet the in-game difference between the rich and the poor can be managed in ways that are impossible out-of-game. The wealthy people in the game can be shown how very wealthy they are, while the poverty of the poor can be hidden from the poor themselves. The comparisons of rich and poor, and the emotions and self-esteem of them all, can be managed by the game developers. Thus while the outer world will provide nothing but glaring contrasts, the in-game world will be everything to everyone.

As developer Gordon Walton once said, "Everybody wants to be a hero" (Castronova, 2013, 15:25). In the play-for-hire game, everyone will still be a hero. But some of them will be paying a lot of real money to be a truly amazing hero, while others will be receiving a very modest amount of real money to be a pretty good hero. They will all be happy. And by this means, the vast wealth produced by automation will trickle down from the owners and the operators to the low-skilled. For it will turn out that the low-skilled were not all that low-skilled after all. They all have something that every human has; an ability to entertain other humans, just by being human. Every player of a game provides entertainment for every other player. It stands to reason that some people, even many people, could get paid to do this entertaining. And the job of being an in-game entertainer will become ever more attractive as the jobs available in the outside world steadily decay.

6. Concluding discussion

6.1. A timeline

The developments discussed in this chapter are already underway. I am told that developers have been aware of the option to pay players but have held back because of legal and regulatory barriers. And yet, as we have seen, young low-education people are already exiting the workforce in much larger numbers. Games are already taking advantage of distinctions between time-rich, skill-rich and money-rich players. The net cost of playing games is becoming negative for some players.

6.2. Within five years

These trends will result in the implementation of a new business model in the game industry. At least one game will innovate around current regulations and set up a system in which most players are given financial incentives to play. The incentives will be indirect: Points on a card, virtual items that can be sold for money, etc. These will be known at first in the industry as "player-retention incentives". They will probably enter discussions as a way to lower churn in the player base, to keep lukewarm players around who might otherwise leave. The game will have a strong pay-to-win element, and will earn huge revenues from spenders. Compared to these revenues, the costs of "player retention incentives" will seem trivial.

The next game in the cycle, still within five years, will go a step further and offer "player retention incentives" to all players from the beginning. This model will also succeed, and the industry will become aware of the new model. After that, most new games will try to develop innovative ways of paying their players.

6.3. Within ten years

These innovations will be complete. Game companies will have created streamlined and efficient ways to channel purchasing power from richer players to poorer ones. A share of the revenues provided by whales will be spent efficiently to generate launch buzz and possibly to keep the overall player numbers up. This will be a standard tactic in the industry.

At the same time, in the outside world, automation will continue. Job prospects for low-skill workers will keeping eroding until it reaches the point that the reservation wage – the minimum wage for which a worker is willing to take a job – will have fallen well below the legal minimum wage. This means that low-skill workers will have greatly increased their willingness to work in games. For many young people, the fantastic work conditions will more than offset the low pay that games offer. Working in games will be more attractive than any job available in the real labor market.

Nonetheless, at this point the game workers will still not generally see themselves as such. They will see themselves as unemployed and unemployable. Yet they will say that they are keeping themselves afloat by going into games and "playing smart", bringing out whatever purchasing power they can. It will at first go almost completely unnoticed that that purchasing power is becoming large enough to cover the meager costs of living which most low-skill workers have come to see as normal. Low-skill people will have become accustomed to living far away from urban centers, with no car, no entertainment, minimal clothing, basic housing, no health care and poor food. Their lives will be lived on the internet. And on the internet, through pay-for-hire games, they will earn enough to support their minimal bodily needs.

6.4. Within 20 years

These trends will have changed the relationship of low-skill people to the labor market. Among those who have no technical abilities and no ownership shares of capital, many if not most will earn their living within multi-player online environments. There will be millions of play-for-hire players all across the globe. If populations in China and India participate, the numbers will be in the billions. For-hire players will have become conscious of this social role; they will understand that playing is their contribution to society. They will express pride at the quality of their play. Indeed, those who do not "pay-to-win" will be quite proud of their hard-earned accomplishments relative to the wealthy who buy their way forward. Players will see themselves as experts in their game of choice and will think of that virtual place as an important social and even political entity. Players will be organized into pressure groups. There will be uprisings and strikes and boycotts in games where compensation is deemed unfair. Meanwhile the wealthy will jump from game to game, always seeking the very best experience for their money. The sad history of inequality and politics will gain a number of new chapters, this time online.

Of course, there are a number of trends that go in a different direction. Fertility is falling in the developed world, which undercuts any labor-surplus argument. In the developing world, incomes are rising, which should increase demand for goods from developed countries. Half the population – women – do not seem to fall into intensive game play in the way this argument assumes. Any prediction could be wrong.

Yet there is a trump card here: The argument here is ultimately driven by technologies of AI, robotics and virtual reality; and technology, at this state of evolution, is beginning to dominate every other trend. It is not unreasonable to assert that tech is becoming the only trend that matters. Technological change has become so powerful and rapid that it is now fair to say that tech does not react to society, tech creates society. That includes everything: Law, gender norms, employment and, of course, games. Even climate change, that great problem of our time, will fade as a public issue as technology remakes how we live. How much energy is required to keep five billion unemployed people happy in their comfy, VR-enabled chairs?

If technology is truly a dominant force, then we can expect technologies of entertainment and automation to create a vast labor market for low-skill game players. It is a new service sector that will thrive in decades to come.

Acknowledgments

I am grateful to Mike Sellers, Joshua Fairfield and Isaac Knowles for helpful comments.

Notes

1 Industry insiders say that PTW motivates spending far more than the other two. At the same time, in US culture at least, PTW generally outrages the player base. As a result, the PTW angle is less used in the United States. In other cultures, PTW is considered acceptable. Game designers try to design PTW in such a way that it makes money on net, that is, so that the added revenue from purchases offsets the lost revenue from players who quit in anger.

2 In China, the game currency QQ Coin was being used as street money until the Chinese government quashed it. However, quashing currencies is a tough task. Black markets always survive.

References

Castronova, E., 2013. *Games and Culture*. [online] YouTube. Available at: <www.youtube.com/watch?v=9qMzj2lNUYA> [Accessed 17 Sep. 2019].

Forrester, 2016. *Robots, AI Will Replace 7% of US Jobs By 2025*. [online] Available at: <https://go.forrester.com/press-newsroom/robots-ai-will-replace-7-of-us-jobs-by-2025> [Accessed 17 Sep. 2019].

Lehdonvirta, V., Hjorth, I., Graham, M. and Barnard, H., 2015. Online labour markets and the persistence of personal networks: Evidence from workers in Southeast Asia. In: *ASA Annual Meeting 2015 Session on the Changing Nature of Work, 22–25 August, 2015*. Chicago, IL, pp. 1–25.

Rose, M., 2014. *Half of All F2P Mobile Game Revenue Comes from 0.22% of Players – Report*. [online] Gamasutra. Available at: <www.gamasutra.com/view/news/215129/Half_of_all_F2P_mobile_game_revenue_comes_from_022_of_players__report.php> [Accessed 17 Sep. 2019].

Swanson, A., 2016. *Why Amazing Video Games Could be Causing a Big Problem for America*. [online] The Washington Post. Available at: <www.washingtonpost.com/news/wonk/wp/2016/09/23/why-amazing-video-games-could-be-causing-a-big-problem-for-america> [Accessed 17 Sep. 2019].

University of Oxford, 2016. *Rise of Online Work Captured in the First Online Labour Index*. [online] Phys.org. Available at: <https://phys.org/news/2016-09-online-captured-labour-index.html> [Accessed 17 Sep. 2019].

13 The global gig economy

Toward a planetary labor market[1]

Mark Graham and Mohammad Amir Anwar

1. Introduction

> *The more production comes to rest on exchange value, hence on exchange, the more important do the physical conditions of exchange – the means of communication and transport – become for the costs of circulation. Capital by its nature drives beyond every spatial barrier. Thus, the creation of the physical conditions of exchange – of the means of communication and transport – the annihilation of space by time – becomes an extraordinary necessity for it.*
>
> – (Marx, 1973, p. 524).

Our world of work is seemingly in crisis. Everywhere we look there are bold proclamations about the ways that technologies are expected to destroy, move and deskill jobs.[2] This chapter seeks to begin from these visions of a rapidly changing world of work, and argue that we are witnessing the emergence of a "planetary labor market" in digital work. By changing the geography of work, a planetary labor market introduces some serious concerns for the livelihoods and structural power of workers. Now more than ever is David Harvey's (1990) famous maxim about the relative power of capital over labor relevant.[3] This chapter points to a need for a reinvigorated program of research and activism to tackle this fact.

Today's online outsourcing platforms host all manner of jobs: ranging from click-work to the training of machine-learning systems to transcription to live personal assistance. These online outsourcing platforms, by becoming key intermediaries in the labor process of outsourced work (Casilli and Posada, 2019), potentially augur a radical shift in the scales at which capital can interact with labor.[4] Upwork, a platform that boasts of having 12 million registered workers, explains the advantages to clients with the following text on their website: "online work can happen wherever there is a reliable internet connection – an office, home, café, or rooftop. This also means you can choose who you work with, among a larger pool of people from around the globe" (Upwork, 2019, para.2). One of the world's largest online outsourcing platforms, Freelancer.com (2019), display their logo and the statement "25 million lives changed" over a map of the planet (noting that their location is "everywhere"). Similarly, Appen, a platform company with workers who train machine-learning systems in 180 countries, explicitly advertise their "global crowd" of workers on their website (Ruby, 2019). The increasing digitalization of work and recent advancements in automation and communication technologies do not just augment the labor process with digital data, digital processes and machines; they also embed it in stretched-out networks of production: with tasks quickly passed in complex assemblages from person to person, person to machine, and machine to machine.

While these arguments are better covered elsewhere, this chapter instead seeks to build on them to make an argument about the spatial implications of these changes.

We will use online outsourcing/platform work as a key case of "digital work" in the rest of the study. Work, in other words, done over the wires and mediated through a platform; work that does not necessarily require proximity between the worker, the work itself and the site of the immediate delivery of the work. The relative lack of fixed organizational infrastructure needed for the online outsourcing sector means that it can be characterized by a broad geographic spread. Unlike traditional forms of employment, companies that outsource digital work, and platforms that mediate those relationships, tend to avoid any formal employment of workers and do not need to share proximity to workers. Jobs are instead listed on digital platforms that allow workers to bid for them. These jobs might take anything from minutes (e.g., click-work or image-tagging) to months (large writing tasks or web design) to complete. According to Heeks (2017), there are about 70 million registered platform workers globally, in the market for work that the World Bank estimates will grow to USD15–USD20 billion by 2020 (Kuek et al., 2015). The economist Guy Standing (2016), meanwhile, predicts that by 2025, platforms will mediate one-third of all labor transactions. The scale and scope that some of these platforms can achieve is in part driven by the development of planetary-scale infrastructures of computation (Bratton, 2016).

Because of the rapid rise of digital work around the world, we ask in this chapter whether we are seeing the emergence of a "planetary labor market" in digital work. To answer this question, we outline the scalar and spatial changes that have been occurring in labor markets, review their implications for the balance of power between labor and capital, and advance some possible responses to ensure that we do not get trapped in a global race to the bottom in which there are constant downwards pressures on wages and working conditions.

The argument that we make here is largely conceptual. However, we illustrate our argument with examples from a five-year (2014–2018) study of digital work in some of the world's economic margins. We conducted semi-structured interviews with 65 online platform workers in South Africa, Kenya, Nigeria, Ghana and Uganda, recruited from one of the world's biggest online labor platforms, Upwork. We sought maximum diversity in our sample, and our respondents were characterized by a range of different attributes, such as number of hours worked on the platform, different types of work activities and income earned. Most workers in our sample had multiple accounts on various platforms such as Freelancer.com, Fiverr.com and Peopleperhour.com. We also recruited Upwork workers through social media (Facebook and LinkedIn) and snowballing. The primary sampling goal was to ensure a diversity of worker experiences. As such, this chapter presents selected cases that indicate the existence of activities, issues and concerns rather than a representative view. Through the interviews we sought to understand the socioeconomic background of workers, the nature and types of work done by these workers, career prospects, livelihood challenges, income, worker-worker and worker-client interactions, strategies to win bids, to stay competitive, to demand higher wages and negotiate working hours and actions to avoid the various risks inherent to platform work. All the worker names have been changed.

2. Moving beyond local labor markets

To describe what is meant by a planetary labor market, it is first useful to describe what labor markets are. A nineteenth-century hiring fair, such as the one described by Thomas Hardy in *Far From the Madding Crowd*, is used by Fevre (1992), in his book about the sociology of labor markets, as a way of illustrating an abstract definition of labor markets through five key distinct processes. These are: informing employers (employers learning about availability and skills of workers), informing workers (workers learning about jobs), screening workers (employers obtaining enough information about workers to decide if they could be hired for a job), screening employers (workers learning about their employers) and offers to buy or sell labor (the actual negotiations and pitches made by workers and employers). Labor markets, in other words, are a way of describing a convergence of workers and employers in specific places and times. While scholars as far back as Karl Marx posited that this convergence in competitive labor markets is a fundamental characteristic of capitalist society, various planned economies in the late twentieth century likewise relied on the concept of a labor market to govern the management and distribution of the labor force (Brown, 1970).

In Hardy's hiring fair, the spatial and temporal co-presence of agricultural workers and employers allowed the five previously-mentioned processes to converge. However, while co-presence has traditionally been a necessary condition for most of these conditions, it has not been a sufficient one. Kalleberg and Sorensen (1979, p. 351) define labor markets as "the arenas in which workers exchange their labor power in return for wages, status, and other job rewards. The concept, therefore, refers broadly to the institutions and practices that govern the purchase, sale, and pricing of labor services. These structures include the means by which workers are distributed among jobs and the rules that govern employment, mobility, the acquisition of skills and training, and the distribution of wages and other rewards obtained contingent upon participation in the economic system." But, ultimately, those institutions and practices still require some level of space-time convergence between employers and workers.

It is important here to distinguish between the way that labor markets have been conceived in orthodox classical economics, and their actual characteristics. Orthodox conceptions put forward a perfectly competitive market that can provide both firms (buyers) and workers (sellers) with perfect information. Wages are set by the relationship between supply and demand, and "workers can move freely in response to changes in supply and demand in different parts of the market" (Kalleberg and Sorensen, 1979, p. 354). Kalleberg and Sorensen (1979) give examples of such markets as the migrant labor market in California, and the 1970s labor market in Afghanistan. In both cases, wages were relatively uniform and institutional forces only had a small influence.

In practice, it is rare for labor markets to fit these sorts of perfect property, instead, labor markets function in imperfect and uneven ways. Workers comprise different classes, genders, races, nationalities and other groups that can get segmented into different functions in labor markets. These markets are further built on, and performed through imperfect information, irrational social behaviors, politics, institutional arrangements and practices, customs and prejudices. As Peck (1996, p. 5) has argued, labor markets are "socially constructed and politically mediated" arenas, "structured by institutional forces and power relations". Thus, we get segmented labor markets functioning at multiple scales and spaces to produce variegated outcomes for workers (Grimshaw et al., 2017). In these segmented or split markets, workers have little opportunity to cross into other groups and are thus constrained to a limited set of outcomes: with factors like gender or race influencing segmentation (with, for instance, women earning lower wages than men) (Bonacich, 1972; Reich et al., 1973).[5]

The takeaway point here is that labor markets function in complex, imperfect, exclusionary ways. When speaking about a physical meeting place, like a hiring fair, the very concept serves as a multi-scalar abstraction. We use the idea of national or regional labor markets not to imply that everyone in those nations or regions have equal opportunities to read or access the market; but rather as a way of indicating that there are distinct economic, social and political enablers and constraints that put rough, porous, but still real geographic boundaries around Fevre's five processes. This is not to say that workers are not enrolled into global-scale associations and production networks. Indeed, workers in many economic sectors have been for centuries (Hunt, 2002). But, as Fevre (1992, p. 14) notes, "Labour markets need have neither a fixed time nor a fixed place, but they must have some sort of time and place otherwise how could people use them? If they do not know when and where, workers cannot find jobs and employers cannot hire workers".

Much of this discussion assumes a located place of work – a farm, factory or office that a worker needs to be physically present in, in order to perform their duties. But, as the nature of work changes, so too must our conceptions of the boundedness of labor markets. Here it is useful to draw on the concept borrowed from geography of a relational understanding of space (Massey, 2005). Rather than only thinking of space as a canvas, it is rather something that can also emerge from social relations (Hudson, 2001). This vision of space as relational and emergent, rather than pre-existing, is useful because it offers a productive way of understanding the impact of digital technologies on labor markets. Stephen Graham (1998, p. 174), in an influential piece on the intersections between technology and space, builds a relational understanding, noting that "such a perspective reveals how new technologies become enrolled into complex, contingent and subtle blendings of human actors and technical artefacts, to form actor-networks". He continues: "new information technologies, in short, actually resonate with, and are bound up in the active construction of space and place, rather than making it somehow redundant".

If we apply these sorts of understandings to the contexts of work, the boundedness of earlier visions of labor markets evaporate. Building on an actor-network understanding of work as constituted through a broad range of associations with objects, Jones (2008, p. 12) further argues that "working practices, the experience of work, the nature of workplaces and the power relations in which people's working lives are entangled require a theoretical understanding of global-scale interrelationships if they are to be properly understood". Describing how work is increasingly performed through global networks of human and non-human objects, he adds: "Contemporary work is becoming less constituted through localized, physically-proximate relations and increasingly constituted through distanciated relations. These multiple spatial associations increasingly extend to the planetary scale" (p. 14). This starting point – moving beyond an understanding of work as inherently local – allows Jones (2008, p. 15) to then build his "global work" thesis:

> Rather than understanding work as a practice undertaken by social actors located in discrete material spaces and framed in a linear chronology, work is reconceptualised as a complex set of spatialised practices involving humans and non-humans . . ., and which is constituted in relational space with a disjunctive, non-linear chronology. . . . This is 'global' work because this reconfigured concept captures the qualitative degree to which all work practices are constituted through distanciated . . . socio-material relations.

As the places of work move beyond single locations, this offers us a pathway for thinking through the impacts of globalization on workers.

3. Toward a planetary labor market

Although the "global work" thesis is useful for providing a framework that allows us to carefully think through the impacts of globalization on workers and the ways that the places of work move beyond single locations, in the rest of this chapter we will argue that it is important to think about the relationships between employers and workers as more than simply distanciated social relations. Using the idea of a "planetary labor market" allows us to show that not just work can be highly (globally) connected, but rather temporary states of co-presence between workers and employers can be brought into being. Like Jones (2008), we build our understanding of a planetary labor market on a relational understanding of space. Specifically, we draw from Doreen Massey (1993, p. 61) who argued that:

> Different social groups and different individuals are placed in very distinct ways in relation to . . . flows and interconnections. This point concerns not

> merely the issue of who moves and who doesn't, although that is an important element of it; it is also about power in relation to the flows and the movement. Different social groups have distinct relationships to this anyway-differentiated mobility: some are more in charge of it than others; some initiate flows and movement, others don't; some are more on the receiving end of it than others; some are effectively imprisoned by it.

As such, the moments of co-presence that will be described later in this chapter rarely fit either the orthodox idea of labor markets or resemble Hardy's nineteenth-century hiring fair. While digital-work platforms have enabled the potential coming together of employers and workers on a planetary scale, the labor market for digital work that is developing is characterized by both asymmetrical scalar relationships and uneven spatial ones: with workers and employers having very different possibilities to read and participate in the labor market. In other words, the argument put forward in this study is that a planetary labor market is not simply a "global" extension of Hardy's hiring fair. It facilitates coming-togethers that can transcend the spatial boundaries that constrained the convergence of employers and workers, but remained shaped and characterized by multi-scalar and asymmetrical technological, political, social, cultural and institutional factors.

3.1. Applying Fevre's characteristics

This section returns to Fevre's (1992) five characteristics of labor markets ("employers learn about workers, workers learn about jobs, employers obtain information about workers, workers obtain information about employers and offers to buy and sell labor transpire") and asks how they apply to the planetary labor market brought into being through online outsourcing. Within each of the following sections, we outline a range of concerns that relate to the structural power of labor *vis-à-vis* that of capital. This strategy is neither intended to imply that these are the only concerns, or that there are not benefits (such as flexibility) to workers at the individual scale.

3.1.1. Employers learn about workers

In the case of online outsourcing, employers (i.e., "clients") have genuine planetary reach when learning about workers. Employers list requirements needed from their workers on online labor platforms, and workers from around the world then bid on those jobs – allowing employers to collect certain information they need about any potential worker. The fact that online outsourcing platforms tend to have a massive oversupply of workers on them (Graham et al., 2017b) means that workers are eager to supply any information that potential employers require. This will typically include location, ratings, reviews, previous clients' feedback, but may also include work history, previous experience, number of hours worked, education and a host of interpersonal skills. When we compare the ways in which employers learn about workers to the ways that workers learn about jobs, the scalar differences in how workers and employers can read the planetary labor market become apparent.

Figure 13.1 shows where workers on the world's largest online labor platforms are based (Ojanperä, 2019). It demonstrates that Upwork and Freelancer's claims to host workers from almost everywhere on the planet are true. However, it is also obvious that there are distinct patterns to the supply of online labor power with large concentrations of workers in a few countries. Employers can join these platforms to either find workers in specific places (for instance when language skills are needed) or to put a job out to competition from workers that can be located anywhere.

This huge number of people who sign up to look for jobs ends up creating a huge oversupply of labor. In Table 13.1, we present data collected from Upwork on a single day in October 2018, to estimate at the potential oversupply of labor on the platform. The table compares the number of people signed up on the platform by country, with the number of workers who have ever earned at least USD1 or worked at least one hour on the platform. Even with such a low threshold of what constitutes work, we see a massive oversupply in the sample of countries in Table 13.1. Globally, less than seven percent of people who register for jobs are ever able to secure one.[6]

While the geography of online labor is far from equally spread around the world, the relative ubiquity of digital connectivity, and the affordances that digital labor platforms provide, mean that employers can now find new workers on the other side of the world in minutes, as long as workers have relevant ICT tools and internet connectivity. However, for workers, the combination of the global market and the oversupply of labor power (or at least the perception of the oversupply of labor power) is experienced as something that significantly depresses the wages they are able to command (see also Graham et al. (2017a) and Wood et al. (2019) for more on this point). Adele, a data-entry worker in South Africa described how this situation played out on the platform Upwork: "You go apply for a job and somebody else will come and apply for less than dollar. Other people are bidding too low and it was people from the Philippines and India. I was angry because they bid too little and . . . Yet they are happy. Yeah, I was quite pissed off there; I was like no way are they doing this!"

3.1.2. Workers learn about jobs

Workers on online outsourcing platforms naturally have a geographically expanded pool of jobs to bid for, compared with the jobs available in their local labor markets. Most platforms allow workers to bid for jobs from anywhere. However, this differs from a simple state of co-presence for two reasons. First, while workers can learn about task vacancies on platforms, clients often reveal relatively little about themselves. Second, these platforms tend to facilitate vertical communication rather than horizontal communication (between workers), thus limiting the associational power of workers.

On the first point, the ability for workers to learn basic information about the jobs, but relatively little about their bosses is particularly pronounced for workers

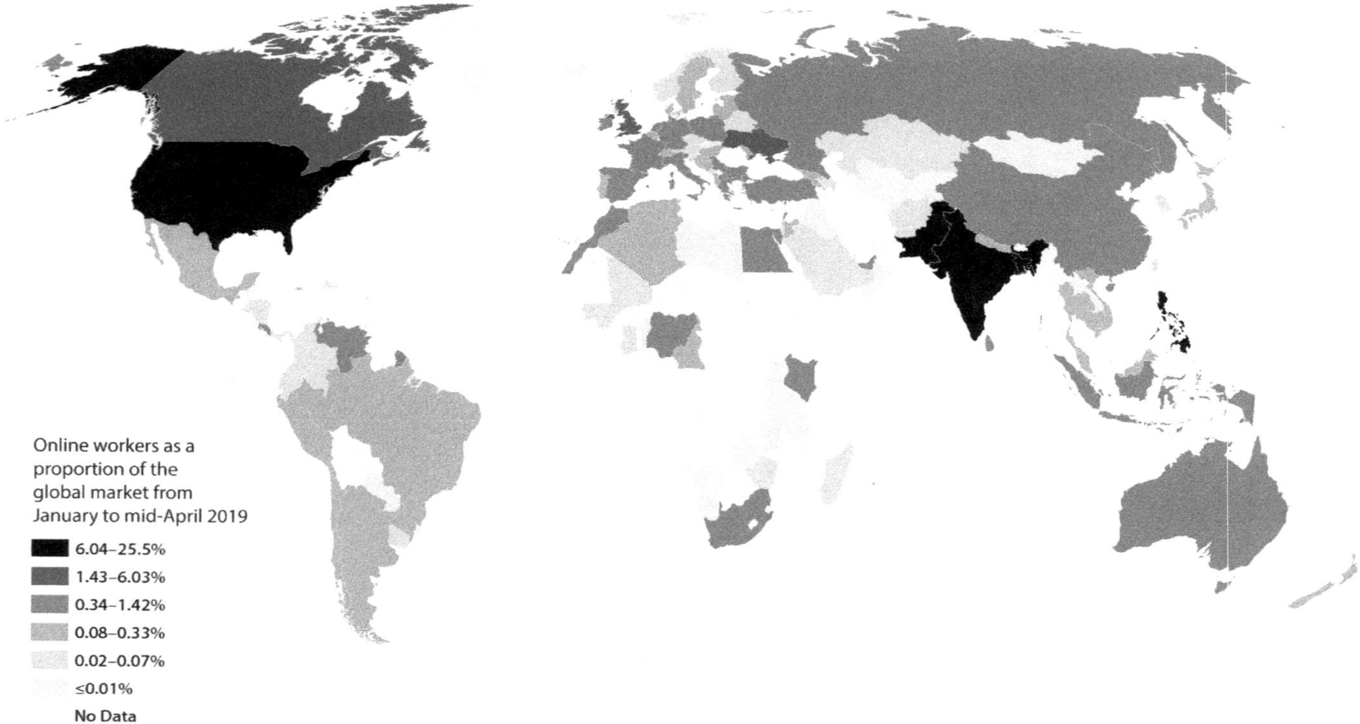

Figure 13.1 The availability of online workers (Ojanperä, 2019).

Source: Reproduced from Graham, M., and Anwar, M. A. 2019. The Global Gig Economy: Towards a Planetary Labour Market? *First Monday*. 24(4). doi.org/10.5210/fm.v24i4.9913 under Creative Commons license.

Table 13.1 Oversupply of labor on Upwork.com.

Country	*Potential workforce*[a]	*Successful workers*[b]	*Over-supply*[c]	*Over-supply percentage (%)*
Global	1,891,648	128,259	1,763,389	93.2
United States	581,717	23,845	557,872	95.9
India	249,698	22,772	226,926	90.8
Philippines	164,757	18,869	145,888	88.5
Pakistan	66,681	6,032	60,649	90.9
United Kingdom	56,644	2,924	53,270	94.0
Ukraine	55,604	8,506	47,098	84.7
Egypt	35,299	1,295	34,004	96.3
Kenya	18,508	898	17,610	95.1
Malaysia	13,385	317	13,068	97.6
South Africa	12,723	593	12,130	95.3
Nigeria	8,032	297	7,735	96.3
Vietnam	7,574	669	6,905	91.1
Ghana	1,656	50	1,606	96.9
Uganda	1,176	31	1,145	97.3

a. Total searchable worker profiles.

b. Worker with at least USD1 earned.

c. Potential workforce minus successful workers.

Source: Data for October 24, 2018, collected and analyzed by the authors.

doing short-term and fixed-price jobs such as document conversion, transcription and writing jobs. Some longer-term jobs such as web-chat support, digital marketing and virtual assistants should in theory allow workers to learn more about clients and their businesses over the course of time and therefore build a relationship of trust with them. Yet, even with these longer-term jobs, many workers struggle to get to know their clients. A Kenyan data-entry worker, Eidi, noted that despite working on a content generation project for over a year, she only knows her line manager who sits in Uganda and has no idea who the main client is, or the owner of the project. Some clients do not tell workers in detail what the job actually requires them to do (see Figure 13.2, simply advertising "repetitive" work).

Here it important to remember that the affordances of online outsourcing platforms are designed for workers and clients to connect with one another, rather than for workers to connect with each other. Historically, the inability for workers to have any effective virtual co-presence has severely limited associational power (Wood et al., 2018). While Fordism enhanced workplace bargaining power (based on the ability of workers to threaten to stop the entire production chains) by uniting workers at the point of production (i.e., physical factories), it decreased marketplace bargaining power (based on the possession of scarce skills and low levels of general unemployment) by bringing a global reserve army of labor under capital's control (Silver, 2003). The point is if by fixing labor in a place often gives it power, it can also be undermined by multiple spatio-temporal fixes created by capital since the crises of the 1970s (1973 oil price crisis, 1973–1974 stock market crash, the fall of the Bretton Woods System) (Harvey, 2011, 2001).

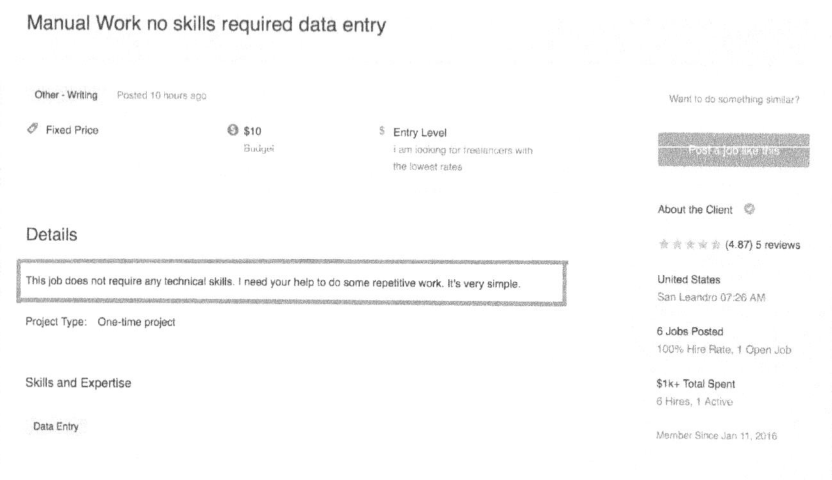

Figure 13.2 Job advert on an online outsourcing platform, revealing little detail of the task being advertised.

Source: Authors' own depiction.

Put differently, relocation of production gave new entrant labor forces a sense of class identity and bargaining power at the workplace, but the ease with which production can be relocated meant an undermining of marketplace bargaining power and threats of job losses. The mobility of capital through the reorganization of production techniques (fragmentation and relocation of production) has tended to weaken the associational power of workers due to the incorporation of a mass of unemployed and unorganized workers who are hard to unionize. Workers lack a sense of collective identity as a working class and a weak state regulatory framework delegitimizes trade unions, making it incredibly hard for such organizations to deliver benefits to workers (Silver, 2003).

These trends have continued with the emergence of digital work that can be performed by a global pool of unorganized workers separated by large physical distances, and workers lacking common linguistic and cultural characteristics. The inability for platform workers to have any effective virtual co-presence severely limits their associational power (see also Wood et al., 2018). This largely relates to the nature of digital work, the technical structure of platforms, the transaction of digital work through the internet, and a global pool of workers who are fragmented and commoditized.

The demand by clients for work to be completed before a set deadline forces workers to confine themselves to their workplaces (usually their rooms), working long hours with high work-intensity to avoid losing wages. Mukondi, in Kenya, was doing internet research for a US-based company dealing with sales of secondhand and end-of-life mobile phones. She was working close to 80 hours a week and as a result, did not have enough time to meet other online workers in the locality or socialize with them. We asked all workers in our sample if they know anyone in their locality and if they meet with them regularly. While some workers knew other digital workers in their cities, they usually find it hard to socialize with them. One of Mukondi's coworkers on this job contract went to the same university in Nairobi, and they never met. Instead, workers tend to utilize whatever time they have to find new work, instead of trying to establish connections with workers either through the internet or locally. As another Kenyan worker, Isa, who does search engine optimization, said: "when you're busy you have no time to go look for another guy". Since there is intense competition between workers on a global scale on these outsourcing platforms, it is understandable that workers will want to prioritize continuously looking for work instead of developing capacities for collective organization. The extreme physical separation of digital workers also makes any collective organization or physical co-presence unlikely. We found a few local networks of platform workers in Ghana, Nigeria and Kenya, where multiple small groups of workers (two to three workers in a group) have developed close working collaboration. Workers also develop networks through social media, although the utility of such networks to transform worker power is debatable.

Figure 13.1 indicates that a range of underlying economic, social and political factors end up bringing into being particular geographies of work. While work can in theory be done from anywhere, myriad factors end up influencing concentrated economic geographies of jobs. Irrespective of its actual geographies, digital work is sufficiently mobile for workers and clients to feel that the marketplace they are operating in is truly global. The result is that workers can lose a sense of any collective organization and feel replaceable, while clients exploit this lack of associational power of workers to exert their demands on workers (also see point 5). Ben, a virtual assistant in Kenya, explains the feeling of being replaceable. He said, "basically I can do what I want but there is always that feeling like what happens tomorrow if the company can't afford me, do I have to cut my salary or what happens if I wake up in the morning and there is this email from Upwork, contract ended. That's the email I fear the most".

3.1.3. Employers obtain information about workers

The way platform work is designed and transacted over the internet reveals stark asymmetries in the ways in which employers and workers obtain information about each other and what they do with that information. While the bidding process enables labor power to be bought as a commodity in the market, a real sense of this commodification comes from the nature of the digital work and types of job contracts offered on digital platforms. Due to the digitally-intensive nature of work activities traded on these platforms, these work tasks can be broken down into simpler tasks (or "gigs") which can be completed by individual workers in a matter of minutes or hours from their homes. Since these tasks can finish quickly, they have to repeat the same bidding process in order to secure new jobs (though some experienced and top-rated workers may get repeat clients who offer work to them directly). There are relatively few jobs advertised that allow some form of trust or working relationship to develop between workers and the employers – usually hourly contract jobs such as digital marketing, social media management, and virtual assistant. But even here there is a problem. Some platforms give clients the power to pay workers only if they are satisfied with their work, and as a result, some workers do not get paid even after they submit their work to clients.

This level of fragmentation of work and commoditization of labor power means the employer-employee relations become contingent (Barker and Christensen, 1998; Barley and Kunda, 2006), and employers are under no obligation to help workers build long-term careers on platforms. The potential for workers to gain experience and build up knowledge and skills for future career development is also constrained, meaning less scope for workers to upgrade to high-skilled and high-income jobs on platforms. A data-entry worker with no formal education and training is highly unlikely to go on to software development tasks or graphic design. Clients do not usually provide/offer training for their platform workforce to upskill them.[7]

Employers are able to demand any information they want prior to the job offer and workers are obliged to supply that information in a codified and quantified form (ratings, hours worked and wage rates), allowing clients to assess the workers' quality and ability to work. Since only workers bid for jobs posted by employers, they signal (or give information to) employers on platforms about their quality which employers use to screen workers and make an informed decision who to award the contract. Employers, with all the information about bidding workers at their disposal, are able to choose the workers they want to work with, which might be the worker with the lowest hourly rate, or the top-rated worker irrespective of their rate. In other words, clients have the ability to access all the information they need before awarding the job contract to workers, who usually know little about their clients (see point 4). The technical infrastructure of the platforms generates and amplifies an information asymmetry between buyers and sellers of labor – in order to favor the buyers (see Graham et al. (2017a) for more on this point).

One of the most significant tools that employers use to learn about workers is the rating system (Wood et al., 2019). As one worker, Mukasa, in Uganda, told us, no client is willing to work with new freelancers with no ratings or feedback, making it difficult for newcomers to land a job easily on platforms. During a group discussion with five platform workers in Abuja, Nigeria, they told us that they had to spend months searching for their first job due to an initial lack of ratings. Clients sometimes use this power asymmetry to exploit workers, by offering extremely low-paid work in return for good reviews and high ratings. Adele, in South Africa, told us that she did her data-entry work for a client at less than one dollar per hour for about a week (usually eight hours a day). She said "it was quite tough and I'm like okay, but at least he gave me that shot. He gave me that because after he did give you a good review and feedback". Onochie, a virtual assistant in Nigeria, explained the importance of ratings for his profile and said:

> I will say my secret is, every client that I work with, I try to leave the best type of impression. Even if the job is not great, I can actually offer to give the

> client a refund. Not that it was my fault that the whole thing went wrong. . . . So, I paid him back and I told please I do not want a review from you. I do not want a [negative] feedback. That is why I decided to give your money back. So, for every client I work with, I try to do the best possible job that I can, so you can give me the best possible feedback.

3.1.4. Workers obtain information about employers

While the information that employers gain about workers can be used as a form of control, some information about clients can also be visible to workers, such as location, whether or not their payment method has been verified and an overall feedback score from other workers at the time a bid is placed. However, some specific platforms like Freelancer.com do not allow information on clients hiring history, such as total money spent on hiring workers or average hiring rates, to be made available at the time of bidding. Such information would be useful to workers during the bidding process.

As already discussed in point two previously, many workers do not know the identity of their clients or even the nature of their business when placing bids for specific jobs. Referring to her client, Adele in South Africa explained: "She said just do the job and then send it to her. I don't know what she is using these for". Similarly, Kobi in Ghana did data-entry work for an American client by sorting 5,000 questions into different subject categories. He said, "It is like a high school website where students post questions and then they get tutors to answer for them. I think I was doing some kind of back-end work, I'm not too sure what I was doing, but I know that there were questions that people needed to answer". The fact that workers can usually only obtain the type of information employers want to release about themselves makes it hard for these workers to upgrade into new job types.

Workers we spoke to told us that they would often find that the person hiring them on a platform is actually an intermediary working for a client who is located elsewhere (a finding replicated in Graham et al. [2017a]). Since digital work can be transferred easily from one location to the other, multiple levels of intermediation can take place, which can effectively obscure knowledge about the source client. This inhibits the ability of workers to take action against their clients in the events of threats of unfair dismissal and non-payment of wages. Since workers and clients are usually separated by large distances and often based in different legal and regulatory landscapes/setting, it is hard for workers to imagine how they would hold clients to account through courts.

Workers are ultimately given just enough information about clients in order to allow transactions to take place. But the fact that workers often cannot see much about the production networks that they are embedded into, or learn much about their clients, limits their bargaining power. For instance, a worker who does not even know what industry they are working in would have a hard time offering knowledge they learnt on the job to other potential employers.

3.1.5. Offers to buy and sell labor transpire

The four previous characteristics all allow a planetary labor market to be brought into being,[8] supporting offers to buy and sell labor across the world. According to Horton et al. (2017), 90% of transactions on Upwork.com are across international borders.

The offers that occur in this international market are characterized by a huge power imbalance between workers and clients. The high-level of individualization and commoditization of labor power, the planetary scale of the labor supply for the platform work and an intense competition between workers means that workers are both left to fend for themselves and compete against one another. The international nature of the transactions that occur leave many workers with an understanding that local labor regulations are of little use in protecting them against some of the worst problems they experience in the platform economy. Some workers we spoke to in places as varied as Kenya, Uganda and Nigeria earned USD1 per hour or less for some of their jobs. However, it was generally understood that not only do their clients have no sense of what a local minimum wage is in their jurisdiction, but also that such considerations would be unenforceable due to the contract types and the fact that local regulators would have little power over foreign clients. Not only are wages often bad, but so too are working conditions. Zain in Ghana explained: "Yes, there's been days that I've stayed up for two days of no sleep at all, not even 30 minutes of sleep because there's a project, I have to get it done and the pay is shit, but you have to get it done". Again, the international nature of the market leaves the sorts of relationships that are set up to encourage Zain to work for days without sleep entirely unregulated.

Workers' lack of bargaining power is also exemplified by the fact that clients can decide to end the contract at any time (without payment of wages), if they are not satisfied with the work submitted by workers. This is particularly pronounced for workers doing fixed-rate or "piecework". Several article writers (mostly paid per article) we interviewed reported that they did not get paid for their article after they submitted their work to their employers because the employers did not like it. One platform worker (editor, web research and data entry) in Ghana, Quinn, edited a book for an American client and clocked 40 hours for a total of USD400, but only got paid USD200 and, despite several complaints made to the platform, she was not paid at the time of the interview.

The lack of structural power for platform workers both manifests in, and is manifested through, the inability of workers to collectively bargain. Offers to buy and sell labor happen at the individual level, making it hard or impossible for workers to take advantage of collective bargaining agreements, or use their collective power to withdraw labor. Dabiku in Kenya was of the opinion that collective bargaining through unions is a good idea. But he remarked, "locally it is unfortunate guys do not trust each other that is one thing. So actually, even [setting up] meetings is always a problem". He was also of the view that while unions would be of help for local work, nothing can be done when clients are located in the US or Canada.

As the millions of offers to buy and sell labor transpire on digital labor platforms, the asymmetries of information and pre-existing asymmetries of power are put into practice by clients. In other words, while people buy and sell labor in a spatially-unbounded way, there is a scalar mismatch in reach, mobility and information that severely limits the bargaining power of workers.

3.2. Planetary labor futures

Following the call by Strauss (2018, p. 626) for "sustained critical attention to what is distinctly spatial about the processes that are of interest, how place matters to those processes, and how scale is relationally constructed . . . and experienced in the production of precarious work situations", this study has explored the ways in which inequality is structured into online labor markets when they are scaled up to the planetary level. We have seen, following Fevre (1992), that employers can learn about workers, workers learn about jobs, both parties learn about each other and transactions take place in ways that seemingly ignore some of the traditional limitations of time and distance. Employers and workers, through the affordances of digital technologies, can seek each out on a genuinely world-spanning scale, escaping some of the constraints that previously bound them exclusively to their local labor markets. Most importantly, many previously bounded labor markets were both transactionally and discursively insulated from a global reserve army of labor and the downward pressure on wages and working conditions that it brings about (Huws, 2003).[9] A market that is planetary in scale will cease to have any of those brakes on the erosion of working conditions. Yet, while all of these interactions occur between economic actors in different parts of the world from one another, what we see is not just Hardy's hiring fair scaled-up to a global-level or scaled-down onto the head of a pin. Instead of seeing the space of the labor market through a Euclidean lens in which geography is a pre-existing canvas on which economic relationships can be formed, the spaces of labor markets are instead relational and emergent.

It is this understanding of space that we seek to bring to discussions about digital work. The discussion in this chapter should encourage us to move away from thinking about labor markets as bounded spaces that you could draw on a map. In a planetary labor market, everything does not happen everywhere. But, key spatial constraints (e.g., the need for commuting, to leave the house and to obtain visas and permits) can be circumvented. This forces us not to imagine away the always-existing economic geographies of work, but to ask questions about how they will shape and be shaped by the potentials for planetary-scale interactions.

Thomas Friedman (2005, p. 110) famously pointed to a globalized world that would allow for "the sharing of knowledge and work – in real time, without regard to geography, distance, or, in the near future, even language". But, as much as some firms and clients might want it to, a planetary market does not do away with geography; it rather exists to take advantage of it. Platforms use uneven geographies to facilitate labor arbitrage, cross-border competition and are able to foster what Peck (2017, p. 42) refers to as an "offshore consciousness". To be clear, references to local labor markets, national labor markets and planetary labor markets should never be made to ignore the myriad ways that those labor markets are brought into being by multi-scalar exogenous factors which, in turn, lead them to be socially and spatially segmented and fragmented. The spatial and scalar prefixes (urban-, local-, national-, etc.) that we add to labor markets instead are intended to signify enablers and constraints that serve to cluster coming-togethers of employers and workers within particular economic geographies. These enablers and constraints can be technological barriers (e.g. transportation costs and the availability of broadband), political (e.g., the availability of visas and work permits), social (e.g., availability of skills and language fluencies), economic (e.g., local reserve wages) and spatial (e.g., work and workers that inherently have to be in particular places).

Underlying material economic geographies of workers and clients are therefore never fully transcended, work is never fully commoditized, and there remain national and regional practices and institutions which govern the purchase, sale and pricing of labor on platforms. But none of those national and regional practices determine the shape of the market as a whole. This is not to say that labor markets in non-digital contexts do not have similar ways of empowering and disempowering different groups. The material architectures, norms, laws and relationships in traditional labor markets all bring particular power dynamics into being. However, what is different in the digital context is that co-presence and the transitory proximity that platforms bring into being is illusory. While workers can, in theory, connect from anywhere, they lose the ability to control a key part of their agency that they otherwise have in any other context: their control over space, and their ability to bring into being labor geographies that are at least in part on their own terms (see e.g. Herod [2001]). Because employers and workers have significantly different abilities to control space, the planetary labor market is a context that serves to further undermine the structural power of labor vis-à-vis that of capital.

This manifests in six key ways. First, mass global connectivity is bringing onstream a massive *oversupply of labor power, mainly from lower-income segments of the world population.* There are far fewer digital jobs than there are people able and willing to do them. The "elemental rationale" of offshoring has always been to cut and suppress costs (Peck, 2017, p. 10), it should therefore come as no surprises that online outsourcing continues the trend. Second, many workers seeking jobs in a planetary labor market are *replaceable and interchangeable.* This is not just due to the oversupply of labor power, but also to the fact that production networks can be footloose while workers are tethered to place. Third, workers mostly interact as *competitors rather than collaborators*. For digital workers, this situation arises primarily because there are few physical sites at which to assemble with coworkers and because the highly commodified nature of their jobs can lead to competition on price above all else. Digital platforms deliberately limit the amount of horizontal information that workers can glean about one another, and the distributed nature of work means that workers have few opportunities to engage in collective action afforded by spatial proximity.

Fourth, there is a *lack of transparency*. Although workers can use digital tools to find jobs on the other side of the planet, the networks and platforms used to mediate those jobs can also conceal much about the nature of those activities. In other words, workers often know little about the production networks that they are embedded into and are offered few opportunities to economically upgrade skills or value-chain positions. Fifth, there is a relative *lack of agency* amongst workers to have their voice in, or shape, their labor conditions. The affordances of most types of digital work tend to be closed to workers – in part because workers rarely have any stake or control in the physical or digital means of production in the digital economy. If we extend spatial metaphors to online labor platforms, they are not public markets; they are rather private spaces. Finally, *workers tend not to be protected by labor* laws. Because the labor market extends well beyond any individual jurisdiction's ability to regulate it, self-regulation tends to be carried out by platforms and clients. Labor laws that exist to protect workers are ignored in some cases, sometimes even willfully.

These issues amplify each other, and all serve to undermine the structural power of workers. And they are all possible because of the specific designs of planetary labor markets that use space against workers. The issues outlined in this chapter paint a picture of a grim future for the balance of power between labor and capital that is likely unsatisfactory to anyone who does not run an outsourcing company. But what can be done if we want to envision and see more equitable outcomes? The solution cannot simply be to turn our backs on innovations in information and communication technologies. We can likely never go back to a world only characterized by local labor markets. Despite the concerns presented here,[10] the digitally mediated relationships presented in this article are far from inevitable. There are two primary reasons why we believe this to be the case. First, all of the digital and virtual infrastructure deployed to bring a planetary labor market into being ultimately depends on material infrastructures, organizations grounded in physical places, and real-world regulation. Current configurations of infrastructure and regulation are thus far from inevitable. Interventions such as platform cooperatives, attempts at cross-border regulation and horizontal organizing amongst workers are more effectively covered elsewhere (Graham and Anwar, 2018; Graham and Woodcock, 2018; Irani and Silberman, 2013; Wood et al., 2018), but the simple point here is that by understanding the spatialities and temporalities of contemporary labor markets, we can better shape them. We can no longer think about labor markets for digital work as being simple shapes on a map.

Second, the ways in which technologies are deployed to produce specific time-spaces and not others do not dictate how we necessarily use, produce or jump over geographies. Indeed, code and algorithms coproduce spaces that are often malleable and hackable (Zook and Graham, 2018). Workers and their advocates have thus far certainly found ways of using technological infrastructures in unintended ways that work in their favor (Wood et al., 2018), and we will need more of this if workers are to exert any significant amount of agency in the labor process. But to build or perform alternatives, we again need to base our efforts on realistic understandings of the relationships between economic actors, technologies and the spaces they bring into being. In Hardy's hiring fair, we would not expect workers to be able to collectively bargain or form a picket if they misread the opportunities and constraints provided by their spatial proximities.

4. Conclusion

This chapter ultimately builds on Doreen Massey's (1994) "global sense of place" – a sense of how distant people, places and processes are always inherently enrolled into any local relationships. We do that by showing that we can use five characteristics of labor markets to think about how online labor platforms create labor markets that are planetary in scope. It has also shown some of the ways in which constructing a planetary labor market changes the balance of power between labor and capital. We have demonstrated not that geography has been eliminated, nor that places have been made irrelevant. No virtual space has been created allowing employers and workers to coexist beyond the confines of the physical realm. Rather what has happened is that digital technologies have been deployed in order to bring into being a labor market that can operate at a planetary scale, and has particular affordances and limitations that rarely bolster both the structural and associational power of workers. Digital technologies that underpin online labor markets help clients operate unboundedly and trans-spatially, and allow them to reconfigure the geography of their production networks for almost zero cost. Workers meanwhile can sell their labor power globally, but still are tethered to the locales in which they go to bed every night.

Acknowledgements

The authors would like thank David Sutcliffe, Jamie Woodcock and Alex Wood for their extensive feedback on earlier drafts and anonymous referees for their comments. We are grateful to the European Research Council (ERC Grant Agreement number 335716), the ESRC (ES/S00081X/1) and The Alan Turing Institute (under the EPSRC grant EP/N510129/1) for funding our research on digital work.

Notes

1 This chapter was originally published as "Graham, M., and Anwar, M. A. 2019. The Global Gig Economy: Towards a Planetary Labour Market? *First Monday*. 24(4). doi.org/10.5210/fm.v24i4.9913". Permission for reprint has been granted by the copyright holder. Thank you to *First Monday* for allowing us to republish this piece.

2 In addition to numerous models that have been created to predict how many jobs artificial intelligence will destroy, it is estimated that up to one-third of all jobs in the United States are offshorable (Peck, 2017).

3 David Harvey (1990, p. 19) noted, "labour power has to go home every night". Capital therefore is always able to take advantage of its relative mobility compared to labor.

4 Large platforms began to emerge after 2008 as a new business model that controlled information in bottlenecks in between digitally-mediated economic, social and political activities. According to Srnicek (2016, p. 48) "Platforms, in sum, are a new type

of firm; they are characterized by providing the infrastructure to intermediate between different user groups, by displaying monopoly tendencies driven by network effects, by employing cross-subsidization to draw in different user groups and by having a designed core architecture that governs the interaction possibilities".

5 The precursor to segmented theory of labor market is the dual labor market theory which describes two distinct sectors in economy that do not have mobility between them. The primary sector is characterized by strong wages, benefits and security, while the secondary sector is characterized by jobs that often require little training, have high-turnover, low wages and insecure contracts (Doeringer and Piore, 1985). The dualist model failed to incorporate the processes of social reproduction and the role of the state in correcting market failures, contributing to workforce reproduction and also regulatory functions, such as enforcement of employment contracts (Peck, 1996).

6 Too much should not be read into the specific oversupply percentages. On one hand, these numbers could be overestimates because it is possible that many people create profiles without having any intention to search for jobs. On the other, it is possible that they are underestimates, because platforms have an interest in keeping pay just above the reserve wage in a variety of industries, there have been time-limited efforts to limit worker sign-ups from some countries.

7 This is not to say that opportunities for workers to learn new skills and earn high income from platform jobs are not present at all. But the point is those who succeed in doing so often come from prosperous family backgrounds and with previous training and education. For example, white South Africans are much more likely to succeed and earn money on platforms than other groups in the country, largely due to their better socio-material conditions. There are 104 workers on Upwork.com in South Africa (on October 25, 2018), who have completed 1000 hours and earned USD10,000 worth of work and all but seven are white.

8 Furthermore, the nature of many digital work platforms, with work done remotely, collaboratively and in real time, means that the "workplace", rather than the just the labor market, could be considered to be planetary (bearing in mind similar geographic caveats discussed in this chapter). This, however, is a topic for another paper.

9 This is not to claim that local labor markets ever reach any sort of equilibrium. Indeed, much important scholarship has taken place refuting such ideas and instead arguing that labor markets are locally constituted (Hanson and Pratt, 1992; Harvey, 1989; Peck, 1989). It is nonetheless clear than many bounded labor markets have been able to avoid an erosion of working conditions through the relative scarcity of labor power and better regulatory frameworks instituted by states.

10 Here it is worth bearing in mind that the planetary scale of the market allows many workers to access jobs and income that they simply would not otherwise have access to. The biggest problem for many potential workers is not that the labor market is full of bad jobs, but rather that they are excluded from those jobs in the first place (as we outline in the section on oversupply). However, the fact that bad jobs are better than no jobs should not stop us from interrogating the conditions that bring these jobs into being.

References

Barker, K. and Christensen, K., 1998. *Contingent Work: American Employment Relations in Transition*. Ithaca, NY: ILR Press.

Barley, S. and Kunda, G., 2006. *Gurus, Hired Guns, and Warm Bodies: Itinerant Experts in a Knowledge Economy*. Princeton, NJ: Princeton University Press.

Bonacich, E., 1972. A theory of ethnic antagonism: The split labor market. *American Sociological Review*, 37(5), pp. 547–559.

Bratton, B., 2016. *The Stack: On Software and Sovereignty*. Cambridge, MA: MIT Press.

Brown, E.C., 1970. Continuity and change in the Soviet labor market. *ILR Review*, 23(2), pp. 171–190.

Casilli, A. and Posada, J., 2019. The platformization of society and its discontents. In: M. Graham and W.H. Dutton, eds., *Society and the Internet, How Networks of Information and Communication Are Changing Our Lives*, 2nd ed. Oxford, UK: Oxford University Press, pp. 293–306.

Doeringer, P. and Piore, M., 1985. *Internal Labor Markets and Manpower Analysis*. Armonk, NY: M.E. Sharpe.

Fevre, R., 1992. *The Sociology of Labour Markets*. New York, NY: Harvester Wheatsheaf.

Freelancer.com, 2019. *Dreaming Bigger*. [online] Available at: <www.freelancer.com/25> [Accessed 16 Sep. 2019].

Friedman, T.L., 2005. *The World is Flat: A Brief History of the Twenty-first Century*. New York, NY: Farrar, Straus and Giroux.

Graham, M. and Anwar, M.A., 2018. Two models for a fairer sharing economy. In: N.M. Davidson, M. Finck and J.J. Infranca, eds., *The Cambridge Handbook of the Law of the Sharing Economy*. Cambridge, UK: Cambridge University Press, pp. 328–340.

Graham, M., Hjorth, I. and Lehdonvirta, V., 2017a. Digital labour and development: Impacts of global digital labour platforms and the gig economy on worker livelihoods. *Transfer: European Review of Labour and Research*, 23(2), pp. 135–162.

Graham, M., Lehdonvirta, V., Wood, A.J., Barnard, H., Hjorth, I. and Simon, D.P., 2017b. *The Risks and Rewards of Online Gig Work At the Global Margins*. [online] Oxford Internet Institute, University of Oxford. Available at: <www.oii.ox.ac.uk/blog/new-report-the-risks-and-rewards-of-online-gig-work-at-the-global-margins> [Accessed 16 Sep. 2019].

Graham, M. and Woodcock, J., 2018. Towards a fairer platform economy: Introducing the fairwork foundation. *Alternate Routes: A Journal of Critical Social Research*, 29, pp. 242–253.

Graham, S., 1998. The end of geography or the explosion of place?: Conceptualizing space, place and information technology. *Progress in Human Geography*, 22(2), pp. 165–185.

Grimshaw, D., Fagan, C., Hebson, G., Tavora, I. (Eds.), 2017. *Making work more equal: A new labour market segmentation approach*. Manchester, UK: Manchester University Press.

Hanson, S. and Pratt, G., 1992. Dynamic dependencies: A geographic investigation of local labor markets. *Economic Geography*, 68(4), pp. 373–405.

Harvey, D., 1989. *The Urban Experience*. Baltimore, MD: Johns Hopkins University Press.

Harvey, D., 1990. *The Condition of Postmodernity: An Enquiry into the Origins of Cultural Change*. Cambridge, MA: Wiley-Blackwell.

Harvey, D., 2001. Globalization and the 'spatial fix'. *Geographische Revue*, 2, pp. 23–30.

Harvey, D., 2011. *The Enigma of Capital and the Crises of Capitalism*. Updated pb ed. London, UK: Profile Books.

Heeks, R., 2017. *Digital Economy and Digital Labour Terminology: Making Sense of the "Gig Economy", "Online Labour", "Crowd Work", "Microwork", "Platform Labour", Etc*. GDI Development Informatics Working Papers, No. 70. Manchester, UK.

Herod, A., 2001. *Labor Geographies: Workers and the Landscapes of Capitalism*. New York, NY: The Guilford Press.

Horton, J., Kerr, W.R. and Stanton, C., 2017. *Digital Labor Markets and Global Talent Flows*. NBER Working Paper Series, No. 23398. Cambridge, MA.

Hudson, R., 2001. *Producing Places*. New York, NY: The Guilford Press.

Hunt, E.S., 2002. *The Medieval Super-Companies: A Study of the Peruzzi Company of Florence*. Updated pb ed. Cambridge, UK: Cambridge University Press.

Huws, U., 2003. *The Making of a Cybertariat: Virtual Work in a Real World.* New York, NY: Monthly Review Press.

Irani, L.C. and Silberman, M.S., 2013. Turkopticon: Interrupting worker invisibility in amazon mechanical turk. In: *CHI '13 Proceedings of the SIGCHI Conference on Human Factors in Computing Systems, April 27 – May 2, 2013, Paris, France*. New York, NY: Association for Computing Machinery, pp. 611–620.

Jones, A., 2008. The rise of global work. *Transactions of the Institute of British Geographers*, 33(1), pp. 12–26.

Kalleberg, A.L. and Sorensen, A.B., 1979. The sociology of labor markets. *Annual Review of Sociology*, 5(1), pp. 351–379.

Kuek, S.C., Paradi-Guilford, C.M., Fayomi, T., Imaizumi, S. and Ipeirotis, P., 2015. *The Global Opportunity in Online Outsourcing*. Washington, DC: World Bank.

Marx, K., 1973. *Grundrisse: Foundations of the Critique of Political Economy*. London, UK: Penguin Classics.

Massey, D., 1993. Power geometry and a progressive sense of place. In: J. Bird, B. Curtis, T. Putnam and L. Tickner, eds., *Mapping the Futures: Local Cultures, Global Change*. London, UK: Routledge, pp. 60–70.

Massey, D., 1994. *Space, Place, and Gender*. Minnesota, MN: University of Minnesota Press.

Massey, D., 2005. *For Space*. London, UK: Sage Publications.

Ojanperä, S., 2019. *Mapping the Availability of Online Labour in 2019*. [online] Oxford Internet Institute, University of Oxford. Available at: <https://geonet.oii.ox.ac.uk/blog/mapping-the-availability-of-online-labour-in-2019> [Accessed 16 Sep. 2019].

Peck, J.A., 1989. Reconceptualizing the local labour market: Space, segmentation and the state. *Progress in Human Geography*, 13(1), pp. 42–61.

Peck, J.A., 1996. *Work-place: The Social Regulation of Labor Markets*. New York, NY: The Guilford Press.

Peck, J.A., 2017. *Offshore: Exploring the Worlds of Global Outsourcing*. Oxford, UK: Oxford University Press.

Reich, M., Gordon, D.M. and Edwards, R.C., 1973. A theory of labor market segmentation. *The American Economic Review*, 63(2), pp. 359–365.

Ruby, J., 2019. *Appen Completes Acquisition of Figure Eight and Achieves Critical Integration Milestone*. [online] Appen.com. Available at: <https://appen.com/blog/appen-completes-acquisition-of-figure-eight-and-achieves-critical-integration-milestone> [Accessed 16 Sep. 2019].

Silver, B.J., 2003. *Forces of Labor: Workers' Movements and Globalization Since 1870*. Cambridge, UK: Cambridge University Press.

Srnicek, N., 2016. *Platform Capitalism*. Malden, MA: Polity Press.

Standing, G., 2016. *The Corruption of Capitalism: Why Rentiers Thrive and Work Does Not Pay*. London, UK: Biteback Publishing.

Strauss, K., 2018. Labour geography 1: Towards a geography of precarity? *Progress in Human Geography*, 42(4), pp. 622–630.

Upwork, 2019. *How It Works: What's the Difference Between Finding Clients Online, Versus Locally?* [online] FAQ. Available at: <www.upwork.com/i/how-it-works/faq> [Accessed 16 Sep. 2019].

Wood, A.J., Graham, M., Lehdonvirta, V. and Hjorth, I., 2019. Good gig, bad gig: Autonomy and algorithmic control in the global Gig economy. *Work, Employment and Society*, 33(1), pp. 56–75.

Wood, A.J., Lehdonvirta, V. and Graham, M., 2018. Workers of the Internet unite? Online freelancer organisation among remote gig economy workers in six Asian and African countries. *New Technology, Work and Employment*, 33(2), pp. 95–112.

Zook, M. and Graham, M., 2018. Hacking code/space: Confounding the code of global capitalism. *Transactions of the Institute of British Geographers*, 43(3), pp. 390–404.

14 Identifying the digital gender divide

How digitalization may affect the future working conditions for women

Anthony Larsson and Yamit Viitaoja

1. Introduction

The digital transformation is rapidly changing how we conduct business and working conditions by and large (Demirkan, Spohrer and Welser, 2016). Technologies that underlie computers, robots and smart equipment are becoming increasingly more advanced, and in turn, transforming organizations much faster nowadays than in the past. Thus, it is easy to gain competitive advantage by being well-versed in digital innovation and transformation. This has led to an increase in the need for future employees to become tech-savvy, as the IT professions can be expected to grow significantly worldwide in the years to come (Vincent, 2017; Kahn, 2017; Winick, 2018). For instance, in the US alone, IT jobs are expected to have grown by 22% from the 2010s through the 2020s (Thibodeau, 2012). However, in spite of significant worldwide growth, women are still under-represented in IT and the tech industry overall (Colby, 2017; Trauth, Nielsen and von Hellens, 2003; Trauth, 2006; Bernhardt, 2014). This, in turn, has dire consequences on women entrepreneurship by and large. For instance, research shows that women hold only 5.5% of all commercialized patents in the United States, while only representing 12% of all US innovators (Hunt et al., 2013; Nager et al., 2016). Also, only 26% of the computing specialists in the Silicon Valley area were made up by women in 2013, a figure that has remained roughly dormant since 1960, with an even bleaker outlook in the engineering profession, where women constituted just 12% of the force in 2013 (Guynn, 2015).

To this end, the average US-born male is roughly nine times more likely to contribute to an innovation as opposed to the average female. Perhaps surprisingly is that in spite of this modest level of female innovators, the US might still outperform Europe. This is even the case in purportedly "progressive" and "liberal" countries that tend to proudly avail themselves of providing "equal opportunities". For instance, Sweden, which is commonly known as being among the most gender equal citizens in the world and boasting the strongest views in favor of gender equality, scored a perfect 100 points in 2019 (the highest possible score) in the World Bank Group's study "Women, Business and the Law index score" (together with Belgium, Denmark, France, Latvia and Luxembourg) (World Bank Group, 2019). However, these figures only tell part of the story and there is still a long way to go. For instance, women hold only six percent of all commercialized patents, i.e., a half percentage point more than the United States (Ädel, 2016). This is in spite of the fact that recent figures from the WIPO has shown an all-time high number of women innovators internationally (WIPO, 2018; Von Hall, 2018). Moreover, in the areas of start-ups and venture capital (VC), the situation is just as, if not even more, dire. Approximately 94–99% of all private and public Swedish VC is awarded to male founders' teams (Dahlgren, Lundeteg and Nord, 2017; Billing, 2015, 2018; Olsson Jeffery, 2019). This is not for a lack of interest, for instance, in 2016, more than half of the start-ups awarded cash by the county's leading incubator were run by women (Savage, 2016). Still, the amount of means awarded through this channel does little to stem the overall disparity and the uneven distribution is apparent also in other areas. For instance, in tech, roughly 82% of all tech cash is given to male entrepreneurs in 2016 (The Local, 2016). A subsequent, more detailed, study showed that this figure had climbed two percentage points to 84% in 2018 with 15% being awarded to mixed teams, and a mere 1% being awarded to teams consisting exclusively of females (Olsson Jeffery, 2019; Wisterberg, 2019).

More so, in Sweden, the participation of women working in the industry is merely 23%, far lower than the EU average at approximately 30% (Edwards, 2017; Roden, 2016; Eurostat, 2018; Fogelqvist, 2016). In fact, Sweden has the third lowest number of women working in the industry (trailing the Netherlands and the United Kingdom respectively) (Fogelqvist, 2016). The situation is somewhat less somber in the information and communication technology (ICT)[1] industry, which has seen a slight increase to 29% women employees in 2017, from 28% the preceding year (IT&Telekomföretagen, 2017; Wisterberg, 2017).

In an unpublished case study by the Swedish women career network International Women Group (IWG Group), it was found that out of 26 women joining a custom-designed 12-month mentorship program called "Empowered Mentoring Program" (EMP), 13 attendees were executive leaders in major Swedish businesses with the remaining 13 women joining as "mentees", i.e., female participants with the aspiration of becoming a future business leader (IWG Group, 2018). The study showed that among the mentees, only one of the members had their program fee (at roughly USD1100) paid for by their employer organization. The study found that there were many different reasons as to why these companies declined to pay for the tuition of the women employees. These arguments included age-related reasons, with the mentee being deemed too young (the ages of the attendees varied between ages 21–35). Other arguments were that that the mentees were not deemed to be experienced enough, or that the company felt no need of establishing a future executive role in their company for that mentee to fill for the foreseeable future. Still, a follow-up evaluation a year after the conclusion of the mentorship program showed that the participating mentees, all active in digital professions, tended to change their career and employer in favor of a better position elsewhere, and in some cases some even started their own company. The mentees posited that attending a mentorship program provided them with a new personal network and instilled them with more confidence, while also boosting their knowledge in digitalization. A recurring contention during the follow-up was that the mentees expressed disappointment with their former employers for not supporting them in their ambition of acquiring leadership insights and skills, which ultimately resulted in them leaving their former workplaces. Hence, it is easy to assume that companies need only to support leadership training programs in order to remedy the problem of the digital gender divide. It is true that this would undoubtedly help women already with an interest in the tech industry, and it is also true that supportive employers are essential to any sense of wellbeing at any workplace. However, a lingering problem is the fact that many women eschew not only the tech industry, but technology at large.

One may contend that the reason why fewer women access and use ICT is that it is not infrequently a direct consequence of their unfavorable conditions in regards to employment, education and wage. Studies have shown that when controlling for these variables, women actually tend to be more active users of digital tools than men (Hilbert, 2011; D'Mello, 2006). The "digital gender divide" may thus become an opportunity. That is, given that women are proficient in, and enjoy using ICT, and that the digital transformation of society provides tools that can improve the human living conditions, ICT represents a real and present opportunity to overturn the challenges of gender inequalities. This includes factors such as improving access to secure employment, wage, education and health services.

Thus, the premise of this study is to provide an analytical commentary based on available research literature and on the authors' "best practice" insights and reflections. In doing so, the aim of this chapter seeks to understand how digitalization and the digital transformation may affect the digital gender divide of business investments and women's overall role in a future digitalized labor market in the Western hemisphere.

2. Discussion

2.1. The digital gender divide

The concept of a "digital divide" refers to the gulf between those who have ready access to computers/smartphones (or equivalent) and the internet, and those who do not (Ragnedda and Muschert, 2013; Selwyn, 2004; Rogers, 2001). Adding to this, there is also a notion of a "digital gender divide", which purports that there is a discrepancy between women's and men's access to IT technology (Cooper, 2006; Cooper and Weaver, 2003; Hilbert, 2011).

According to a 2018 OECD report, there are many root causes as to why there is gender-based digital exclusion (Borgonovi et al., 2018). The most prominent reasons include restricted access to digital tools in the form of affordability, lack of proficiency and/or education and deficiencies in technological literacy in addition to inherent gender biases and socio-cultural norms. While connectivity by and large is a problem for the developing parts of the world, the aforementioned factors affect women all across the world. Worldwide, there are approximately 327 million more men than women who possess a smartphone and can access mobile internet. On average, women are approximately 26% less likely than men to possess a smartphone. In South Asia and Africa these proportions are particularly prominent as they stand at 70% and 34%, respectively (Borgonovi et al., 2018).

2.2. The dwindling number of ICT-education

In Europe, there is currently a decreasing percentage of Europeans with ICT-related education (irrespective of gender), while there is (paradoxically) a growing demand of ICT specialists and digital profiles (Quirós et al., 2018; Wever, 2012). This is a problem also in the US, Asia and other regions as well (Umoh, 2017; Nkhoma, Dang and Lu, 2012). Naturally, there are many reasons the interest in pursuing an ICT-related education is falling. According to Selwyn (2003b), there are a number of factors that have constituted the main reasons as to why ICT has failed in popularity over the past few decades.

2.2.1. Limited access to material resources and economic restrictions

Undoubtedly, access to ICT is contingent on there being available resources in the form of money and materials (Murdock, Hartmann and Gray, 1995). Needless to say, there will always be differences in the ability of the individual ability to purchase and/or lease IT equipment, along with the relevant accompanying services, such as internet, etc. While most people in the Western hemisphere will own some kind of electronic device, the quality thereof may vary. In the event that such a device of adequate standard is not available, the universities and public libraries will often provide some degree of accessibility to IT services available to all at little or no cost (Liff and Steward, 2001; Holley, 2013). However, while universities and libraries will provide access to these facilities and provide people with at least a basic standard of equipment for use, there are indications that these facilities merely reinforce the existing patterns of the students' ICT use in "private" settings (Selwyn, 2003b; Henderson, Selwyn and Aston, 2017; Ramalingam and Kar, 2014). That is to say, universities and libraries do not tend to be as effective in actually widening the level of ICT usage to the category of individuals who were not already using ICT but rather helps increase the levels of use among those individuals who are already using it. In that way, there is a distinction between the de facto access to ICT, and the effectiveness it carries (Lim, 2002; Wilson, 2000). More aptly formulated by Rogers (2001, p. 105) publicly available ICT facilities "can provide the public access function, but they need adequate computer facilities, adequate access time per user and help desk facilities which were not [always] available".

2.2.2. Cognitive disparities

Although access to ICT tools is a prerequisite to using them, the difference between using them and not using them is not solely a matter of a so-called "equipment gap" between the "have" and the "have-nots", as the "equipment gap" has been steadily shrinking in the past few decades (Krieg, 1995; Ottestad and Quale, 2009).

There has been some scientific debate regarding people's cognitive proclivity to use technology (Bain and Rice, 2006; Mitzner et al., 2010; Colley and Comber, 2003; Heflin, Shewmaker and Nguyen, 2017; Cai, Fan and Du, 2017). Naturally, possessing an interest and knowledge in how to use a certain type of technological equipment are obvious factors, albeit underpinned by the individual's experience of, and attitudes toward, using technological devices (Durndell, Macleod and Siann, 1987; Kang and Yoon, 2008). Naturally, one could argue that that technological experience and technological disposition are inherently intertwined, as the more a certain type of behavior is carried out, the more one's attitude about said behavior is formed and reinforced. For instance, Todman and Monaghan (1994) contended that if one has a positive initial experiences with, for instance, a computer, one is prone to display more positive affective attitudes toward computer interaction in the future.

To this end, psychologists have highlighted a range of cognitive and affective factors that act as important determinants that affect a person's interaction with technology. This includes perceived ease of use, perceived behavioral control, self-efficacy, and perceived ability (He, Chen and Kitkuakul, 2018; Ajzen, 2002; Wu, Hsia and Tennyson, 2011). Though it is important to emphasize that there is indeed a complex relationship between attitudes and casual factors and the way these are affected by other psycho-social factors. As in the case of computers, there is historical linkage between mathematical ability and level of interest toward computers (Schumacher, Morahan-Martin and Olinsky, 1993; Selwyn, 1999; Miller and Varma, 1994; Shashaani, 1995). Yet, in other contexts, there is research to indicate that attitudes are significantly correlated with people's creativity level (Kuśpit, 2016; Offir, Golub and Cohen-Fridel, 1993), learning and social self-image (Katz, 1993; Demo, 1985; Blascovich and Tomaka, 1991) and locus of control (Tomer and Eliason, 2000; Woodrow, 1990).

2.2.3. Technophobia

A different psychological component concerns various forms of "phobia" of various digital objects and technology, commonly referred to as "technophobia". This signifies individuals who harbor negative opinions and/or anxiety toward the use of technology and consequently tend to use it less often, even when it poses no real or immediate threat (Brosnan, 1998; Nimrod, 2018; Selwyn, 2003b). In this way, technophobia serves to obfuscate the individual's perception of a particular type of technology inasmuch that they will believe that it is "not for them".

Much of the research carried out in this area has focused on the avoidance of ICT in various settings, such as one's workplace or home (Khasawneh, 2018; Shashaani, 1993; Colley, Gale and Harris, 1994). The premise has often been that the aversion toward ICTs is brought on by either ignorance, lack of experience or an apparent deep-rooted adverse reaction stemming from the belief that the ICT is somehow posing a threat to the user (Rosen, Sears and Weil, 1993; Felt, Schumann and Schwarz, 2015; Brosnan and Davidson, 1994; Holzer, 2015; Oliveira and Jerónimo, 2016). This approach has historically often departed from the notion that an individual's aversion to ICT constitutes a transitory frame of mind that can somehow be "conquered" or at least "treated" (Rosen, Sears and Weil, 1993; Kennewell, 1992; Richard, 1997). Research from the late twentieth century would often contend that technophobia differed little from the anxieties surrounding the introduction of any kind of technology throughout history and that anxieties regarding the use of ICT would erode when they became more commonplace (Selwyn, 2003b). However, research indicates that technophobia is still just as relevant and present in this day and age as ICT usage tends to create new situations in which not even teachers/instructors always possess the needed skills or credentials to deal with them optimally, which makes it difficult for people to develop a sense of in-depth competence (Fernández-Cruz and Fernández-Díaz, 2016; Revilla Muñoz et al., 2017). Moreover, it is known that technophobia is a particularly recurrent and prominent phenomenon among senior individuals (Hou, Wu and Harrell, 2017; Rosen and Weil, 1995; Johnson, 2012). However, technophobia may also correlate with other individual characteristics, although there is contention as to whether or not gender is such a factor as results have varied (Trauth, Quesenberry and Huang, 2010; Hogan, 2009; dos Santos and Santana, 2018).

2.2.4. Ideological refusal

Another reason why some people make less use of ICT is because of ideological reasons (Van Dijk, 2012). That is to say "want nots" rather than "have nots" who refuse to engage with ICT for ideological reasons despite being able to do so in practice. This practice often carries a non-conformist angle that purportedly conveys an act of opposition against technology, more often than not by individuals who can afford to do so (Hesselberth, 2018; Selwyn, 2003a). As expressed by Bauer (1995, p. 19): "The resistance against information technology is mostly local and a matter of 'intellectuals'; it is mainly informal, individual and passive, such as a refusal to work with computers".

There are, of course, many different reasons for why one would take on such an ideological stance, but Norman (1993) theorized that there is an apparent clash between the "hard" nature of machines and the "soft" nature of the humans that are expected to use them. That is to say, humans are inherently good at the aspects that are intrinsic to human nature, such as creativity, invention, empathy and emotion. People adhering to these views while eschewing the use of ICT can therefore be regarded as deliberate non-users of technology (Selwyn, 2003b).

2.2.5. Diffusion theory

Diffusion theory believes in a recurring "s-curve" of expansion of technology use in society (Rogers, 2003). This relates from initial groups of "early adopters" all the way through to the majority of the population, who adopt the technology at a later stage (Mohr, Sengupta and Slater, 2010). According to this theory, those who do not pick up on technological advancements, in due course, are considered "laggards", and that societal use of an innovation is hastened by its relative advantage (e.g., the degree of benefit it provides as opposed to what it purports to replace) (Rogers, 2003). The internet, for instance, is considered to possess a high degree of relative advantage, as expressed by Rogers (2001, p. 97): "Compared to postal mail, email via the internet is faster, cheaper and quicker. Compared to books or other sources of information, the World Wide Web is a more convenient means of searching for information (that is, if an individual has access to a computer and modem)".

2.2.6. Total cost of ownership

Another reason, on a business level, that might act as a deterrent toward engaging oneself in ICT is the total costs of ownership (TCO). This is particularly the case for entrepreneurs and small and medium-sized businesses (SME) (Lochner, 2005; Kirche and Srivastava, 2017). That is to say, the initial purchase cost of ICTs may be misleading, as it in fact only constitutes a smaller sum of what one ends up having to pay in the end. Associated costs, such as new software, installation, transition costs, employee training, security costs, disaster-recovery planning, ongoing support and future upgrades actually account for the largest economic drain over time (Kling, 1999; Kagan, 2018). While the cost of a computer for business (desktop or laptop) can range from approximately USD400 for a basic model with limited storage space to USD3500 for a top of the line computer, the TCO amounts to much more than this, both in actual cost and in time expenditure (Thompson, 2017). This in turn, may act as an incentive of seeking oneself to lesser technology-oriented ventures.

According to Power (2004), 20 computers costing USD1000 a piece (i.e., with a total cost of USD20,000) would after three years cost approximately USD38,240 in TCO (including the initial purchase cost) when accounting for the cost of maintenance, supplies and electricity, etc. To this end, Figure 14.1 gives a rough example of what some of the associated TCO are for entrepreneurs and SMEs (Lazar, 2016; Betts, 2004).

Benchmark	*Amount*
Average annual revenue spent by SMBs on IT expenses.	6.4%
Average sum spent by trade associations on hardware and software every year.	USD74,000
Number of IT costs occurring after the initial purchase.	80%
Annual average cost per unmanaged PC.	USD5000
Number of minutes spent each week by employees per week trying to fix PC problems or helping a coworker.	30
Average amount spent each month by firms when all IT expenses are factored.	USD700

Figure 14.1 Examples of figures included in ICT total cost of ownership (TCO) for entrepreneurs and small- and medium-sized businesses.

2.3. Digital opportunities and barriers for women on the labor market

While the aforementioned factors are inherently valid for both men and women, they tend to have a greater impact on women, given the fact that women already have historical disadvantage of taking up ICT-related jobs and educations (Quirós et al., 2018; Nsibirano, Kabonesa and Madanda, 2012). On the other hand, digitalization and the digital transformation offer a variety of opportunities for female empowerment and for a greater degree of female participation in, for instance, labor markets, financial markets and entrepreneurship. In theory, digitalization would appear to favor the female labor force, as women, on average, face lower risk of being replaced by machines, as compared to men (AlphaBeta, 2017; Hanrahan and Evlin, 2017; Simonton, 2006). This is often on account that women are more likely to work in occupations that need social, interpersonal, creative and decision-making skills (Hanrahan and Evlin, 2017). According to studies, women also tend to outperform men on most measures of educational attainment worldwide (indubitably in itself also a form of "gap", but nonetheless a topic for a different debate) (Bidwell, 2018; Bilton, 2018; Van Hek, Kraaykamp and Wolbers, 2016). In theory, possessing high levels of social skills complemented with higher educational attainment and advanced digital literacy, would account for a competitive advantage on the labor market. Moreover, digitalization would, at least in principle, make it easier for workplaces to implement a greater degree of flexibility (Ang et al., 2018). This would make it easier to combine paid work with various forms of caring responsibilities since these tasks are still generally carried out by women (OECD, 2018; Graham, 1993).

While women may benefit from such a potential increased flexibility in work, this flexibility may come at a cost of the conventional type of "fixed contract" employment, in favor of less transparent and more insecure and unclear types of arrangements, such as turning more toward a freelance-based format, with "gig-contracts", contractors, self-employed consultants, "zero-hour" contracts, internships, volunteers, etc. (UK Department for Business, 2017). This in turn, may lead to cynical and unscrupulous use of new, atypical, work arrangements that seeks to exploit low-skilled and/or low-paid labor, which in turn erodes the quality of the working conditions (OECD, 2017; UK Department for Business, 2017). This is but one example of the many existing barriers that serves to obstruct this potential advantage, preventing women from acquiring many beneficial opportunities, including employment and entrepreneurial ventures (Krieger-Boden and Sorgner, 2018).

In addition, it is not solely the potential working conditions that may pose a problem. As mentioned previously, male-dominated jobs have been (and are still) at the greatest risk of becoming automatized, and we have hitherto mostly seen this development occurring in such sectors as agriculture and manufacturing (Gordon, 2018; Fraser and Charlebois, 2016; Belton, 2016). However, in the future, automation is expected to spread to other (if not all) sectors as well, albeit to different degrees (Ford, 2015; World Economic Forum, 2018). This, in turn, includes sectors harboring professions traditionally dominated by women, such as retail trade, food and beverage services.

Moreover, jobs are expected to grow in various sectors that are traditionally female-dominated, such as business services, health, education and social services (Yate, 2019; OECD, 2017). At the same time, there are still huge gender differences in the educational system throughout the various fields of studies (OECD, 2017). This means that, if perpetuated, women run the risk of benefitting less from new business opportunities in various STEM[2]-related occupations (OECD, 2017).

2.4. *Finding a way forward*

The issue concerning the digital gender divide is indeed complex and deeply rooted, and involves many different aspects in various settings. As such, there is no universal "fix-all" solution. However, one prudent place to begin is at school. At the age of 15, the digital gender gap is not ostensible. While girls tend to underperform boys in specific digital-related activities and skills, they tend to outperform boys in other areas that are valued by employers, such as collaborative problem-solving skills (Borgonovi et al., 2018). Girls, by and large, also tend to have greater literacy skills, while boys often have better numeracy skills, although by how much continues to be a topic of contention (Lindberg et al., 2010; Henry, Lagos and Berndt, 2012; Borgonovi et al., 2018). Curiously, however, the literacy gap is bridged at the age of 27 for the average man, while men's lead in numeracy skills tend to increase with age (Stoll and Notter, 2000; Borgonovi et al., 2018; Feinstein, Vorhaus and Sabates, 2010). This affects women's opportunities in the aforementioned STEM professions and for this reason there is a pressing need to provide the possibility for adults to upgrade their skills in various areas, and for women to be given the chance to strengthen their skills in areas pertinent to STEM subjects. In this regard, the digital era provides for flexible solutions in removing barriers to adult education, but this necessitates coordination across various institutions and actors, such as education and training institutions, employers and social-policy institutions (Borgonovi et al., 2018). This is exemplified by the fact that women are less likely than men to engage in massive open online courses (MOOCs), which are often given as free courses and cover broad ranges of different topics (Zhenghao et al., 2015; Davis et al., 2017).

A major component here is the attitudinal issues. According to the OECD, only 0.5% of 15-year old girls wish to pursue a career as ICT professionals, as opposed to 5% of boys in the same age group, while twice the number of boys expect to become future engineers, scientists or architects (Borgonovi et al., 2018). In this regard, female role models in STEM are important from a signaling perspective and for encouraging girls to enter STEM at an early age. However, success at increasing the number of girls/women studying STEM does little to overcome any problems if women continue to be faced with unchanged biases in the workplace. For that reason, there is also a need for actions that seek to address the systemic bias perceptions that in turn serve to perpetuate the digital gender divide. For instance, as mentioned previously, there still exists a problem with the overall lack of representation among female entrepreneurs and innovation team. While there are some favorable developments in terms of an increase of overall female participation in patenting activities over the past decade (with ICTs increasing relatively more than other technological domains), the increase is occurring at such a slow pace that it will be the year 2080 before women are involved in half of all patented inventions within the five largest IP offices (IP5) (Borgonovi et al., 2018; Romei, 2018; Burk, 2018; Shaw and Hess, 2018). Notwithstanding, there is potential for betterment. Data shows that VC firms with at least one female partner are more than twice as likely to invest in a venture with female representation in its management team, and three times as likely to invest in female CEOs (STEMconnector, 2017; Stengel, 2017; Kerpen, 2018). To this extent, there is a need for coordination among different initiatives that seek to bridge the digital gender divide in addition to changes that make it possible for corporations to facilitate "top-down" investment in female leadership.

That is to say, gender equality is a global issue that needs to be supported not primarily by legislation but by changed modes of business practices and conventions. In this context, "bottom-up" initiatives such as disruptor firms have preliminarily shown to have a greater impact on closing the gap on the digital gender divide. In 2017, female-founded companies comprised 4.4% of all VC deals in the US, which despite its low-sounding figure, is actually the largest percentage since 2006 (Boorstin, 2018).

3. Conclusion

The aim of this chapter was to understand how digitalization and the digital transformation may affect the digital gender divide of business investments and the role of women in a future digitalized labor market. The results of this study indicates that there is still a digital gender divide present, particularly in regards to the low numbers of women entering STEM-educations and professions, and the challenges of women entrepreneurship in ICT-related ventures (often coupled with the difficulty for women founder's teams in securing VC). However, the study concludes that many of the aspects that specifically act to the detriment of women in STEM and ICT professions carry attitudinal connotations. That is to say, changes in attitudes, behavior and habits would have more fundamental impact than changes in policy and regulation. To this extent, female role models in STEM and promotional campaigns will have an important role to inspire and motivate girls and women to seek out a career in these areas. Also, it is essential to find ways to combat the systemic bias perceptions against women by increasing and improving the coordination among the various initiatives that seek to bridge the digital gender divide. Moreover, organizational/strategic changes that make it possible for corporations to facilitate "top-down" investment in female leadership should be encouraged and incentivized.

However, this chapter has also shown that digitalization process, in and of itself, may have various ramifications on women's working conditions on the future labor market. It is true that the emergent automatization will likely start to harvest some professions that have traditionally been women-dominated professions. In this case, a system for reskilling the women who risk being made redundant must be set in place in good time before the automatization of their jobs have been completed.

It is also true that the digital transformation may lead to an improvement of securing more flexible working conditions in ways that benefit women. However, while this development is not expressly geared toward women specifically, there is also a risk of more atypical and cynical work arrangements emerging in professions that are overrepresented by women. For that reason, there is a need for lawmakers to ensure that there is an adequate system of checks and balances in place that makes it impossible to make cynical use of various legal loopholes in the labor laws. Also important is that there is increased transparency and awareness among the social partners that seek to cooperate in securing working relationships and mutually agreed upon goals among employers and employees.

Acknowledgements

The authors would like to extend their sincerest gratitude to Dana Icikzone for her insights and feedback during the planning stage of this chapter. The authors are also deeply grateful to Marie Ekström for her assistance in collecting data and compiling the IWG case study referenced in this study. A special acknowledgment is also extended to Ellen Broström for her support and inspiration in the making of this chapter.

Notes

1 Information and communication technology (ICT) refers to all technical means used to handle information and aid communication. This includes computer and network hardware, as well as any associated software.

2 STEM = Science, technology, engineering and mathematics.

References

Ädel, D., 2016. *Fortfarande få kvinnor söker patent i Sverige [Still Few Women to Seek Patent in Sweden]*. [online] Sveriges Radio. Available at: <https://sverigesradio.se/sida/artikel.aspx?programid=83&artikel=6384638> [Accessed 18 Sep. 2019].

Ajzen, I., 2002. Perceived behavioral control, self-efficacy, locus of control, and the theory of planned behavior. *Journal of Applied Social Psychology*, 32(4), pp. 665–683.

AlphaBeta, 2017. *The Automation Advantage*. [online] Available at: <www.alphabeta.com/our-research/the-automation-advantage> [Accessed 18 Sep. 2019].

Ang, S.S., Orozco, M., Gijbels, D. and Van den Bossche, P., 2018. Learning in the context of work in a digital age: The use of digital media in informal and formal learning contexts. In: C. Harteis, ed., *The Impact of Digitalization in the Workplace: An Educational View*. Cham, Switzerland: Springer, pp. 87–101.

Bain, C.D. and Rice, M.L., 2006. The influence of gender on attitudes, perceptions, and uses of technology. *Journal of Research on Technology in Education*, 39(2), pp. 119–132.

Bauer, M., 1995. Resistance to new technology and its effects on nuclear power, information technology and biotechnology. In: M. Bauer, ed., *Resistance to New Technology:*

Nuclear Power, Information Technology and Biotechnology. Cambridge, UK: Cambridge University Press, pp. 1–44.

Belton, P., 2016. *In the Future, Will Farming be Fully Automated?* [online] BBC News. Available at: <www.bbc.com/news/business-38089984> [Accessed 18 Sep. 2019].

Bernhardt, S., 2014. *Women in IT in the New Social Era: A Critical Evidence-Based Review of Gender Inequality and the Potential for Change*. Hershey, PA: Business Science Reference.

Betts, M., 2004. *Think Tank*. [online] Computerworld. Available at: <www.computerworld.com/article/2567737/it-management/think-tank.html> [Accessed 18 Sep. 2019].

Bidwell, A., 2018. *Report: Women Outperform Men in Educational Attainment, But Still Earn Less*. [online] NASFAA. Available at: <www.nasfaa.org/news-item/14514/Report_Women_Outperform_Men_in_Educational_Attainment_But_Still_Earn_Less> [Accessed 18 Sep. 2019].

Billing, M., 2015. *Nio av tio riskkapitalinvesteringar går till en man [Nine Out of Ten Venture Capital Investments are Given to a Man]*. [online] Dagens Industri. Available at: <https://digital.di.se/artikel/nio-av-tio-riskkapitalinvesteringar-gar-till-en-man> [Accessed 18 Sep. 2019].

Billing, M., 2018. *Granskning: Kvinnliga entreprenörer får mindre än 1 procent av riskkapitalet [Review: Female entrepreneurs Receive Less Than 1 Percent of the Venture Capital]*. [online] Dagens Industri. Available at: <https://digital.di.se/artikel/granskning-kvinnliga-entreprenorer-far-mindre-an-1-procent-av-riskkapitalet> [Accessed 18 Sep. 2019].

Bilton, I., 2018. *Women Are Outnumbering Men At a Record High In Universities Worldwide*. [online] StudyInternational. Available at: <www.studyinternational.com/news/record-high-numbers-women-outnumbering-men-university-globally> [Accessed 18 Sep. 2019].

Blascovich, J. and Tomaka, J., 1991. Measures of self-esteem. In: J.P. Robinson, P.R. Shaver and L.S. Wrightsman, eds., *Measures of Personality and Social Psychological Attitudes: Measures of Social Psychological Attitudes*. San Diego, CA: Academic Press, pp. 115–160.

Boorstin, J., 2018. *Silicon Valley's Gender Gap: Here's Where Powerful Women Are Moving Fast to Close It*. [online] CNBC. Available at: <www.cnbc.com/2018/05/22/silicon-valley-gender-gap-heres-where-powerful-women-are-closing-it.html> [Accessed 18 Sep. 2019].

Borgonovi, F., Centurelli, R., Dernis, H., Grundke, R., Horvát, P., Jamet, S., Keese, M., Liebender, A.S., Marcolin, L., Rosenfeld, D. and Squicciarini, M., 2018. *Bridging the Digital Gender Divide: Include, Upskill, Innovate*. [online] OECD. Available at: <www.oecd.org/newsroom/more-needs-to-be-done-to-bridge-the-digital-gender-divide.htm> [Accessed 18 Sep. 2019].

Brosnan, M.J., 1998. *Technophobia: The Psychological Impact of Information Technology*. London, UK: Routledge.

Brosnan, M.J. and Davidson, M., 1994. Computerphobia: Is it a particularly female phenomenon? *The Psychologist*, 7(2), pp. 73–78.

Burk, D.L., 2018. *Bridging the Gender gap in Intellectual Property*. [online] WIPO Magazine. Available at: <www.wipo.int/wipo_magazine/en/2018/02/article_0001.html> [Accessed 18 Sep. 2019].

Cai, Z., Fan, X. and Du, J., 2017. Gender and attitudes toward technology use: A meta-analysis. *Computers & Education*, 105, pp. 1–13.

Colby, L., 2017. *Quicktake: Women and Tech*. [online] Bloomberg. Available at: <www.bloomberg.com/quicktake/women-are-underrepresented-in-the-high-tech-industry-globally> [Accessed 18 Sep. 2019].

Colley, A. and Comber, C., 2003. Age and gender differences in computer use and attitudes among secondary school students: What has changed? *Educational Research*, 45(2), pp. 155–165.

Colley, A.M., Gale, M.T. and Harris, T.A., 1994. Effects of gender role identity and experience on computer attitude components. *Journal of Educational Computing Research*, 10(2), pp. 129–137.

Cooper, J., 2006. The digital divide: The special case of gender. *Journal of Computer Assisted Learning*, 22(5), pp. 320–334.

Cooper, J. and Weaver, K.D., 2003. *Gender and Computers: Understanding the Digital Divide*. Mahwah, NJ: Lawrence Erlbaum Associates, Inc.

D'Mello, M., 2006. Gendered selves and identities of information technology professionals in global software organizations in India. *Information Technology for Development*, 12(2), pp. 131–158.

Dahlgren, T., Lundeteg, A. and Nord, C., 2017. *Riskkapitalet trängs vid skampålen [The Venture Capital is Crowding at the Pillory]*. [online] Allbright. Available at: <www.allbright.se/nyheter/2017/3/6/riskkapitalet-trngs-vid-skamplen> [Accessed 18 Sep. 2019].

Davis, D., Jivet, I., Kizilcec, R.F., Chen, G., Hauff, C. and Houben, G.J., 2017. Follow the successful crowd: Raising MOOC completion rates through social comparison at scale. In: *Proceedings of the Seventh International Learning Analytics & Knowledge Conference on – LAK '17, March 13–17*. Vancouver, Canada: ACM Press, pp. 454–463.

Demirkan, H., Spohrer, J.C. and Welser, J.J., 2016. Digital innovation and strategic transformation. *IT Professional*, 18(6), pp. 14–18.

Demo, D.H., 1985. The measurement of self-esteem: Refining our methods. *Journal of Personality and Social Psychology*, 48(6), pp. 1490–1502.

dos Santos, T.D. and Santana, V.F. de, 2018. Computer anxiety and interaction. In: *Proceedings of the Internet of Accessible Things, W4A '18, April 23–25*. Lyon, France: ACM Press, pp. 1–10.

Durndell, A., Macleod, H. and Siann, G., 1987. A survey of attitudes to, knowledge about and experience of computers. *Computers & Education*, 11(3), pp. 167–175.

Edwards, C., 2017. *Swedes Hold Strongest Views on Gender Equality in EU*. [online] The Local. Available at: <www.thelocal.se/20171122/how-swedes-feel-about-gender-equality-study> [Accessed 18 Sep. 2019].

Eurostat, 2018. *ICT Specialists – Statistics on Hard-to-Fill Vacancies in Enterprises*. [online] Statistics Explained. Available at: <http://ec.europa.eu/eurostat/statistics-explained/index.php/ICT_specialists_-_statistics_on_hard-to-fill_vacancies_in_enterprises> [Accessed 18 Sep. 2019].

Feinstein, L., Vorhaus, J. and Sabates, R., 2010. Learning through life: Future challenges. In: C.L. Cooper, J. Field, U. Goswami, R. Jenkins and B.J. Sahakian, eds., *Mental Capital and Wellbeing*. Chichester, UK: Wiley-Blackwell, pp. 307–342.

Felt, U., Schumann, S. and Schwarz, C.G., 2015. (Re)assembling natures, cultures, and (nano)technologies in public engagement. *Science as Culture*, 24(4), pp. 458–483.

Fernández-Cruz, F.J. and Fernández-Díaz, M.J., 2016. Generation Z's teachers and their digital skills. *Comunicar*, 24(46), pp. 97–105.

Fogelqvist, J., 2016. *Få kvinnor inom industrin – Sverige under EU-snittet [Few Women Working in the Industry – Sweden Under EU-average*. [online] Arbetsmiljöforum. Available at: <www.arbetsmiljoforum.se/nyheter/faa-kvinnor-inom-industrin-sverige-under-eu-snittet> [Accessed 18 Sep. 2019].

Ford, M., 2015. *Rise of the Robots: Technology and the Threat of a Jobless Future*. New York, NY: Basic Books.

Fraser, E. and Charlebois, S., 2016. *Automated Farming: Good News for Food Security, Bad News for Job Security?* [online] The Guardian. Available at: <www.theguardian.com/sustainable-business/2016/feb/18/automated-farming-food-security-rural-jobs-unemployment-technology> [Accessed 18 Sep. 2019].

Gordon, A., 2018. *What the Future of Manufacturing Automation Could Look Like*. [online] Forbes. Available at: <www.forbes.com/sites/forbestechcouncil/2018/02/01/what-the-future-of-manufacturing-automation-could-look-like> [Accessed 18 Sep. 2019].

Graham, H., 1993. *Hardship & Health Womens Lives*. London, UK: Routledge.

Guynn, J., 2015. *Silicon Valley Gender Gap is Widening*. [online] USA Today. Available at: <https://usatoday.com/story/tech/2015/03/26/silicon-valley-gender-gap-widening/70444276> [Accessed 18 Sep. 2019].

Hanrahan, C. and Evlin, L., 2017. *Artificial Intelligence: Men's Jobs Face Higher Risk of Automation than Women, Low-paid Workers Also At Risk*. [online] ABC News. Available at: <www.abc.net.au/news/2017-08-09/ai-automation-men-and-lower-paid-workers/8741518> [Accessed 18 Sep. 2019].

He, Y., Chen, Q. and Kitkuakul, S., 2018. Regulatory focus and technology acceptance: Perceived ease of use and usefulness as efficacy. *Cogent Business & Management*, 5(1).

Heflin, H., Shewmaker, J. and Nguyen, J., 2017. Impact of mobile technology on student attitudes, engagement, and learning. *Computers & Education*, 107, pp. 91–99.

Henderson, M., Selwyn, N. and Aston, R., 2017. What works and why? Student perceptions of 'useful' digital technology in university teaching and learning. *Studies in Higher Education*, 42(8), pp. 1567–1579.

Henry, K., Lagos, A. and Berndt, F., 2012. Scholarship-in-Practice Bridging the literacy gap between boys and girls: An opportunity for the National Year of Reading 2012. *The Australian Library Journal*, 61(2), pp. 143–150.

Hesselberth, P., 2018. Discourses on disconnectivity and the right to disconnect. *New Media & Society*, 20(5), pp. 1994–2010.

Hilbert, M., 2011. Digital gender divide or technologically empowered women in developing countries? A typical case of lies, damned lies, and statistics. *Women's Studies International Forum*, 34(6), pp. 479–489.

Hogan, M., 2009. Age difference in technophobia: An Irish study. In: C. Barry, K. Conboy, M. Lang, G. Wojtkowski and W. Wojtkowski, eds., *Information Systems Development: Challenges in Practice, Theory, and Education, Volume 1*. New York, NY: Springer, pp. 117–130.

Holley, R.P., 2013. Library space and technology. In: G. Walton and G. Matthews, eds., *University Libraries and Space in the Digital World*. Farnham, UK: Ashgate Publishing Limited, pp. 51–70.

Holzer, D., 2015. *The BIM Manager's Handbook: Guidance for Professionals in Architecture, Engineering, and Construction*. Chichester, UK: John Wiley & Sons.

Hou, J., Wu, Y. and Harrell, E., 2017. Reading on paper and screen among senior adults: Cognitive map and technophobia. *Front. Psychol.*, 8, p. 2225, pp. 1–10.

Hunt, J., Garant, J.P., Herman, H. and Munroe, D.J., 2013. Why are women underrepresented amongst patentees? *Research Policy*, 42(4), pp. 831–843.

IT&Telekomföretagen, 2017. *Statistik kvinnor och män [Statistics Women and Men]*. [online] Available at: <www.itot.se/om-oss/statistik/statistik-kvinnor-och-man> [Accessed 18 Sep. 2019].

IWG Group, 2018. *Empowered Mentoring Program (EMP) Evaluation 2018 [Unpublished Case Study]*. Stockholm, Sweden.

Johnson, V., 2012. The gender divide: Attitudinal issues inhibiting access. In: R. Pande and T.P. Van der Welde, eds., *Globalization, Technology Diffusion and Gender Disparity: Social Impacts of ICTs: Social Impacts of ICTs*. Hershey, PA: IGI Global, pp. 110–119.

Kagan, J., 2018. *Total Cost Of Ownership – TCO*. [online] Investopedia. Available at: <www.investopedia.com/terms/t/totalcostofownership.asp> [Accessed 18 Sep. 2019].

Kahn, J., 2017. *Demand for AI Talent Turns Once-Staid Conference Into Draft Day*. [online] Bloomberg. Available at: <www.bloomberg.com/news/articles/2017-12-06/demand-for-ai-talent-turns-once-staid-conference-into-draft-day> [Accessed 18 Sep. 2019].

Kang, N.E. and Yoon, W.C., 2008. Age- and experience-related user behavior differences in the use of complicated electronic devices. *International Journal of Human-Computer Studies*, 66(6), pp. 425–437.

Katz, Y.J., 1993. Self-image, locus of control and computer-related attitudes. In: *Proceedings of the IFIP TC3/WG3.3 Working Conference on Lessons from Learning, September 6–8*. Amsterdam, Netherlands: North-Holland Publishing Co, pp. 105–109.

Kennewell, S., 1992. Computing for the terrified. *Computers & Education*, 18(1–3), pp. 195–200.

Kerpen, C., 2018. *How Women Entrepreneurs Are Closing the Venture Capital Gap*. [online] Forbes. Available at: <www.forbes.com/sites/carriekerpen/2018/04/09/how-women-entrepreneurs-are-closing-the-venture-capital-gap> [Accessed 18 Sep. 2019].

Khasawneh, O.Y., 2018. Technophobia without boarders: The influence of technophobia and emotional intelligence on technology acceptance and the moderating influence of organizational climate. *Computers in Human Behavior*, 88, pp. 210–218.

Kirche, E. and Srivastava, R., 2017. A staged strategy for understanding organizational requirements in the acquisition of information and communication for SMEs. In: R. Bogaschewsky, M. Eßig, R. Lasch and W. Stölzle, eds., *Supply Management Research: Aktuelle Forschungsergebnisse 2016*. Wiesbaden, Germany: Springer, pp. 3–28.

Kling, R., 1999. Can the 'next-generation Internet' effectively support 'ordinary citizens'? *The Information Society*, 15(1), pp. 57–63.

Krieg, R.M., 1995. Information technology and low-income, inner-city communities. *Journal of Urban Technology*, 3(1), pp. 1–17.

Krieger-Boden, C. and Sorgner, A., 2018. Labor market opportunities for women in the digital age. *Economics*, 12(2018–28), pp. 1–8.

Kuśpit, M., 2016. Creative attitude and understanding of emotions by artistically-gifted students. *Polish Journal of Applied Psychology*, 15(1), pp. 59–78.

Lazar, M., 2016. *Here's Why Small Business IT Consulting Is a Wise Prerequisite*. [online] Insight. Available at: <www.insight.com/en_US/learn/content/2016/09292016-service-provider-security-it-consulting.html> [Accessed 18 Sep. 2019].

Liff, S. and Steward, F., 2001. Community e-gateways: Locating networks and learning for social inclusion. *Information, Communication & Society*, 4(3), pp. 317–340.

Lim, C.P., 2002. A theoretical framework for the study of ICT in schools: A proposal. *British Journal of Educational Technology*, 33(4), pp. 411–421.

Lindberg, S.M., Hyde, J.S., Petersen, J.L. and Linn, M.C., 2010. New trends in gender and mathematics performance: A meta-analysis. *Psychological Bulletin*, 136(6), pp. 1123–1135.

The Local, 2016. *82 Percent of Tech Cash Goes to Male Entrepreneurs*. [online] The Local.se. Available at: <www.thelocal.se/20161128/82-percent-of-tech-cash-goes-to-male-entrepreneurs> [Accessed 18 Sep. 2019].

Lochner, F.C., 2005. A cost maturity model for community informatics projects in the developing world. *The Journal of Community Informatics*, 1(2), pp. 116–133.

Miller, F. and Varma, N., 1994. The effects of psychosocial factors on Indian children's attitudes toward computers. *Journal of Educational Computing Research*, 10(3), pp. 223–238.

Mitzner, T.L., Boron, J.B., Fausset, C.B., Adams, A.E., Charness, N., Czaja, S.J., Dijkstra, K., Fisk, A.D., Rogers, W.A. and Sharit, J., 2010. Older adults talk technology: Technology usage and attitudes. *Computers in Human Behavior*, 26(6), pp. 1710–1721.

Mohr, J.J., Sengupta, S. and Slater, S.F., 2010. *Marketing of High-technology Products and Innovations*. 3rd ed. Upper Saddle River, NJ: Prentice Hall.

Murdock, G., Hartmann, P. and Gray, P., 1995. Conceptualising home computing: Resources and practices. In: N. Heap, R. Thomas, G. Einon, R. Mason and H. Mackay, eds., *Information Technology and Society: A Reader*. London, UK: Sage Publications, pp. 269–283.

Nager, A., Hart, D.M., Ezell, S. and Atkinson, R.D., 2016. *The Demographics of Innovation in the United States*. [online] ITIF. Available at: <https://itif.org/publications/2016/02/24/demographics-innovation-united-states> [Accessed 18 Sep. 2019].

Nimrod, G., 2018. Technophobia among older Internet users. *Educational Gerontology*, 44(2–3), pp. 148–162.

Nkhoma, M.Z., Dang, D.P.T. and Lu, T.H., 2012. Towards an understanding of implementation and benefits of ICT in education: Review of issues to be considered by developing countries. In: *Proceedings of the 3rd International Conference on Society and Information Technologies (ICSIT 2012), March 25th – 28th*. Orlando, Florida, USA: International Institute of Informatics and Systemics, pp. 31–35.

Norman, D.A., 1993. *Things That Make Us Smart: Defending Human Attributes in the Age of the Machine*. New York, NY: Basic Books.

Nsibirano, R., Kabonesa, C. and Madanda, A., 2012. Gender symbolism and technology uptake: A literature review. In: R. Pande and T.P. Van der Welde, eds., *Globalization, Technology Diffusion and Gender Disparity: Social Impacts of ICTs*. Hershey, PA: IGI Global, pp. 120–128.

OECD, 2017. *Going Digital: The Future of Work for Women*. [online] Policy brief on the future of work. Available at: <www.oecd.org/employment/Going-Digital-the-Future-of-Work-for-Women.pdf> [Accessed 18 Sep. 2019].

OECD, 2018. *Entrenched Social Norms Prevent the Equal Distribution of Caring Responsibilities Between Men and Women*. [online] OECD. Available at: <www.oecd.org/gender/data/entrenched-social-norms-prevent-the-equal-distribution-of-caring-responsibilities-between-men-and-women.htm> [Accessed 18 Sep. 2019].

Offir, B., Golub, M.R. and Cohen-Fridel, S., 1993. Attitudes toward courseware as a function of high school students' creativity level. In: *Proceedings of the IFIP TC3/WG3.1/WG3.5 Open Conference on Informatics and Changes in Learning, June 7–11*. Amsterdam, Netherlands: North-Holland Publishing Co, pp. 211–216.

Oliveira, C.S. and Jerónimo, N.A., 2016. Small but focused: Women (self) empowerment in a rural village. In: J. Wilson and N.D. Gapsiso, eds., *Overcoming Gender Inequalities through Technology Integration*. Hershey, PA: IGI Global, pp. 93–122.

Olsson Jeffery, M., 2019. *Bolag grundade av kvinnor får 1 procent av riskkapitalet [Corporations Founded by Women Are Awarded 1 Percent of the Venture Capital]*. [online] Dagens Industri. Available at: <https://digital.di.se/artikel/mannens-techbolag-far-99-procent-av-allt-riskkapital> [Accessed 18 Sep. 2019].

Ottestad, G. and Quale, A., 2009. Trends in instructional ICT infrastructure. In: T. Plomp, R.E. Anderson, N. Law and A. Quale, eds., *Crossnational Information and Communication: Technology Policy and Practices in Education*, 2nd ed. Charlotte, NC: Information Age Publishing, pp. 41–66.

Power, T., 2004. ICT and teacher education in the global south: Costing the benefits of learning. In: *Third Pan Commonwealth Forum on Open Learning, 4–8 July*. Dunedin, New Zealand: Commonwealth of Learning, pp. 1–13.

Quirós, C.T., Morales, E.G., Pastor, R.R., Carmona, A.F., Ibáñez, M.S. and Herrera, U.M., 2018. *Women in the Digital Age*. [online] European Commission & Iclaves. Available at: <https://ec.europa.eu/digital-single-market/en/news/increase-gender-gap-digital-sector-study-women-digital-age> [Accessed 18 Sep. 2019].

Ragnedda, M. and Muschert, G.W., 2013. Introduction. In: M. Ragnedda and G.W. Muschert, eds., *The Digital Divide: The Internet and Social Inequality in International Perspective*. Oxon, UK: Routledge, pp. 1–14.

Ramalingam, A. and Kar, S.S., 2014. Is there a digital divide among school students? An exploratory study from Puducherry. *Journal of Education and Health Promotion*, 3, p. 30, pp. 1–4.

Revilla Muñoz, O., Alpiste Penalba, F., Fernández Sánchez, J. and Santos, O.C., 2017. Reducing techno-anxiety in high school teachers by improving their ICT problem-solving skills. *Behaviour & Information Technology*, 36(3), pp. 255–268.

Richard, P.L., 1997. Conquering technophobia: Preparing faculty for today. *Studies in Health Technology and Informatics*, 46, pp. 437–441.

Roden, L., 2016. *Swedes Named Fourth Most Gender Equal in the World*. [online] The Local. Available at: <www.thelocal.se/20161026/swedes-named-fourth-most-gender-equal-in-the-world> [Accessed 18 Sep. 2019].

Rogers, E.M., 2001. The digital divide. *Convergence*, 7(4), pp. 96–111.

Rogers, E.M., 2003. *Diffusion of Innovations*. 5th ed. New York, NY: Free Press.

Romei, V., 2018. *Lack of Women 'STEM' Students Has Led to Gender 'Digital Divide'*. [online] Financial Times. Available at: <www.ft.com/content/4554e816-d872-11e8-a854-33d6f82e62f8> [Accessed 18 Sep. 2019].

Rosen, L.D., Sears, D.C. and Weil, M.M., 1993. Treating technophobia: A longitudinal evaluation of the computerphobia reduction program. *Computers in Human Behavior*, 9(1), pp. 27–50.

Rosen, L.D. and Weil, M.M., 1995. Computer availability, computer experience and technophobia among public school teachers. *Computers in Human Behavior*, 11(1), pp. 9–31.

Savage, M., 2016. *Women Zoom into Swedish Startup Accelerator Spots*. [online] The Local.se. Available at: <www.thelocal.se/20160218/women-zoom-into-top-swedish-startup-accelerator-programme> [Accessed 18 Sep. 2019].

Schumacher, P., Morahan-Martin, J. and Olinsky, A., 1993. Computer experiences, attitudes, computer and mathematical anxiety, and grades of MBA students. *Collegiate Microcomputer*, 11(3), pp. 183–193.

Selwyn, N., 1999. Student's attitudes towards computers in sixteen to nineteen education. *Education and Information Technologies*, 4(2), pp. 129–141.

Selwyn, N., 2003a. Apart from technology: Understanding people's non-use of information and communication technologies in everyday life. *Technology in Society*, 25(1), pp. 99–116.

Selwyn, N., 2003b. *Why Students Do (and do not) Make Use of ICT in University*. [online] Available at: <www.leeds.ac.uk/educol/documents/00003130.htm> [Accessed 18 Sep. 2019].

Selwyn, N., 2004. Reconsidering political and popular understandings of the digital divide. *New Media & Society*, 6(3), pp. 341–362.

Shashaani, L., 1993. Gender-based differences in attitudes toward computers. *Computers & Education*, 20(2), pp. 169–181.

Shashaani, L., 1995. Gender differences in mathematics experience and attitude and their relation to computer attitude. *Educational Technology*, 35(3), pp. 32–39.

Shaw, E. and Hess, C., 2018. *Closing the Gender Gap in Patenting, Innovation, and Commercialization: Programs Promoting Equity and Inclusion*. [online] Institute for

Women's Policy Research. Available at: <https://iwpr.org/publications/gender-diversity-patenting-program-scan> [Accessed 18 Sep. 2019].

Simonton, D., 2006. Women workers: Working Women. In: D. Simonton, ed., *The Routledge History of Women in Europe Since 1700*. London, UK: Routledge, pp. 134–176.

STEMconnector, 2017. *Women's Quick Facts: Compelling Data on Why Women Matter*. New York, NY: Morgan James Publishing.

Stengel, G., 2017. *Women Get It Done: Fixing the Broken Venture Capital System*. [online] Forbes. Available at: <www.forbes.com/sites/geristengel/2017/11/08/women-get-it-done-fixing-the-broken-venture-capital-system> [Accessed 18 Sep. 2019].

Stoll, F. and Notter, P., 2000. Domain-specific control beliefs in literacy and numeracy versus actual performance among adults. In: W.J. Perrig and A. Grob, eds., *Control of Human Behavior, Mental Processes, and Consciousness: Essays in Honor of the 60th Birthday of August Flammer*. Mahwah, NJ: Lawrence Erlbaum Associates, Inc., pp. 466–479.

Thibodeau, P., 2012. *IT Jobs Will Grow 22% Through 2020, Says U.S.* [online] Computerworld. Available at: <https://computerworld.com/article/2502348/it-management/it-jobs-will-grow-22--through-2020--says-u-s-.html> [Accessed 18 Sep. 2019].

Thompson, E., 2017. *How Much Should a Business Computer Cost?* [online] Business.org. Available at: <www.business.org/finance/cost-management/much-computer-cost> [Accessed 18 Sep. 2019].

Todman, J. and Monaghan, E., 1994. Qualitative differences in computer experience, computer anxiety, and students' use of computers: A path model. *Computers in Human Behavior*, 10(4), pp. 529–539.

Tomer, A. and Eliason, G., 2000. Beliefs about self, life and death: Testing aspects of a comprehensive model of death anxiety and death attitudes. In: A. Tomer, ed., *Death Attitudes and the Older Adult: Theories Concepts and Applications*. London, UK: Routledge, pp. 137–156.

Trauth, E.M., 2006. An agenda for research on gender diversity in the global information economy. In: E.M. Trauth, ed., *Encyclopedia of Gender and Information Technology*. Hershey, PA: Idea Group Publishing, pp. xxiv – xxxiii.

Trauth, E.M., Nielsen, S.H. and von Hellens, L.A., 2003. Explaining the IT gender gap: Australian stories for the new millennium. *Journal of Research and Practice in Information Technology*, 35(1), pp. 7–20.

Trauth, E.M., Quesenberry, J.L. and Huang, H., 2010. Factors influencing career choice for women in the global information technology workforce. In: M.G. Hunter and F. Tan, eds., *Technological Advancement in Developed and Developing Countries: Discoveries in Global Information Management*. Hershey, PA: IGI Global, pp. 23–48.

UK Department for Business, E. and I.S., 2017. *Employment Status Review (2015)*. [online] GOV.UK. Available at: <www.gov.uk/government/publications/employment-status-review-2015> [Accessed 18 Sep. 2019].

Umoh, R., 2017. *The US has a Shortage of Tech Workers. Here's How Kids and Schools Can Solve the Problem*. [online] CNBC. Available at: <www.cnbc.com/2017/08/23/why-we-have-a-shortage-of-tech-workers-in-the-u-s.html> [Accessed 18 Sep. 2019].

Van Dijk, J.A.G.M., 2012. *The Network Society*. 3rd ed. London, UK: Sage Publications.

Van Hek, M., Kraaykamp, G. and Wolbers, M.H.J., 2016. Comparing the gender gap in educational attainment: The impact of emancipatory contexts in 33 cohorts across 33 countries. *Educational Research and Evaluation*, 22(5–6), pp. 260–282.

Vincent, J., 2017. *Tencent Says There are Only 300,000 AI Engineers Worldwide, but Millions are Needed.* [online] The Verge. Available at: <www.theverge.com/2017/12/5/16737224/global-ai-talent-shortfall-tencent-report> [Accessed 18 Sep. 2019].

Von Hall, G., 2018. *Rekordmånga patent tas av kvinnor [Record Number of Patents Taken by Women].* [online] SvD. Available at: <www.svd.se/rekordmanga-patent-tas-av-kvinnor> [Accessed 18 Sep. 2019].

Wever, C., 2012. *Europe Is in Need of Its Own Highly E-skilled Professionals.* [online] Neth-ER. Available at: <www.neth-er.eu/en/news/Europe-need-its-own-highly-e-skilled-professionals> [Accessed 18 Sep. 2019].

Wilson, E.J., 2000. *Closing the Digital Divide: An Initial Review. Briefing the President.* Washington, DC: Internet Policy Institute.

Winick, E., 2018. *Five Jobs That Are Set to Grow in 2018.* [online] MIT Technology Review. Available at: <www.technologyreview.com/s/609644/five-jobs-that-are-set-to-grow-in-2018> [Accessed 18 Sep. 2019].

WIPO, 2018. *World Intellectual Property Day 2018 Celebrates Women's Accomplishments: New WIPO Figures Show Highest-Ever Rate of Women Inventors, but Gender Gap Persists.* [online] World Intellectual Property Organization: Press Releases. Available at: <www.wipo.int/pressroom/en/articles/2018/article_0003.html> [Accessed 18 Sep. 2019].

Wisterberg, E., 2017. *Kartläggning: Så mansdominerade är de svenska techjättarna [Survey: This is How Male-dominated the Swedish Tech-gigants are].* [online] Breakit. Available at: <www.breakit.se/artikel/9393/hur-mansdominerade-ar-de-svenska-tech jattarna-egentligen> [Accessed 18 Sep. 2019].

Wisterberg, E., 2019. *84 procent av riskkapitalet i Sverige går till männens bolag [84 Percent of the Venture Capital is Awarded to Male-run Corporations].* [online] Breakit. Availableat:<www.breakit.se/artikel/19336/kartlaggning-mannen-far-84-procent-av-allt-riskkapital-i-sverige> [Accessed 18 Sep. 2019].

Woodrow, J.E.J., 1990. Locus of control and student teacher computer attitudes. *Computers & Education*, 14(5), pp. 421–432.

World Bank Group, 2019. *Women, Business and the Law 2019: A Decade of Reform.* [online] Available at: <http://hdl.handle.net/10986/31327> [Accessed 18 Sep. 2019].

World Economic Forum, 2018. *The Future of Jobs Report 2018.* [online] World Economic Forum. Available at: <www.weforum.org/reports/the-future-of-jobs-report-2018> [Accessed 18 Sep. 2019].

Wu, J.H., Hsia, T.L. and Tennyson, R.D., 2011. Design strategies for improved online instructional systems. In: P. Ordóñez de Pablos, M. Lytras, W. Karwowski and R.W.B. Lee, eds., *Electronic Globalized Business and Sustainable Development Through IT Management: Strategies and Perspectives: Strategies and Perspectives.* Hershey, PA: IGI Global, pp. 222–236.

Yate, M., 2019. *6 Fastest Growing Jobs in Education and Social Services.* [online] Monster. Available at: <www.monster.com/career-advice/article/most-stable-professional-sectors-education-social-services> [Accessed 18 Sep. 2019].

Zhenghao, C., Alcorn, B., Christensen, G., Eriksson, N., Koller, D. and Emanuel, E.J., 2015. *Who's Benefiting from MOOCs, and Why.* [online] Harvard Business Review. Available at: <https://hbr.org/2015/09/whos-benefiting-from-moocs-and-why> [Accessed 18 Sep. 2019].

15 Consulting in the digital era?

The role of tomorrow's management consultants

Anthony Larsson, Nicole Andersson, Peter Markowski, Malin Nilsson and Ivy Mayor

1. Introduction

The term "consultant" can indeed take on many different forms. At bedrock, it refers to a professional who provides expert advice within a specific, specialized area (Oxford Dictionaries, 2018; Tordoir, 1995). Consultants are commonly differentiated as being either "internal" or "external" consultants, depending on what function they serve or to whom they provide consulting advice. An internal consultant typically refers to someone operating within an organization. They may be consulted on their area of expertise by others within the same organization. An external consultant, on the other hand, typically refers to an externally employed expert who provides assistance or advice to an actor on a temporary basis, usually in exchange for a fee (Armbrüster, 2006; O'Mahoney and Markham, 2013).

While the two categories are similar inasmuch that they both adhere to issues concerning confidentiality, risk project, project termination, etc., there are several practical differences between them as well. For instance, internal consultants are usually contracted in a rather informal manner as opposed to external consultants, and tend to be considerably cheaper to contract. They also tend to have a better knowledge about the organization from the outset than an external consultant. However, their strong tie to the organization carries the innate risk of them becoming overly cautious and/or apprehensive in taking or suggesting an action that would risk upsetting someone with the ability of influencing the internal consultant's career in either direction. They may also lack certain skills in facilitating organizational change (Cummings and Worley, 2013; Burtonshaw-Gunn, 2010).

External consultants, on the other hand, are often able to select their clients according to their own criteria and/or profile. They are generally looked upon as being more prestigious, which in turn elevates the organizations expectations for them to achieve their goal. This, by extension, enables the consultants to probe difficult issues and assess the organization in a more objective manner, devoid of any personal attachments and without fear of reprisals from the manager (Cummings and Worley, 2013; Scott and Barnes, 2011).

Moreover, consulting firms range in size from sole proprietorships, consisting of a single consultant, and small businesses consisting of a small number of consultants, to mid- to large consulting firms. The latter of which may in some cases be multinational corporations. This type of consultant generally engages with multiple and changing clients, which are typically companies, nonprofit organizations or governments.

While a plethora of specific types of consultants exists, this chapter will primarily focus on management consultants, as this is one of the most common, and among recently graduated university students, most popularly sought after types of consultancies (Wickham and Wilcock, 2016; White, 2011; Hope, 2016). One of the reasons for this is that management consulting is known to generate high streams of revenue, both for the individual consultant, as well as for the consultancy firm, with some recent university graduates receiving offers from top firms with a remuneration approaching or even exceeding USD90,000 in their first year (Nisen, 2013; Harvard Business School, 2018; Management Consulted, 2018).

Management consultants are typically external consultants who provide the client management with strategic and/or operational advice (data driven). The reason why companies hire management consultants is explained by Greiner and Metzger (1983, p. 7):

> Management consulting is an advisory service contracted for and provided to organizations by specially trained and qualified persons who assist, in an objective and independent manner, the client organization to identify management problems, analyze such problems, and help, when requested, in the implementation of solutions.

As the digital transformation continues to make its way through various businesses, the consultancy profession is no exception, as pointed out by Christensen, Wang and van Bever (2013). Digital transformation aims to increase efficiency, competitiveness and accessibility of consultants by transitioning much of their businesses to digital technology. However, there is currently a lack of research on how digital transformation affects the role of management consultancy in the future, as there is confusion as to how consultants should structure their digital business (Marriage, 2018). There is also a pressing issue in regards to whether or not the consultants as we know them today are likely to look the same tomorrow, given the technological advancements (Czerniawska, 1999).

Thus, the overarching research questions are:

> RQ 1: *How may digitalization influence the consultant's role of tomorrow?*
> RQ 2: *How may the profile of the typical consultant change in the future?*

As a theoretical/speculative study, this chapter seeks to draw upon some of the available literature and the authors' own best-practice experiences in exploring some of the most pressing issues of the digitalization of consulting of today, with an anticipation of how the role of consultants may come to develop in the near future (Kim, Sefcik and Bradway, 2017; Cooper and Endacott, 2007; Elliott and Timulak, 2005; Murphy and Dingwall, 1998).

2. The background of traditional consultancy

A recurring point throughout the years has been contention that consultants receive vast amounts of money for their services and that much of this money is spent on impractical data and poorly implemented recommendations (Turner, 1982). Thus, in order to reduce waste, there is a need for potential customers to better understand what consulting assignments can accomplish. Historically, the traditional role of a consultant has been "to advise and assist the client in carrying out the project definition and contracting process, as well as with the management and execution of design, plus administration, supervision and quality control of the . . . contracts" (Harrison and Lock, 2004, p. 85). Typically, the consultant carries out a lead role in a given project, but falls short of overall project management and/or integration inasmuch that they are generally not accountable for, or in charge of, all parts of the project (Harrison and Lock, 2004).

The years following World War II are often described as the "emergent period of management consulting" (Srinivasan, 2014, p. 259). During this period, consulting entrepreneurs would highlight the significant contrasts between the status quo and broad cultural logics and use insights from outside their professional field to suggest solutions to problems. Moreover, they would emphasize the larger societal benefits of the proposed solutions, establish the uniqueness of their profession by establishing social codes, and establish relationships with prominent actors outside their professional field in order to legitimate their problem-solving models (David, Sine and Haveman, 2013). This evolution would eventually lead to an industry consisting of various actors and firms that are conceptually similar, but yet markedly differently positioned (Srinivasan, 2014).

In later years, various corporations have begun making increased use of titles that include "consultant" (Srinivasan, 2014). These staff members are effectively "internal consultants" (as described earlier in this chapter). These consultants provide the company with specialized expertise, but as "internal" consultants they are an integral part of the organization. As such, they do not generally bring in the "outside" perspective that clients often seek (Srinivasan, 2014). Arguably, the external perspective has traditionally been of key importance as Fincham, Mohe, and Seidl (2013, p. 6) identify management consulting as including "any activity that has as its apparent justification the provision of some kind of support in identifying or dealing with management problems, provided by individuals, groups, or organizations that are external to the particular management domain and which are contracted by the management on a temporary basis". The added value that external consultants bring to their clients is that the consultants are able to provide them with unique expertise, innovation and/or swiftness not readily available to the client (Momani, 2013; Srinivasan, 2014). To this end, a vital component of management consulting has also been the ability of providing advisory services by specialists who can assist the client in an objective and independent fashion in identifying management problems, analyzing problems, proving suggested courses of actions and in some cases, even assist in the implementation of solutions (Greiner and Metzger, 1983).

In time, however, the value proposition of the consultancies have gradually shifted from providing specialists to solve clients' business problems to granting clients the ability to tap into the consultancy's knowledge base, as many clients and consultancies have similar access to the resource pools for hiring new recruits, i.e., promising graduates from top business schools (Sarvary, 1999). This means that consulting firms in the past couple of decades have had to emphasize the power of its collective knowledge rather than the individual expertise among its staff (Srinivasan, 2014).

The term "consultant" has shifted meaning from solely pertaining to expert advice during a limited amount of time, to also including concepts such as staffing consultants, or contractors (Hyman, 2016; Berry and Oakley, 1994; Turner, 1982). Some companies have employed a strategy of hiring consultants rather than employing staff, as it enables them to quickly cut back on staffing costs whenever recession looms (Banks and Coutu, 2008; Baumann, 2009).

3. The four phases of consulting

Prior to implementing solutions, the solutions in question need to be devised and clearly articulated in the upcoming implementation plan. This is typically done along with the consultants during a phase called "solutions design" (or something to that effect) (Griffin, 2017). These are executed by either the management consultants or the organization itself.

In an oversimplified manner, consulting can be expressed as consisting of four different phases: (1) the pre-analysis phase, (2) the problem-identification phase, (3) the analysis phase and (4) the implementation phase (as depicted in Figure 15.1). These four phases each carry their own potential issues.

3.1. Pre-analysis phase

Initially, there is the pre-analysis phase that seeks to answer the "why" of what needs to be accomplished. In a strictly oversimplified and theoretical world, this phase can be omitted and a consultant would be able to dive right in to deal with the problem at hand. However, in practice this is rarely, if ever, possible, due to

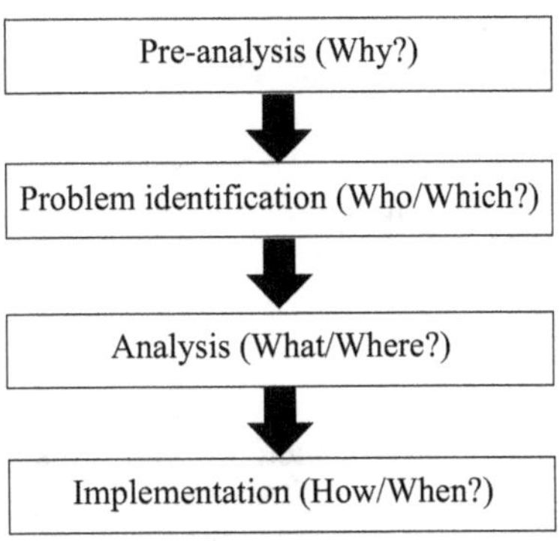

Figure 15.1 The four phases of consulting (authors' own depiction).

the fact that the management consultancy services are ever so often subject to various forms of organizational politics and demands from other hierarchical levels within the organization (Verlander, 2012; Hodges, 2017). Moreover, the local executive in charge of contracting the consultant ever so often lacks the insight in the actual (or perceived) problem at hand and often needs someone to guide them in taking the next steps (Cummings, 2010; Keuning, 2007; Cohen, 2016). This problem was highlighted in a 1989 landmark study, in which Yoshida (1989) coined the famed expression, "the Iceberg of Ignorance". This pertained to the realization that only 4% of an organization's frontline problems are known by top-management, 9% are known by middle-management, 74% by managers and 100% by employees.

By and large, the issue of "the Iceberg of Ignorance" remains a problem to this day and age (Jankowski, 2017; Corey and Elliott, 2018; Ray, 2016; Albert, 2018). In practice, this means that the management consultant during the initial phase is often tasked with greeting staff across the hierarchy (i.e., not only executives), in order to acquire a level of empathy and gaining a better understanding of their situation, why it is important, and how to go about helping the organization achieve its aspirations (Gourguechon, 2017; Poulfelt and Paynee, 1994; Senge and Krahnke, 2014).

3.2. Problem-identification phase

The problem-identification phase that seeks to answer the "who" or "which" that lies at the root of the client's problems (Heiser and Farah, 2018; Benn, Jones and Rosenfield, 2008; Schmidt, 2017). This phase is a critical part in the management consultant's work as it seeks to establish the problem as identified not only by the client but also by the consultant. That is to say, the way problems are defined affects the ability to solve them (Kubr, 2002; Ashkenas, 2012; Conoley, Conoley and Gumm, 1992). For instance, a company might find itself struggling with declining revenues/profits, or with increasing costs. They might lose market shares but fail to understand why.

In order to be able to identify root causes and solve these problems, the consultants typically start by gathering quantitative and qualitative data, mainly from internal data sources but also external if needed (Newton, 2010; Andler, 2016). Examples of internal data can be financial data, company annual reports, interviews, surveys, etc. External data sources that may be used are e.g., competitor annual reports, interviews with external experts and customer surveys. The internal and external data is used to build a solid foundation in order to understand the industry context as well as the internal starting point from where to undertake the analysis.

Analyzing the collected data to understand root causes is instrumental to every project. Management consultants often operate from a hypothesis-driven structure when developing the right choice of methods and tools for their tasks (Liedtka, 2006; Rasiel, 1999; Garrette, Phelps and Sibony, 2018). This means that the management consultant will depart from their best, educated guess of an answer to a given problem. It should be stressed that a hypothesis-driven deduction is not "a shot in the dark", based on conjecture or personal opinion, but rather on background information, preliminary data analyses and input from various experts in the field and actors in the organization (Hamann, 2012; Weiss, 2011). While there may not be too many concrete facts upon which to base the hypothesis during the early stages of the engagement, more facts emerge the further a consultant delves into the client's engagements, meaning that the hypothesis development is the result of a highly organic and evolving process. This approach allows the consultant to quickly gain a grasp of the organization and finding a hypothesis that can either be supported or rejected rather than having to start from a blank sheet.

Nevertheless, this phase places a lot of demands on the consultant as this is a phase in which several mistakes are prone to be made. According to Kubr (2002, p. 186), common issues during this phase include:

- Mistaking symptoms for problems
- Having preconceived notions about the causes of the problems
- Looking at problems only from one sole technical viewpoint
- Disregarding how the problem is perceived in other parts/sections of the organization
- Miscalculating the sense of urgency of the problem
- Incomplete/deficient problem diagnosis
- Failure to clearly articulate the focus purpose

It is thus the role of the management consultant to help the organization avoid these common pitfalls by bringing in an objective perspective. Organizations may face difficulty in trying to avoid these kinds of mistakes due to internal forces of power and politics (Mintzberg, 1983). Although the consultant may also face such difficulties, they are often in a better position to assume an objective/neutral standpoint (Greiner and Metzger, 1983).

3.3. Analysis phase

Following the problem-identification phase, is the analysis phase, which seeks to answer the question concerning "what" and possibly even "where" something needs to be addressed. The analysis phase may actually consist of several substages, depending on the analysis and research methods carried out (Biggs, 2010). During this phase, the consultant(s) will carry out an in-depth diagnosis of the problem, while assessing the type change the organization will have to undergo in order to achieve the purposes of the assignment while also assessing the client's performance, resources, needs and aspirations (Harrison, 2005; Kubr, 2002). The consultant(s) will at this stage determine the client's attitude toward change and if the client is likely to carry out suggested changes without much ado or if they need further convincing before taking action (Kubr, 2002). During this phase the consultant(s) will be able to see some possible solution emerging from the data processed. Nevertheless, a lingering issue with this phase is that fact-finding often receives the least amount of attention (Kubr, 2002). At the same time, decisions regarding what data to look for and what data to disregard predetermines the relevance and quality of the proposed solution(s).

Another problem is that by manipulating processes by, for instance, collecting data and talking to people, the consultant may effectively yield the potential to influence the client's firm, even if on a micro-political level (Armbrüster, 2006). This may lead to altered behaviors among the client's staff as a direct result of the consultant's presence, through what is known as the "Heisenberg effect" (Verlander, 2012). This carries certain similarities to the "Hawthorne effect" and refers to a phenomenon in which the presence of a consultant/researcher affects what is being researched (Simonton, 2010). However, the role of the consultant goes further than that, as a central part of succeeding with actionable analyses is getting the organization to internalize the results of the analysis. The consultant typically helps the organization through the problem formulation where they together contextualize and articulate the problems, thus opening up for the possibility of launching concrete initiatives to address the identified problems (Baer, Dirks and Nickerson, 2013).

3.4. Implementation phase

Finally, there is the implementation phase, which seeks to answer the question concerning "how" a proposed solution could be enacted and integrated into the firm's operations. This phase marks the culmination of the consultant and the client's collaboration. If no implementation occurs, the consultant's efforts are at best incomplete or at worst have failed (Kubr, 2002). It is important for consultants to also be part of the implementation phase if they wish to influence, monitor or oversee the changes being put into practice (Baaij, 2014; Kubr, 2002). It is, however, not the consultant's prerogative to opt whether or not they should take part in the implementation.

Oftentimes, the clients believe they have the necessary skills and capacity to run the implementation by themselves, even though they, more often than not, actually lack the necessary skills. Alternatively, they may lack the finances or the interest needed to fund the implementation phase. In other cases, it is a combination of both of these reasons. In these events, it is difficult, if not to say wrong, to blame consultants for unsuccessful implementations. Consultants whose sole focus lies on a specific area of expertise and who need not concern themselves with the regular business routines of their clients would indeed have more time at their disposal, meaning that they can implement solutions at a far more rapid pace. Moreover, they also possess the required skills and knowledge to carry out these implementations, often drawing upon insights gained from past projects. However, it is important to bear in mind that the effects of consultancy may sometimes materialize after some time has passed after the completion of the project. Likewise, it is not possible to measure the firm's performance had they not chosen to enlist a consultant or vice versa. Hence, it may in some cases be difficult to estimate the causality between consultancy and firm performance (Baaij, 2014). However, as previously discussed, management consultants are often hired to aid top management in assessing a situation and suggesting possible routes forward, rather than actually implementing a solution. In such cases, the possibility of informed choice, rather than an implemented solution, is what the organization gains from hiring consultants.

4. The digitalization of consultancy services

Management consultants of today are devoting much of their time to conducting analysis, possibly in vain, since they do not always know the extent to which their work is actually going to be implemented, or even if it will ultimately remedy the problem at hand (Kubr, 2002; Srinivasan, 2014). A consequence of this is that consultants may find it tempting to opt for the "low-hanging fruits" in the interest of achieving quick results rather than spending time on more profound and complex problems (Chase and Kumar, 2010). Thus, a salient issue that has been subject to much debate is to what extent, if at all, consultants are solving the "right" problem (Spradlin, 2012). That is to say, clients will still continue to experience the need for consultancy services, but there is an increased need for the clients to reduce their risk while still ensuring that they receive sustainable solutions that address the core of their problems (Newton, 2010). What distinguishes the best management consulting firms is their ability to go beyond the quick and sometimes simple solutions to solve the complex problems and achieve real and sustainable change. By being able to do so, they create a reputation for themselves that leads to repeat business.

With digitalization shaping the business environment, we see increasing data availability and ranges of analytics tools, leading to larger datasets to analyze and more data to navigate (Sivarajah et al., 2017; Newman, 2015). Digitalization and the digital transformation is changing the way companies do business and the problems they face, and thus also the consultant's role (Bieler, 2014). This chapter will take a closer look at what happens when the consultant is challenged to adapt to the changing market conditions to stay relevant to their clients.

The consultant's role is heavily influenced by data. Consultants tend to be data-driven in the sense that they often use different data sources in their work, using experience to bridge the gap between data sources (Curuksu, 2018). Clients hire consultants for expertise they cannot get in-house. Will digitalization make these skills available to everyone? If data access and analysis is facilitated by digitalization and made possible for everyone to learn and excel at, then what need is there for consultants at all? Several identified factors seem to decrease the need for external management consultants in the future, such as increasing data availability and increasing availability of analytics tools (Davenport, 2017). These factors, outlined next, might also change the way we look at a consultant project, leading to more modularization of the business.

4.1. Increasing data availability

Once it is simpler for organizations to gather data (internal as well as external data), they do not need to hire consultants to find data that was previously difficult to find. However, consultants will still have important functions to fill, even if gathering such data will be done to much less extent. To this end, datasets still need to be interpreted and consultants are able to infer personal experiences when making these interpretations. That is to say, consultants use business judgment and experience to bridge the gap in poor datasets. Examples of external data sources often used by consultants are industry reports, expert networks with industry experts and research and surveys done by large strategy houses. With increased access to these sources, now most of them are only a Google search or a phone call away, it is no longer necessary to bring in consultants to piece together information.

With higher quality of internal data and ways of gathering data getting better, the need for relying on external input and data sources will likely also decrease. This will decrease the need to rely on management consultants to present facts. However, management consultants also, to a large extent, provide tacit, experience-based knowledge. This allows for rapid diagnosis of situations based on heuristics, which is (barring other instructions) a set of default, go-to rules that have been developed over an extended period of time following a process of repeatedly having to address similar problems in other organizations (Newell, Shaw and Simon, 1959). As contended by Baer, Dirks and Nickerson (2013), problem formulation is a central and complex part of organizational development. For this reason, consultants who possess the ability to do this in a swift manner will likely continue to be in high demand, as opposed to the articulated knowledge of different management solutions, which will become widely accessible through networks of shared data.

4.2. Automation of organizational processes

Organizational processes are invariably complex systems, meaning that they consist of a network of several highly interactive and interrelated elements, with each of these performing its own function (Gino, 2002; Langlois, 2002). In regards to the aforementioned "four phases of consulting" (as mentioned in Section 3), digitalization impacts the first pre-analysis phase in that complex organizational processes will likely become more automatized in the future due to the advancements of various robots and AI-algorithms (Daugherty and Wilson, 2018; Davenport and Ronanki, 2018; Manyika and Sneader, 2018). Consequently, the traditional setup of most organizations of today may very well change in the near future as robots may come to take on increasingly more complicated challenges, requiring no human involvement.

The automation process that follows the digital transformation also entails that organizations are able to operate in a more agile manner, while reducing lead-times (since machines operate faster than humans). This means that it will become more difficult, perhaps even futile, for the consultant to establish relationships with the staff with the intent of gaining insights into how the business operates. That is to say that the hypothesis-driven approach may yield less information if used in the same manner as it has been used hitherto. Moreover, digital systems and algorithms can amass vast quantities of data, meaning that the future consultants will be less likely to contribute "hard facts" that are not already known to the organization and its AI system. Hence, the consultant's "know-what" will become less important in the future, and rather the emphasis will come to rely on the consultant's "know-how", in terms of their ability of tethering out information from complex systems.

4.3. Increased availability of data-analytics tools

Increased data accuracy and higher quality of internal data, combined with an increasingly advanced analytics tool readily available to the public, makes it easier for companies to set up in-house analytics teams (Isson and Harriott, 2016; Bell and Zaric, 2013). This is already in progress with companies such as Walmart, IBM and FedEx, as they rely on analytics teams in order to gain a competitive advantage over their rivals (Bell, 2015; Mochari, 2015).

Reverting back to the four phases of consulting (found in Section 3), the second phase of "problem identification" becomes a salient issue at this stage. During this stage, the consultant devotes much time toward identifying the management problems, analyzing data to understand the root causes of these problems and attempting to devise solutions to these problems. We have seen from companies using in-house analytics teams that such arrangements are especially helpful during the third phase of a project, namely during the data analysis. It is possible that this third phase may in the future be transferred from external management consultants in favor of having it handled by the organization's in-house analytics teams.

4.4. Complex analytics tools

The access to advanced analytics tools will also increase the speed and quality of data analysis, as machines can detect patterns in big data better than humans and are not prone to the same risks of making subjective and arbitrary interpretations as humans. However, many, if not most, of these complex tools will require a sizable amount of training in order to become fully versed in them. This can in turn affect the consultant's work in different ways. One might be that consulting projects will become "modularized", where the client might request a team versed in using a certain tool or skill set. It can also lead to the internal analytics team becoming considerably streamlined, meaning that consultants will have very few sets of skills outside the designated analytics tool, which in turn may lead to a diminishing need for consultants.

While the aforementioned trends would seem to decrease the need for external management consultants in the future, there are also some other factors that work in the consultants' favor. Specifically, one such factor is the consultant's prerogative to ask the right questions, as this is often contingent on the consultant's experience of the subject matter, as well as their decision-making abilities. Another factor working in the consultants' favor in the future is the project-based business model. These factors are discussed in greater detail next.

4.5. Asking the right questions

As previously discussed, the most important factor of a consulting project is the ability to define it accurately. Thus, it is crucial to understand the business objective to delimit the scope of what it is supposed to achieve (Hanna, 2016). Trying to find patterns in large datasets without knowing what to look for will undoubtedly lead to valuable time being wasted. As the need to have someone who knows how to ask the right questions is such an important factor, it is quite possible that there may even be an increase in the need of consultants in the future. Especially with the increased amounts of data, consultants will likely be needed to navigate the data to an even larger extent than today.

A research study conducted by the management consulting firm McKinsey and Company interviewed executives in data-driven businesses (Barton and Court, 2013; Díaz, Rowshankish and Saleh, 2018). The executives agreed that the business objective was crucial. While access to data and tools may increase the speed of the analysis and the possibility to analyze more things than in the past, it is still critical to understand the desired outcome and what problems there are to resolve. This is becoming even more important, since the quantity of data seemingly continues to grow in numbers. The external consultants have an additional advantage from conducting multiple projects within certain functional capabilities, leaving them with experience the client may lack. They also offer an outside in perspective, to look at the business from an external perspective which might be valuable. This is clearly explained by Curuksu (2018, p. 19):

> Predictive analytics may be used to identify risks and opportunities such as economic forecasts, cross-sell/up-sell targets and credit scoring. But the type of intuition that consultants develop to ask questions, pose hypotheses and drive executive decisions is still the realm of science fiction, not existing computer programs. Hence, the arrival of data scientists and big data analytics does not eliminate the need for traditional business professionals.

4.6. Big data does not mean accurate data

While data availability increases, it does not necessarily mean that the data accuracy is high (Delgado, 2015; Schuck, 2018). There are several studies showing the contrary, for example a study conducted by Deloitte (Lucker, Hogan and Trevor, 2017). The data might be from a limited sample, respondents might not answer accurately and so on. Making decisions based on inaccurate data may be even worse than making decisions based on experience combined with data. This will likely keep the demand for management consultants in the future at a stable level.

4.7. Updating extant business models

In adapting to the changes in the business environment following the digital transformation process, some consultancies have opted to capitalize on big data and advanced analytics by extending their service offerings to this category as well, providing niched and specialized services to customers needing assistance in these areas specifically. BCG Gamma and McKinsey Analytics are a couple of examples working in this direction (Curuksu, 2018; Duranton, 2019; McKinsey and Company, 2019). That is to say, rather than losing this market by leaving it up to the client's in-house analytics teams to handle, the consultancies have expanded their service offerings to better accommodate for this type of demand. Boston Consulting Group has adopted an approach where data scientists from BCG Gamma work together with the management consultants to solve the issues clients face (Duranton, 2019; AI Multiple, 2019).

Digital transformation has become a salient part of management consulting, as this transformation constitutes a major change required for their clients to survive in a digitalized world. Management consultants become key players in this regard, as the transition to a digital environment is more about management than it is about technology. Putting digital on the top management agenda, introducing agile working methods and enabling for experimentation are parts of becoming a digitally mature organization (Snow, Fjeldstad and Langer, 2017). To enable digital transformation, management consultants thus aid the organization in designing and adjusting routines in tandem with the introduction of new technologies, making it possible for the organization to use these new technologies to achieve a new, digitally-enabled, state of business. For this reason, management consultants play a pivotal role – while new technologies and analytics are a key component of becoming digitally mature, these tools are of no value unless combined with the relevant management principles. In this sense, digitalization, although often mistakenly regarded as an end state, is in essence an implementation of technological tools, which, combined with appropriate management practices, enable organizations to function in a digitalized world.

4.8. Combining management consultants with data scientists

As previously discussed, the advanced analytics tools will require an extensive amount of training from the consultants' part in order to gain proficiency in them (Consultancy.uk, 2018b). Additionally, these tools will require proficiency in statistics (Tong, Kumar and Huang, 2011). One possible way for management consultants to retain their strong market position would thus be to collaborate with data scientists, who possess knowledge of both statistics and the advanced analytics tools (Flinn, 2018; Granville, 2014). This way, the benefits of the management consultants, such as business intuition, decision-making abilities and the sense for detecting the right questions to ask, may be combined with the technical expertise of the data scientists. This, in turn, leads to strong analytical capabilities. By basing at least part of the analysis carried out by the management consultant on solid data science would in all likelihood improve competitiveness as society moves further into the fourth industrial revolution (4IR).[1] The aforementioned project-based business models would likely facilitate a transition to teams consisting of both management consultants and data scientists. An example of this approach is the previously mentioned Gamma team at Boston Consulting Group.

4.9. The project-based business model and the project processes

Management consultants are often employed on a "need basis". While digitalization improves the convenience of having an on-going business support that provides the organization access to data so that they may make informed decisions based on the available data, there will always be fluctuations in an organization's workload and there will be times when the available staff will not have the ability or the resources to be able to solve the organization's challenges. For shorter and/or irregularly occurring projects with occasional spikes in workload, it will (even in the future) likely be easier to temporarily enlist the services of a trained task force than hiring new people with the right skills and talent.

In truth, we have already witnessed part of the digital transformation of the consultancy industry in the form of cloud-based *Kanban* boards,[2] provided by e.g., Trello, Waffle (GitHub), etc. (Błaś, 2016; Swartout, 2018). There are also more advanced cloud-based project-planning tools that view the whole process in a flow-like manner, where it is possible to zoom into the small parts of the project and add information and comments on the right granularity level. This enables real-time follow-up of the consultants' work, as they update the progress. Commenting the posts with thoughts and questions to be asked, may also facilitate the communication between the consultant and the client. There are possibilities to connect Kanban boards or project-planning tools to communication platforms such as Slack, so that the client is instantly notified when the consultant makes a comment. One may expect these features to develop even more so in the future in order to make the notification scheme even more seamless while upholding fast communication routines.

4.10. Opportunity for scalability, growth and flexibility

Of particular interest to the consulting industry is the strong potential of digitalized business models for scalability. Contrary to traditional consulting, where the number of projects and growth are limited by human resources, technology-based consulting allows scalability and growth without raising cost to a similar level (Werth, Zimmermann and Greff, 2016; Stampfl, Prügl and Osterloh, 2013). Earlier in this chapter we have identified and discussed a number of key areas and strategies that could enable technology-cognizant consultants to gain competitive advantages in the future digitalized economy. By facilitating certain processes to become more technology and customer based, consultants gain a possibility to focus on their primary decision-supporting competences, therefore consulting services can be provided in a more flexible, more individualized and more cost-efficient manner.

Deploying digital technologies will enable routine information-based tasks of consulting to become increasingly more automated and outsourceable, therefore boosting the effectiveness of physical resources, potentially reducing expenses and time invested by consultants in the services. Additionally, the non-routine essential value-generating and business-operating factors will be further enhanced. The digital transformation of consultancy also allows consultants to become more flexible both in terms of time and space, as they are no longer bound by the constraints of having to travel to the client at any one particular location (Nissen and Seifert, 2015). By integrating customers and potential third-party actors, whose help consultants may enlist for specific processes (e.g., statisticians, programmers, interface designers, etc.), into a digital interface, it is possible to facilitate a service process that is more efficient in acquiring and storing information, while providing more economical and individualized solutions. The changing roles and activities of the consultants as they evolve from traditional consulting to digital consulting, should in theory lead to increased scalability, higher growth and flexibility. Latecomers, who are too slow to recognize the potential and to embrace the power of digital technology in consulting, could soon find their services becoming obsolete or too cost-inefficient in order to provide meaningful services in the future world of consulting.

4.11. Further opportunities, risks and implications of digital consulting

The degree of success a consulting service can expect to reap through transforming conventional processes into digital ones is primarily dependent on the consultants' ability to cater for the changing needs of their customers in the technology-oriented market. While digital consulting carries many benefits over traditional consulting (such as greater flexibility, faster lead-times, more cost-efficiency and better catchment area), there is still a discernible resistance to the digital transformation of consulting among a great portion of clients as well as consultants. Indeed, digital consulting has made great headway, with many newer innovations such as web-based file-hosting systems (e.g., Dropbox) becoming more commonplace in everyday consulting use. However, due to a general lack of knowledge and trust in new technologies and their capabilities, many people tend to be skeptical and cautious of using them, at least initially.

The disparity between the standards and practices used in digital solutions as well as in consultancies themselves, is also the cause of significant barriers for widespread implementation of digital consulting. By establishing international and national standards for the services provided via digital consulting, it would be possible to make the future consulting practice more compatible with pre-existing consulting practices, meaning that already established conventions could unequivocally also be part of the new digital consulting practice. In 2017, the *ISO 20700:2017 Guidelines for Management Consultancy Services* were developed as a guideline for people or organizations for the effective management of management consulting services (ISO, 2017). By drawing upon research and experience from a wide array of management consultancies around the world, the *ISO 20700:2017* seeks to increase transparency and effectiveness for clients as well as consultancies and aims to provide practical guidelines based on outcome while emphasizing the importance of understanding the clients' needs (Boler, 2017). To this end, a practical first step could be to update the *ISO 20700:2017* guidelines to establish a set of recognized standards that better reflect the aspects relating to digital consulting. In doing so, it would be possible to further strengthen and increase trust for and acceptance of these types of services.

To ensure maximum benefits at minimal loss for consulting providers, it is essential to clarify and to further discuss some novel opportunities and risks from perspective of both the consultant and the client. Over the past decade, the digitalization process has allowed for reduced direct face-to-face interaction in specific stages of the consulting project (in some cases the interaction may be exclusively digital on a remote basis). Using this virtual approach enables consultants to deliver customized solutions anytime and anywhere while optimizing the workload to gain a sustainable competitive advantage. Besides financial benefits and the improved flexibility of consulting services, such new type of interaction is advantageous for the client as the availability of consulting is not contingent on arranging physical meetings. Moreover, this digital type of interacting reduces much of the waiting times associated with arranging physical meetings, which in turn helps expedite the consulting project. That is not to say that the potential lack of physical meetings is without concern. It is known that face-to-face meetings help strengthen the bond of trust between the consultant and clients (Taylor, Daymond and Willard, 2018; Goman, 2016). While physical consulting meetings take a back-seat, the clients continue placing higher demands on the quality of their consultancy services (Bryder, Malmborg-Hager and Söderlind, 2016; Nissen, 2018). To that end, there is a risk that the reduced direct client-consultant interaction incurs added communication difficulties, a sense of deindividualization and weaker client-advisor relationship.

Another risk consultants must beware of is the fact that digitalization and automation of processes make consulting services increasingly prone to cyberattacks and fraud. Responsible dealing with data and adequate stability of the infrastructure are essential for successful digitalization of consulting services. Moreover, the protection of personal data as well as business data needs to be guaranteed (Schuster, 2005). When developing solutions utilizing digital technology, consultants need both to uphold the client's trust and to offer legally valid data security (Nissen and Seifert, 2015).

The lack of common practices, standards and regulatory framework in information security is an impediment to the implementation of digitalization in the consulting industry. Legal ambiguities are of particular concern, since consulting services are based on large amounts of complex data from various sources. For example, when a consultant working on the behalf of their client, extracts information about consumers from market data and then processes this information using an analytical application, an additional data-privacy approval may or may not be needed depending on the context and legal framework of the country of the client being serviced. Thus, added consideration should be given to relevant security technologies and concepts so that the consultants are completely familiarized with all the intricacies of data security and privacy in an international setting.

All of the aforementioned factors may have damaging effects on the client-consultant trust, which is in and of itself an integral component of consulting (Glückler and Armbrüster, 2003). From a strategic point of view, it is important to establish a feeling of cohesion between the involved parties beyond the limitations of digitalization, consequently it is highly desirable for clients to feel secure about the privacy of their data and to have the continuous support and access to a consultant through personal contact if need-be.

Research studies have shown that rising degree of digitalization of consulting services lead to an observable shift in the clients' expectation and service quality criteria (Nissen, Seifert and Blumenstein, 2015). The personal client-consultant relationship decreases in significance from the client's perspective, whereas factors such as support availability, privacy and data security, reaction capability, efficiency, aesthetics and compensation rise in importance. Given the growing number of clients wishing to have a combination of digital-consulting services and conventional personal consulting, it is essential for consultants to continue to accommodate the client's wishes rather than coerce them into a style that panders to the consultant's convenience at the expense of the client's trust. To this end, it is vital that consultants ensure that they have a secure and stable digital platform and analytics infrastructure, so that the designed digital-consulting products serve to strengthen the trust and relationship with their clients. Nevertheless, a great part of the challenge for future consultants is to ensure that the quality and balance of traditional/digital services live up to the satisfaction of the ever-changing demands of their customers.

5. Conclusion

The premise of this chapter was to explore the future role of management consulting following digitalization and the digital transformation. The chapter set itself out to explore the following two research questions:

RQ 1: *How may digitalization influence the consultant's role of tomorrow?*
RQ 2: *How may the profile of the typical consultant change in the future?*

In doing this, this chapter drafted up a model outlining the four phases of consulting, consisting of the pre-analysis phase, problem identification phase, the analysis phase and the implementation phase (illustrated in Figure 15.1).

In response to RQ 1, this chapter concludes that data analytics tools will play a central role in the future. Above and beyond, it is primarily phase 3, i.e., the analysis phase, that will see the greatest benefits of digitalization. As such, the overall digitalization (and digital transformation) may decrease the perceived need for (external) management consultants in the future, as various forms of analytics tools, AIs, algorithms, scripts, etc. may become available on the market that purports to enable for organizations to take ownership of their own optimization process.

There will undoubtedly surface companies whose business model seeks to capitalize on the advancements of digital tools in order to sell various iterations of customized package solutions to organizations in order for them to optimize their own business performances in the belief that they are saving on consultancy costs. Hence, companies may find it tempting to outsource this task to their in-house data scientists. This, in turn, may have disastrous effects as part of the consultant's role is to help the client contextualize/articulate the problem, something which the client's invariably lack the insight to do on their own accord. Management consultants also possess "tacit knowledge", which means that no matter how much data/information that is made readily available on the open market, the consultants have their own set of heuristics and knowledge of how to facilitate groups, handle organizational politics, stakeholders, etc. This also provides management consultants with the advantage of being able to swiftly assess any given situation based on their own experiences and know-how, while identifying solutions that will work well within a given particular context.

This is not to belittle the future role of data scientists by any means. On the contrary, data scientists possess valuable knowledge of statistics as well as proficiency in how to best use and interpret the advanced analytics tools. To this end, digitalization may actually serve to prompt a more integrated, multidisciplinary arrangement of management consultants and data scientists working in tandem to solve complex organizational problems.

To this end, while the fourth stage, the implementation phase, is where the whole endeavor comes to fruition, it is important to stress that implementation is not always everything. A suggestion brought forth by a consultant that is not implemented is not necessarily tantamount to failure. Sometimes the chief gain from consultancy can be that one becomes aware of one's situation and having all possible scenarios and outcomes presented to oneself and being given a sense of agency to choose one's own direction going forward.

In regards to RQ 2, the role of the typical consultant may change inasmuch that there is an added need for consultants to at least familiarize themselves with the workings of digital tools and what they can accomplish. There will also be a need for consultants to learn to work in closer collaboration with other professions, chiefly data scientists, which will place greater emphasis on the consultants' ability to be "team players". Traditionally, business students have constituted the natural selection of management consultants (Curran and Greenwald, 2006). However, with the digital age emerging, students of more data-oriented and/or technological disciplines can be expected to make a foray into management consulting (Kubr, 2002; Wright and Kipping, 2012). Thus, the importance of multidisciplinary approaches and the ability to communicate across educational backgrounds will become even more important in the digital age.

As digital technology becomes an integrated part of organizational processes, management consultants may, to a larger extent, aid organizations in working with data, rather than trying to reduce latency in manual processes. While management consultants with long-standing practical experience will continue to be a sought-after commodity even in the future, old consultants will eventually retire and new consultants will need to earn practical experience on fresh merits. Thus, in the future, it is likely that management consulting will not only be about being able to know one's way around people, but also (if not more) about knowing one's way around "4IR" technology.

Management consultants will need to work with new technologies in a new digital and innovation-driven economy where clients will want to know how their enterprise can benefit from such digital advancements as blockchain, smart contracts[3] and algorithms (Corrales, Fenwick and Haapio, 2019). Specifically, management consultants will need to offer value that exceeds what digital technology will soon purport itself to do of its own accord (Kelley, 2016; Martin, 2009). Clients will therefore need to enlist consultants that are knowledgeable in these types of technologies in order to provide strategic advice and those consultants who are not competent enough in this area may risk losing their customers to another consultancy. Hence, being tech-savvy will in a way become quintessential in securing the customers' "brand loyalty" to the consulting firm (Corrales, Fenwick and Haapio, 2019). For this reason, fluency in digital will be a central part of core consulting skills, just as integrating systems will be a natural part of organization design and process development. However, this is not to say that the future consultants should forgo their ability to interact with humans and only be hired on the basis of possessing the necessary technological expertise (Erikson and Markuson, 2001). Rather, complexity will increase as today's distinction between human and technology processes will become less obvious, and interfaces between humans and technology will become more sophisticated and less rigid. This will require management consultants to be comfortable in interacting with both people and technology in fast-paced business processes and offer clients contextual insights and proficiency that a mere algorithm cannot. This is yet another argument favoring collaboration between management consultants and data scientists along with other professions of a heavier-set technical background.

With clients wanting advice on how to benefit from digital advancements, one could easily envision a process in which the management consultant is initially hired in order to evaluate the business needs and suggest various technical solutions, such as algorithms for e.g., predictive maintenance, or other types of predictive analysis. Following this example, the management consultant would then engage data scientists or algorithm developers/programmers in order to implement the suggested actions. Following a close working relationship between the management consultants and the technical experts, the clients would have favorable odds of being able to implement cutting-edge technology and reap its rewards, while the management consultants would deepen their knowledge and insight of the technical possibilities without losing sight of their tacit knowledge as previously discussed.

Consequently, rather than launching large-scale business transformation programs involving prolonged change-management efforts, consulting will become more agile as the result and output of change efforts may be instantaneous, making experimentation and iterative problem-solving in short time frames into the standard practice of management consulting. Management consultant profiles may gravitate toward skills within iterative experimental methods in order to fit with the agility of the digital business.

Most essential value-generating and business-operating factors have the potential to be enhanced or automated using digital technologies. However, in order to gain a competitive advantage, these factors need to be constantly attuned to the changes in the wants and needs of the clients, as well as to the market and the technological development. Moreover the digitalization of the consulting industry offers a number of economic advantages. One example is the scalability of virtual (remote) consulting services. Another example is the cost-savings and time-efficiency brought on by automatization of analytics as well as the decreased traveling activities. This, in turn, could open up new market shares and for a new type of client that was previously unable to afford the costly services rendered via conventional face-to-face consulting. Nevertheless, future consultants should take caution of the limitations of digitalization and take as many precautionary measures as necessary in order to preempt and counteract the risks associated with over reliance on digital technologies.

Of course, consultants who are early adopters of digital technology will likely continue to have a head start over those consultants who do not, especially the early adopters who are able to add value through their own creative input. The latter category entails that they have the ability to put their own touch on things and are able to infer unpredictable, but accurate conclusions in a way that induces the same Eureka effect that a machine cannot (Hull, 2002). In this sense, (and tying into the previously answered RQ 1), human consultants will continue to be indispensable to the consultancy profession even in a future where AI has advanced beyond the Turing test[4] (Christian, 2011).

Admittedly, many management consultants of today would already define their work style as "agile" and it is true that the word "agile" has become something of a buzzword that has permeated the consultancy industry for many years to describe a sense of being fashionable and up-to-date with how to implement processes, projects and products (Rigby, Sutherland and Takeuchi, 2016; Fuchs and Golenhofen, 2019; Consultancy.uk, 2018a). However, agile methods will become even more accentuated in a digitalized age and will in many cases form a building block of the consultants' work. This will in turn affect the scope of projects (end-to-end), the consultants' skills and/or team setup, as well as the cost-revenue-structure of the project controlling (Krüger and Teuteberg, 2018).

Nevertheless, the digital age may prompt the consultancy organizations to take on a more agile profile. This is in particular regard to those organizations that deal with large-scale and far-reaching transformations that have hitherto not had the capacity to conduct their work in a faster manner.

Acknowledgements

A special acknowledgement is extended to Vendela Klint for her insights and support in preparation of this chapter.

Notes

1 The fourth industrial revolution (4IR) denotes a fusion of technologies that blurs the lines between the physical, digital and biological spheres via technological breakthroughs in different fields, such as robotization, automatization, Internet of Things (IoT), artificial intelligence, 3D printing, etc.
2 *Kanban* (Japanese: *看板*) is a lean method to manage and improve work across human systems. A *Kanban board* is an agile project-management tool designed to help visualize work, limit work-in-progress and/or maximize efficiency or flow.
3 A smart contract consists of a computer protocol that seeks to digitally facilitate, verify or enforce the negotiation or performance of a contract. These types of contracts allow credible transactions to take place without the need of involving third parties as these types of transactions are trackable and irreversible.
4 The Turing test (named after English mathematician Alan Turing [1912–1954]) denotes a situation in which an AI is able to communicate with a human being via a text-based interface in a way that is indiscernible from another human being.

References

AI Multiple, 2019. *Data Science Consulting & Consultants in 2019: In-depth Guide.* [online] Available at: <https://blog.aimultiple.com/data-science-consulting> [Accessed 24 Sep. 2019].

Albert, B., 2018. *Principled Profits: Outward Success is an Inside Job.* New York, NY: Morgan James Publishing.

Andler, N., 2016. *Tools for Project Management, Workshops and Consulting: A Must-Have Compendium of Essential Tools and Techniques.* 3rd ed. Erlangen, Germany: Publicis Publishing.

Armbrüster, T., 2006. *The Economics and Sociology of Management Consulting.* Cambridge, UK: Cambridge University Press.

Ashkenas, R., 2012. *From Problem Solver to President: What It Takes for a Consultant to Become a Leader.* [online] Forbes. Available at: <www.forbes.com/sites/ronashkenas/2012/09/19/from-problem-solver-to-president-what-it-takes-for-a-consultant-to-become-a-leader> [Accessed 24 Aug. 2019].

Baaij, M.G., 2014. *An Introduction to Management Consultancy.* London, UK: Sage Publications.

Baer, M., Dirks, K.T. and Nickerson, J.A., 2013. Microfoundations of strategic problem formulation. *Strategic Management Journal*, 34(2), pp. 197–214.

Banks, J. and Coutu, D., 2008. *How to Protect Your Job in a Recession.* [online] Harvard Business Review. Available at: <https://hbr.org/2008/09/how-to-protect-your-job-in-a-recession> [Accessed 24 Sep. 2019].

Barton, D. and Court, D., 2013. *Three Keys to Building a Data-driven Strategy.* [online] McKinsey & Company. Available at: <www.mckinsey.com/business-functions/digital-mckinsey/our-insights/three-keys-to-building-a-data-driven-strategy> [Accessed 24 Sep 2019].

Baumann, H.D., 2009. *Building Lean Companies: How to Keep Companies Profitable as They Grow.* Garden City, NY: Morgan James Publishing.

Bell, P.C., 2015. *Sustaining an Analytics Advantage.* [online] MIT Sloan Management Review. Available at: <https://sloanreview.mit.edu/article/sustaining-an-analytics-advantage> [Accessed 24 Sep. 2019].

Bell, P.C. and Zaric, G., 2013. *Analytics for Managers: With Excel.* London, UK: Routledge.

Benn, A.E., Jones, G.W. and Rosenfield, S., 2008. Analysis of instructional consultants' questions and alternatives to questions during the problem identification interview. *Journal of Educational and Psychological Consultation*, 18(1), pp. 54–80.

Berry, A. and Oakley, K., 1994. Consultancies: Agents of organizational development. Part II. *Leadership & Organization Development Journal*, 15(1), pp. 13–21.

Bieler, D., 2014. *Digital Transformation is Changing the Market For Consulting–Observations About Internal Cultural Transformation At PWC*. [online] Forrester. Available at: <https://go.forrester.com/blogs/14-08-15-digital_transformation_is_changing_the_market_for_consulting_observations_about_internal_cultural_trans> [Accessed 24 Sep. 2019].

Biggs, D., 2010. *Management Consulting: A Guide for Students*. Andover, UK: Cengage Learning.

Błaś, R., 2016. Building trust among the clients of cloud computing industry: Case study of Trello, Inc. *Journal of Positive Management*, 7(2), pp. 3–14.

Boler, J., 2017. *New Standard ISO 20700 to Optimize Management Consultancies*. [online] The Auditor. Available at: <www.theauditoronline.com/new-standard-iso-20700-to-optimize-management-consultancies> [Accessed 24 Sep. 2019].

Bryder, K., Malmborg-Hager, A. and Söderlind, E., 2016. *Virtual Business Models: Entrepreneurial Risks and Rewards*. Amsterdam, Netherlands: Woodhead Publishing.

Burtonshaw-Gunn, S.A., 2010. *Essential Tools for Management Consulting: Tools, Models and Approaches for Clients and Consultants*. Chichester, UK: John Wiley & Sons.

Chase, R. and Kumar, R., 2010. Operations management consulting. In: L.E. Greiner and F. Poulfelt, eds., *Management Consulting Today and Tomorrow: Perspectives and Advice from 27 Leading World Experts*. New York, NY: Routledge, pp. 151–172.

Christensen, C.M., Wang, D. and van Bever, D.C.M., 2013. Consulting on the cusp of disruption. *Harvard Business Review*, 91(10), pp. 106–114.

Christian, B., 2011. *The Most Human Human: What Artificial Intelligence Teaches Us About Being Alive*. London, UK: Penguin.

Cohen, W.A., 2016. *Peter Drucker on Consulting: How to Apply Drucker's Principles for Business Success*. New York, NY: LID Publishing Inc.

Conoley, C.W., Conoley, J.C. and Gumm, W.B., 1992. Effects of consultee problem presentation and consultant training on consultant problem definition. *Journal of Counseling & Development*, 71(1), pp. 60–62.

Consultancy.uk, 2018a. *Agile Working is Booming. Five Trends in Agile to Look Out For*. [online] Available at: <www.consultancy.uk/news/17043/agile-working-is-booming-five-trends-in-agile-to-look-out-for> [Accessed 24 Sep. 2019].

Consultancy.uk, 2018b. *How Data analytics are Changing the Consulting Industry*. [online] Consultancy.uk. Available at: <www.consultancy.uk/news/18522/how-data-analytics-are-changing-the-consulting-industry> [Accessed 24 Sep. 2019].

Cooper, S. and Endacott, R., 2007. Generic qualitative research: A design for qualitative research in emergency care? *Emergency Medicine Journal*, 24(12), pp. 816–819.

Corey, D. and Elliott, G., 2018. *Build It: The Rebel Playbook for World-Class Employee Engagement*. Chichester, UK: Wiley.

Corrales, M., Fenwick, M. and Haapio, H., 2019. Digital technologies, legal design and the future of the legal profession. In: M. Corrales, M. Fenwick and H. Haapio, eds., *Legal Tech, Smart Contracts and Blockchain*. Singapore, Singapore: Springer, pp. 1–15.

Cummings, T., 2010. Intervention strategies in management consulting. In: L.E. Greiner and F. Poulfelt, eds., *Management Consulting Today and Tomorrow: Perspectives and Advice from 27 Leading World Experts*. London, UK: Routledge, pp. 279–302.

Cummings, T.G. and Worley, C.G., 2013. *Organization Development and Change*. 10th ed. Stamford, CT: Cengage Learning.

Curran, S. and Greenwald, S., 2006. *Smart Moves for Liberal Arts Grads: Finding a Path to Your Perfect Career*. Berkeley, CA: Ten Speed Press.

Curuksu, J., 2018. *Data Driven: An Introduction to Management Consulting in the 21st Century*. Cham, Switzerland: Springer.

Czerniawska, F., 1999. *Management Consultancy in the 21st Century*. West Lafayette, IN: Purdue University Press.

Daugherty, P.R. and Wilson, H.J., 2018. *Human + Machine: Reimagining Work in the Age of AI*. Boston, MA: Harvard Business Review Press.

Davenport, T.H., 2017. *How Analytics Has Changed in the Last 10 Years (and How It's Stayed the Same)*. [online] Harvard Business Review. Available at: <https://hbr.org/2017/06/how-analytics-has-changed-in-the-last-10-years-and-how-its-stayed-the-same> [Accessed 24 Sep. 2019].

Davenport, T.H. and Ronanki, R., 2018. Artificial intelligence for the real world. *Harvard Business Review,* 96(1), pp. 108–116.

David, R.J., Sine, W.D. and Haveman, H.A., 2013. Seizing opportunity in emerging fields: How institutional entrepreneurs legitimated the professional form of management consulting. *Organization Science*, 24(2), pp. 356–377.

Delgado, R., 2015. *Improving the Accuracy of Big Data Analysis*. [online] Dataconomy. Available at: <https://dataconomy.com/2015/10/improving-the-accuracy-of-big-data-analysis-2> [Accessed 24 Sep. 2019].

Díaz, A., Rowshankish, K. and Saleh, T., 2018. *Why data culture matters*. [online] McKinsey Quarterly. Available at: <www.mckinsey.com/business-functions/mckinsey-analytics/our-insights/why-data-culture-matters> [Accessed 24 Sep. 2019].

Duranton, S., 2019. *Growing Business Through Applied Data Science and Advanced Technology*. [online] BCG Gamma. Available at: <www.bcg.com/beyond-consulting/bcg-gamma/default.aspx> [Accessed 24 Sep. 2019].

Elliott, R. and Timulak, L., 2005. Descriptive and interpretive approaches to qualitative research. In: J. Miles and P. Gilbert, eds., *A Handbook of Research Methods in Clinical and Health Psychology*. Oxford, UK: Oxford University Press, pp. 147–160.

Erikson, R. and Markuson, C., 2001. *Designing a School Library Media Center for the Future*. 1st ed. Chicago, IL: American Library Association.

Fincham, R., Mohe, M. and Seidl, D., 2013. Guest editors' introduction. *International Studies of Management & Organization*, 43(3), pp. 3–10.

Flinn, S., 2018. *Optimizing Data-to-Learning-to-Action: The Modern Approach to Continuous Performance Improvement for Businesses*. New York, NY: Apress.

Fuchs, C. and Golenhofen, F., 2019. *Mastering Disruption and Innovation in Product Management: Connecting the Dots*. Cham, Switzerland: Springer.

Garrette, B., Phelps, C. and Sibony, O., 2018. *Cracked It!: How to Solve Big Problems and Sell Solutions Like Top Strategy Consultants*. Cham, Switzerland: Palgrave Macmillan.

Gino, F., 2002. Complexity measures in decomposable structures. In: *EURAM (European Academy of Management) Conference on "Innovative Research in Management", May 9–11*. Stockholm, Sweden.

Glückler, J. and Armbrüster, T., 2003. Bridging uncertainty in management consulting: The mechanisms of trust and networked reputation. *Organization Studies*, 24(2), pp. 269–297.

Goman, C.K., 2016. *The Immeasurable Importance of Face-To-Face Meetings*. [online] Forbes. Available at: <www.forbes.com/sites/carolkinseygoman/2016/03/11/the-immeasurable-importance-of-face-to-face-meetings> [Accessed 24 Sep. 2019].

Gourguechon, P., 2017. *Empathy Is an Essential Leadership Skill – And There's Nothing Soft About It*. [online] Forbes. Available at: <www.forbes.com/sites/prudygourgue

chon/2017/12/26/empathy-is-an-essential-leadership-skill-and-theres-nothing-soft-about-it> [Accessed 24 Sep. 2019].

Granville, V., 2014. *Developing Analytic Talent: Becoming a Data Scientist*. Indianapolis, IN: John Wiley & Sons.

Greiner, L.E. and Metzger, R.O., 1983. *Consulting to Management*. Englewood Cliffs, NJ: Prentice-Hall.

Griffin, R.W., 2017. *Management*. 12th ed. Boston, MA: Cengage Learning.

Hamann, T.K., 2012. Application of systems thinking in management consulting. In: S.N. Grösser and R. Zeier, eds., *Systemic Management for Intelligent Organizations: Concepts, Models-Based Approaches and Applications*. Heidelberg, Germany: Springer, pp. 195–208.

Hanna, N.K., 2016. Mastering digital transformation. In: E.G. Carayannis and N.K. Hanna, eds., *Mastering Digital Transformation (Innovation, Technology, and Education for Growth)*. Bingley, UK: Emerald Publishing, pp. i–xxvi.

Harrison, F. and Lock, D., 2004. *Advanced Project Management: A Structured Approach*. 4th ed. Aldershot, UK: Gower.

Harrison, M.I., 2005. *Diagnosing Organizations: Methods, Models, and Processes*. 3rd ed. Thousand Oaks, CA: Sage Publications.

Harvard Business School, 2018. *Detailed Charts*. [online] Recruiting. Available at: <www.hbs.edu/recruiting/data/Pages/detailed-charts.aspx> [Accessed 24 Sep. 2019].

Heiser, D.R. and Farah, B.N., 2018. *Problem Solving*. [online] Reference for Business. Available at: <www.referenceforbusiness.com/management/Or-Pr/Problem-Solving.html> [Accessed 24 Sep. 2019].

Hodges, J., 2017. *Consultancy, Organizational Development and Change: A Practical Guide to Delivering Value*. London, UK: Kogan Page.

Hope, K., 2016. *What Does a Management Consultant Do Anyway?* [online] BBC News. Available at: <www.bbc.com/news/business-35220061> [Accessed 24 Sep. 2019].

Hull, J., 2002. *Are Humans Obsolete?: How To Deal With What May Come*. Scotts Valley, CA: CreateSpace.

Hyman, L., 2016. Temps, consultants, and the rise of the precarious economy. *The Hedgehog Review*, 18(1), pp. 18–32.

ISO, 2017. *ISO 20700:2017 Guidelines For Management Consultancy Services*. [online] Available at: <www.iso.org/standard/63501.html> [Accessed 24 Sep. 2019].

Isson, J.P. and Harriott, J., 2016. *People Analytics in the Era of Big Data: Changing the Way You Attract, Acquire, Develop, and Retain Talent*. Hoboken, NJ: Wiley.

Jankowski, A., 2017. *Interactive Granular Computations in Networks and Systems Engineering: A Practical Perspective*. Cham, Switzerland: Springer.

Kelley, B., 2016. *Artificial Intelligence and the Consulting Industry – Innovation Excellence*. [online] Innovation Excellence. Available at: <www.innovationexcellence.com/blog/2016/08/31/artificial-intelligence-and-the-consulting-industry> [Accessed 24 Sep. 2019].

Keuning, D., 2007. *Management: A European Perspective*. 2nd ed. London, UK: Routledge.

Kim, H., Sefcik, J.S. and Bradway, C., 2017. Characteristics of qualitative descriptive studies: A systematic review. *Research in Nursing & Health*, 40(1), pp. 23–42.

Krüger, N. and Teuteberg, F., 2018. Consulting business models in the digital era. In: P. Drews, B. Funk, P. Niemeyer and L. Xie, eds., *Data driven X – Turning Data into Value, Band III, Multikonferenz Wirtschaftsinformatik (MKWI), Leuphana Universität, 6–8 March*. Lüneburg, Germany: Leuphana Universität, pp. 1273–1284.

Kubr, M., 2002. *Management Consulting: A Guide to the Profession*. 4th ed. Geneva, Switzerland: International Labour Office.

Langlois, R.N., 2002. Modularity in technology and organization. *Journal of Economic Behavior & Organization*, 49(1), pp. 19–37.

Liedtka, J., 2006. *Using Hypothesis-Driven Thinking in Strategy Consulting*. Charlottesville, VA: Darden School of Business, University of Virginia.

Lucker, J., Hogan, S.K. and Trevor, B., 2017. Predictably inaccurate: The prevalence and perils of bad big data. *Deloitte Review*, 21, pp. 8–25.

Management Consulted, 2018. *2018 Management Consulting Salaries for Undergraduates, MBAs/PhDs, & Interns*. [online] Available at: <https://managementconsulted.com/consulting-salaries/2018-management-consulting-salaries-for-undergraduates-mbas phds-interns> [Accessed 24 Sep. 2019].

Manyika, J. and Sneader, K., 2018. *AI, Automation, and the Future of Work: Ten Things to Solve for*. [online] McKinsey & Company. Available at: <www.mckinsey.com/featured-insights/future-of-work/ai-automation-and-the-future-of-work-ten-things-to-solve-for> [Accessed 24 Sep. 2019].

Marriage, M., 2018. *Management Consultants Split on How to Make Digital Add Up*. [online] Financial Times. Available at: <www.ft.com/content/6dd91d80-caf8-11e7-8536-d321d0d897a3> [Accessed 24 Sep. 2019].

Martin, R.L., 2009. *The Design of Business: Why Design Thinking is the Next Competitive Advantage*. Boston, MA: Harvard Business Press.

McKinsey & Company, 2019. *About This Practice*. [online] McKinsey Analytics. Available at: <www.mckinsey.com/business-functions/mckinsey-analytics/how-we-help-clients/about-this-practice> [Accessed 24 Sep. 2019].

Mintzberg, H., 1983. *Power in and Around Organizations*. Englewood Cliffs, NJ: Prentice Hall.

Mochari, I., 2015. *5 Ways to Sustain Your Analytics Advantage*. [online] Inc. Available at: <www.inc.com/ilan-mochari/5-ways-to-sustain-your-analytics-advantage.html> [Accessed 24 Sep. 2019].

Momani, B., 2013. Management consultants and the United States' public sector. *Business and Politics*, 15(3), pp. 381–399.

Murphy, E. and Dingwall, R., 1998. Qualitative methods in health services research. In: N. Black, B. Reeves, J. Brazier and R. Fitzpatrick, eds., *Health Services Research Methods: A Guide to Best Practice*. London, UK: BMJ Books, pp. 129–140.

Newell, A., Shaw, J.C. and Simon, H.A., 1959. The Processes of Creative Thinking. In: *Symposium on Creative Thinking, University of Colorado, Boulder, Colorado, May 16*. Santa Monica, CA: RAND Corporation, p. 1320, pp. 1–82.

Newman, D., 2015. *The Role Big Data Plays In Digital Transformation*. [online] Forbes. Available at: <www.forbes.com/sites/danielnewman/2015/12/22/the-role-big-data-plays-in-digital-transformation> [Accessed 24 Sep. 2019].

Newton, R., 2010. *The Management Consultant: Mastering the Art of Consultancy*. London, UK: Financial Times/Prentice Hall.

Nisen, M., 2013. *Why Management Consultants Make So Much*. [online] Business Insider. Available at: <www.businessinsider.com/why-management-consultants-make-so-much-2013-11> [Accessed 24 Sep. 2019].

Nissen, V., 2018. Digital transformation of the consulting industry – Introduction and overview. In: V. Nissen, ed., *Digital Transformation of the Consulting Industry: Extending the Traditional Delivery Model*. Cham, Switzerland: Springer, pp. 1–60.

Nissen, V. and Seifert, H., 2015. Virtualization of Consulting – Benefits, Risks and a Suggested Decision Process. In: P. Pavlou and C. Saunders, eds., *21st Americas Conference on Information Systems (AMCIS 2015), August 13–15, Puerto Rico*. Atlanta, GA: Association for Information Systems, pp. 1380–1391.

Nissen, V., Seifert, H. and Blumenstein, M., 2015. Virtualisierung von Beratungsleistungen: Qualitätsanforderungen, Chancen und Risiken der digitalen Transformation in der Unternehmensberatung aus der Klientenperspektive [Virtualization of consulting services: Quality requirements, opportunities and risks]. In: T. Deelmann and D.M. Ockel, eds., *Handbuch der Unternehmensberatung [Handbok för Management Consulting]*, 25th ed. Berlin, Germany: Erich Schmidt Verlag, pp. 7311–7312.

O'Mahoney, J. and Markham, C., 2013. *Management Consultancy*. 2nd ed. Oxford, UK: Oxford University Press.

Oxford Dictionaries, 2018. *Consultant*. [online] Definition of 'Consultant' in English by Oxford Dictionaries. Available at: <https://en.oxforddictionaries.com/definition/consultant> [Accessed 24 Sep. 2019].

Poulfelt, F. and Paynee, A., 1994. Management consultants: Client and consultant perspectives. *Scandinavian Journal of Management*, 10(4), pp. 421–436.

Rasiel, E.M., 1999. *The McKinsey Way*. New York, NY: McGraw Hill.

Ray, G.D., 2016. Navigating the human landscape in cross-border mergers and acquisitions. *Strategic Direction*, 32(9), pp. 11–14.

Rigby, D.K., Sutherland, J. and Takeuchi, H., 2016. Embracing agile. *Harvard Business Review*, 94(5), pp. 40–50.

Sarvary, M., 1999. Knowledge management and competition in the consulting industry. *California Management Review*, 41(2), pp. 95–107.

Schmidt, A., 2017. *The Phases of Consulting*. [online] Bizfluent. Available at: <https://bizfluent.com/info-8636969-phases-consulting.html> [Accessed 24 Sep. 2019].

Schuck, H., 2018. *Why Data Accuracy Is Critical to the Evolution of Artificial Intelligence In B2B Sales*. [online] Forbes. Available at: <www.forbes.com/sites/forbestechcouncil/2018/05/02/why-data-accuracy-is-critical-to-the-evolution-of-artificial-intelligence-in-b2b-sales> [Accessed 24 Sep. 2019].

Schuster, K., 2005. *E-Consulting: Chancen und Risiken [E-Consulting: Chances and Risks]*. Munich, Germany: Oldenbourg Wissenschaftsverlag.

Scott, B. and Barnes, B.K., 2011. *Consulting on the Inside: An Internal Consultant's Guide to Living and Working Inside Organizations*. Alexandria, VA: ASTD Press.

Senge, P. and Krahnke, K., 2014. Transcendent empathy: The ability to see the larger system. In: K. Pavlovich and K. Krahnke, eds., *Organizing through Empathy*. New York, NY: Routledge, pp. 185–202.

Simonton, D.K., 2010. Heisenberg effect. In: N.J. Salkind, ed., *Encyclopedia of Research Design*. Thousand Oaks, CA: Sage Publications, pp. 563–567.

Sivarajah, U., Kamal, M.M., Irani, Z. and Weerakkody, V., 2017. Critical analysis of Big Data challenges and analytical methods. *Journal of Business Research*, 70, pp. 263–286.

Snow, C.C., Fjeldstad, Ø.D. and Langer, A.M., 2017. Designing the digital organization. *Journal of Organization Design*, 6(1), p. 7, pp. 1–13.

Spradlin, D., 2012. *Are You Solving the Right Problem?* [online] Harvard Business Review. Available at: <https://hbr.org/2012/09/are-you-solving-the-right-problem> [Accessed 24 Sep. 2019].

Srinivasan, R., 2014. The management consulting industry: Growth of consulting services in India: Panel discussion. *IIMB Management Review*, 26(4), pp. 257–270.

Stampfl, G., Prügl, R. and Osterloh, V., 2013. An explorative model of business model scalability. *International Journal of Product Development*, 18(3–4), pp. 226–248.

Swartout, P., 2018. *Continuous Delivery and DevOps – A Quickstart Guide: Start Your Journey to Successful Adoption of CD and DevOps*. 3rd ed. Birmingham, UK: Packt Publishing.

Taylor, M., Daymond, J. and Willard, J., 2018. Building trust across international boundaries: The founder's role in consulting firm's early development. In: M. Zhang, ed., *Trust Building and Boundary Spanning in Cross-Border Management*. New York, NY: Routledge, pp. 86–106.

Tong, H., Kumar, T.K. and Huang, Y., 2011. *Developing Econometrics*. Chichester, UK: John Wiley & Sons.

Tordoir, P.P., 1995. *The Professional Knowledge Economy: The Management and Integration of Professional Services in Business Organizations*. Dordrecht. Netherlands: Springer.

Turner, A.N., 1982. Consulting is more than giving advice. *Harvard Business Review*, 60(5), pp. 120–129.

Verlander, E.G., 2012. *The Practice of Professional Consulting*. San Francisco, CA: Pfeiffer.

Weiss, A.E., 2011. *Key Business Solutions: Essential Problem-solving Tools and Techniques That Every Manager Needs to Know*. Harlow, UK: FT Press.

Werth, D., Zimmermann, P. and Greff, T., 2016. Self-service consulting: Conceiving customer-operated digital IT consulting services. In: *End-user Information Systems, Innovation, and Organizational Change (SIGOSRA), Proceedings of the 22nd Americas Conference on Information Systems (AMCIS 2016), San Diego, CA*. Atlanta, GA: Association for Information Systems, pp. 1–10.

White, A., 2011. *Management Consultancy Careers: What the Experts Said*. [online] The Guardian. Available at: <www.theguardian.com/careers/management-consultancy-advice> [Accessed 24 Sep. 2019].

Wickham, L. and Wilcock, J., 2016. *Management Consulting: Delivering an Effective Project*. 5th ed. Harlow, UK: Pearson.

Wright, C. and Kipping, M., 2012. The engineering origins of the consulting industry and its long shadow. In: M. Kipping and T. Clark, eds., *The Oxford Handbook of Management Consulting*. Oxford, UK, pp. 29–50.

Yoshida, S., 1989. The Iceberg of ignorance. In: *International Quality Symposium*. Mexico City, Mexico.

16 Digitalization, circular economy and the future of labor

How circular economy and digital transformation can affect labor[1]

Anthony Larsson and Linn Lindfred

1. Introduction

Circular economy is an economic model that offers a way of producing and consuming goods while simultaneously allowing mankind to stay within the planetary bounds (European Commission, 2014; Stahel, 2016; Ellen MacArthur Foundation, 2015c). Notwithstanding, changing the global economic system means substantial implications in many, if not all, areas. Specifically, there will be a greater need for human employment in a circular economy and the growth will be based on human capital instead of the extraction of natural resources (Groothuis, 2015). However, the transition to a circular economy relies on tax shifts and reforms in legislation, which in turn also carries an impact on labor (Groothuis, 2015; Wijkman and Skånberg, 2015; Lewandowski, 2018).

While such tax shifts and legislation reforms may seem straightforward in principle, the transformation from a linear to a circular economy is laden with many other obstacles as well, and for that reason it is necessary to discuss the tools available to overcome them. Obstacles range from "hard values", such as policies and laws to such "soft values" as individual behavior, lack of knowledge and lack of data. Digital technology provides a tool that can overcome some of those barriers. In fact, an increasing number of researchers are now discussing the role of digital technology as a catalyst for achieving a more sustainable society and the necessity of using it in order to achieve circular economy (Antikainen, Uusitalo and Kivikytö-Reponen, 2018; Ellen MacArthur Foundation, 2015d; Stuchtey, Enkvist and Zumwinkel, 2016). Many researchers have addressed how digital technologies at-large affect society. Fewer are the numbers of researchers who have investigated how society in general and labor in particular is affected when digital technology is used to propel and enable circular processes and circular economies. Overall, there has been a lack of research in how social sustainability is affected by a circular transition (Schroeder, Anggraeni and Weber, 2019). Thus, this conceptual study seeks to draw upon available literature and research findings with the aim of answering how the labor conditions are affected when digitalization is used to achieve circular businesses in different ways.

2. Linear vs. circular economy

What defines a business model varies greatly and is indeed subject to much debate. However, in its most essential form, it can be said to describe how an organization creates and captures value (Kaplan, 2012). The feature of this model defines the organization's customer value proposition and pricing mechanism. This in turn indicates how the company organizes itself and how it structures its supply chain (Kavadias, Ladas and Loch, 2016; Lahti et al., 2018). An economic model, on the other hand, is a framework that is used to conceptualize a complex process in a comprehensible manner. Of course, there are many different types of economic models, but when discussing societies at large, the economic model that has characterized societies since the beginning of industrialization is called the "linear model" (World Economic Forum, 2014). The linear economic model is based on extracting resources, producing items/products, using them and ultimately disposing of them after their initial use (Stahel, 2016; Ellen MacArthur Foundation, 2015c; Ellen MacArthur Foundation, McKinsey & Co and Google, 2019).

The linear model maximizes throughput without considering the environmental and economic cost of the large quantities of waste, the environmental pollution, the depreciation of value and the depletion of new resources that follows with it (Ellen MacArthur Foundation, 2015d; Stahel, 2016; Wijkman and Skånberg, 2015). In fact, a decisive majority of products' original value today are lost after the first intended use (Wijkman and Skånberg, 2015; Wastling, Charnley and Moreno, 2018). In Europe alone, statistics show that 60% of the discarded materials in 2012 went to landfill or incineration whereas the remaining 40% were recycled or reused as materials (Ellen MacArthur Foundation, 2015c). This means that 95% of the material and energy value was lost after the products' initial use. Thus, only 5% of the raw material value was preserved in material recycling and waste-based energy recovery. Of particular concern is plastic packaging, as indicated in a report by the World Economic Forum (2016, p. 6):

> After a short first-use cycle, 95% of plastic packaging material value, or \$80 billion to \$120 billion annually, is lost to the economy. A staggering 32% of plastic packaging escapes collection systems, generating significant economic costs by reducing the productivity of vital natural systems such as the ocean and clogging urban infrastructure.

As a result of the production and consumption patterns in our world economies, the resources are increasingly scarce and the ecological footprint is growing continuously larger (Fu et al., 2015; Stuchtey, Enkvist and Zumwinkel, 2016; Ellen MacArthur Foundation, McKinsey & Co and Google, 2019). Thus far, 2018 has placed itself as the fourth hottest year in history with record high temperatures all over the globe, with only three other years having been hotter at various places across the globe: 2015, 2016 and 2017 (Levenson and Miller, 2018; NASA, 2018; Sengupta, 2018). The pattern of elevating temperatures indicates that the threat of global warming is clear and present. The negative environmental externalities is foremost a result from handling materials in the linear economy (Maitre-Ekern, 2018; World Economic Forum, 2014; Horvath et al., 2018). More than half of all greenhouse gas emissions is a consequence of materials-management activities and the current trajectory implies that greenhouse-gas emissions related to materials management will more than double in 2060 (OECD, 2019).

At the same time as we are crossing the planet's boundaries for environmentally safe operations each day in order to supply the global consumption demand and keep our industries afloat, the global consuming middle class is expected to increase from 1.8 to 4.9 billion people by 2030, meaning that demand for all commodities will increase from 30% to 80% the same year (Batra, 2017; OECD, 2011a). Today, the use of resources has more than tripled since 1970, resulting in 90% bio-diversity loss and water stress (the lack of sufficient access to potable water), which in turn stems from how we extract resources and produce products (UN environment and International Resource Panel, 2018). Considering the implications that have already been incurred on the planet, there is a pressing need to find a new way to serve the emerging demand while staying within our planetary bounds. The linear economy is reaching the end of its physical limit and it does not proffer a solution to the equation of meeting a growing demand while at the same time lowering environmental pollution (Ellen MacArthur Foundation, 2015d; Bonciu, 2014).

In contrast, the circular business model was formed as a response, and solution, to the increasingly crucial problems and threats to future life on this planet (Murray, Skene and Haynes, 2017; Bettencourt, 2018). At bedrock, the circular business model should be viewed as a means of capturing value while aspiring to achieve an ideal state of resource usage (Lahti et al., 2018). The exact definition of circular economy is still subject to much debate, as the definitions range in the hundreds (Kirchherr, Reike and Hekkert, 2017). Nevertheless, one of the most elaborate and exhaustive definitions thus far has been articulated by the French Environment and Energy Management Agency [*Agence de l'environnement et de la maîtrise de l'énergie*] (ADEME) (Gallaud and Laperche, 2016). The agency defines circular economy as "a system of exchange and production which, at every stage of the product lifecycle (goods and services), is aimed at increasing the efficiency of use of resources and reducing the impact of production activities and consumption on the environment" (Gallaud and Laperche, 2016, p. 8).

Circular economy disrupts and rethinks the system as we know it. The circular economy is based on the mindset of keeping all material at its highest value at all times (Lahti et al., 2018; Hannon, Magnin and Rosenfield, 2016; Linder, 2017). The circular economy concept describes an economy where the products are reused for as long as possible where after they are recycled and taken care of when they can no longer be reused (Lahti et al., 2018; Stahel, 2016). In this way, it is possible to retain as much of the original value as possible for as long as possible and design out, and thus eliminate, waste. When the purpose is to increase the utility rate and the lifetime of the product, the product needs to be designed for wear and tear and to hold the quality for longer. It thus means that the design of the product is central for a circular economy to function (Lund University et al., 2018). Product and concept designer Mäkilä means that in a circular economy it is necessary to design the product's life cycle and not only the product itself (SB Insight, 2019). The products and materials go through one of two life cycles, either the biocycle, where products and materials are returned to the biosphere and regenerated, or the technical cycle, where products and materials are recovered and restored (Ellen MacArthur Foundation, 2015d).

Table 16.1 illustrates that the different ways of capturing value and eliminating waste throughout the product lifecycle are plentiful and differ depending on the stakeholder, perspective and position in the value chain (Lund University et al., 2018).

In this context it should be noted that circular economy advocates performance and/or access over ownership, meaning that the customer does not necessarily need to own the asset in question, but purchases the right to use it. Done right, this leads to an increased utility rate of the asset, a reduction of total asset cost and a potential increase of the retailer's profitability and customer service (Ellen MacArthur Foundation, 2016). It is thus necessary to adopt a mindset that thinks

Table 16.1 Illustration of efforts required to increase circularity at different levels.

At the level of the consumer	*At the level of the product cycle*		*Between different product cycles*
Extending product lifetime *Products that hold the demanded qualities over a long time span.*	**Repair** *Correcting a fault to enable longer lifetime.*	**Material recovery** *Using industrial residues from same production chain.*	**Parts reuse** *Using parts of used products in other production chains.*
Dematerialization *Offering the value without selling a physical product. Using less or no material (i.e., streaming music).*	**Reuse/share** *The multiple use of a product in different time frames (reuse). Or common use of a product during a specific time frame (share).*	**Close the loop** *Valorization of materials after initial use in the same production chain.*	**End-of-life recycling** *Recycle the product, convert into reusable materials to use in other production chains.*
	Refurbish/ Remanufacture *Perform change to used product – to original specifications (remanufacturing) – to satisfactory working conditions (refurbishing).*	**Secondary raw materials** *Materials recycled from other production chains.*	**Recycling of industrial residues** *Using industrial production residues in other production chains.*

Source: Table adapted from "Circular Economy: Sustainable Materials Management" by Lund University et al. (2018).

beyond the conventional "take-make-dispose" model, and open one's mind to new business models that can solve the inherent wastefulness of linear economies. For instance, an electric drill is on average used between 12–20 minutes in a lifetime, and the average car in the Western world is usually used for three months during an average lifetime of 12 years, while some sources would have that the car is unused for approximately 22 hours a day (Lund University et al., 2018; Kessler, 2015; Botsman and Rogers, 2010). In the former case, this gives a utility rate of a car at 0.02%. Still, both the car and the electric drill are items that are presently commonplace possessions in many households.

The need of drilling could be solved in a much more efficient, environmentally friendly and economic way. Offering distribution and access of the drilling machine instead of selling it, would mean that less products are created, resulting in less extraction of new resources and ultimately a reduction of waste. While cars are in and of themselves not inherently environmentally friendly (save for, arguably, electric cars), cars cause a lot more waste and pollution than they need to. That is to say, the aforementioned example of the car could also benefit from adopting a circular practice, by ways of utilizing such examples as carpooling, car sharing and rental services to a much greater extent. In many cases, it is the service we need rather than the product in itself. A circular economy based on distribution and access will in this way create new patterns of interactions between people and necessitate a change in how we own, use, consume and interact with our products (Lund University et al., 2018; Markopolu et al., 2019). In short, this would entail radical changes to the way we currently live our lives. From a business perspective, leasing the electric drill would enable earning more from each product and potentially also offer the service of drilling, which would mean more man-hours and by extension, more job opportunities.

The examples of the electric drill and the car are but two of many that illustrate the inherent wastefulness of the linear economy. It would appear as though the linear economy has ultimately forgotten the purpose of why products are created in the first place i.e., to be used. To name a few more examples, manufacturers in the textile industry often produce two garments in order to be able to sell one of them. This means that one of two garments will end up either in landfill, get incinerate, or will be sold at a discount price due to fluctuating demands. On top of this, the clothes that are actually bought are oftentimes used sporadically at most, while some clothes are never even worn at all. In fact, in Great Britain, 30% of all clothes bought in households are never worn (Lund University et al., 2018). In a similar way, our current food system is wasteful (Ellen MacArthur Foundation, 2019a). In our global economy, 1.3 billion tons of food is wasted yearly, which means that one-third of all food produced globally is wasted (Food and Agriculture Organization of the United Nations, 2019). The Ellen MacArthur Foundation (2019a, p. 8) writes that "Overall, for every dollar spent on food, society pays two dollars in health, environmental and economic costs. Half these costs – totaling USD5.7 trillion each year globally – are due to the way food is produced".

According to Wijkman and Skånberg (2015, p. 5) "Resource constraints as well as increasing volumes of waste and pollution are likely to impose increasing threats to welfare and wellbeing and, from a business point of view, to competitiveness, profits and business continuity". An increasing number of corporations have seconded this statement and have begun acknowledging this present need for a new model. For instance, the CEO of Philips, Frans van Houten, already in 2015 stated that, "Two years ago, we decided to embed circular economy thinking in our strategic vision and mission, both as a competitive necessity and with the conviction that companies solving the problem of resource constraints will have an advantage" (Ellen MacArthur Foundation, 2015a, p. 3). Apple (2019, para.2) has stated, "Ultimately, we want to make products using only renewable resources or recycled material. And we want to return an equivalent amount of material to the market, to be used by us or others. Our ambition is that one day we'll extract nothing from the earth". Also, H&M incorporates circular economy in their long-term strategy by having a defined goal of becoming 100% circular and renewable by 2030 (H&M Group, 2018).

A report by Material Economics (2018) on the plastic, cement, aluminum and steel industry showed that should an ambitious circular-economy scenario be reached for these materials by 2050, their CO_2 emissions would be cut by 56%, saving 250 million tons of CO_2 emissions annually in the EU and 3.6 billion tons annually globally. A key takeaway from the report was that circular economy should have a prominent position in EU climate policy, since utilizing resources and products more efficiently will enable European industrial growth while decreasing industrial pollution.

3. Circular economy and job creation

Transitioning to a circular economy does not only imply a reduced environmental impact and a competitive advantage for businesses, it also generates economic growth and creates jobs (Morgan and Mitchell, 2015; Kalmykova, Sadagopan and Rosado, 2018; Stahel, 2016; Wijkman and Skånberg, 2015). In fact, according to Stahel (2016) the concept of a circular economy arose from the idea of substituting manpower for energy, as a solution to the increasingly high energy prices and high unemployment that characterized the early 1970s. The idea was based on the logic that it requires more labor and fewer resources when refurbishing goods and products than when creating new ones. When concepts such as "reuse", "recycling" and "extended product life" is the goal of the economy (circular economy), it is thus by definition more labor intensive than an economy where products are consumed and wasted (linear economy) (Wijkman and Skånberg, 2015; Stahel, 2016; Aurich, Fuchs and Wagenknecht, 2006). The process of taking care of goods and products (maintaining, remanufacturing and repairing) creates skilled jobs in local businesses. According to the Ellen MacArthur Foundation (2015c, p. 14), the new business services that would require skills and process know-how in ways that generate job opportunities in a circular economy are:

- Collection and reverse-logistics[2] firms that support end-of-life products being reintroduced into the system

- Product remarketers and sales platforms that facilitate longer lives or higher utilization of products
- Parts and component remanufacturing and product refurbishment offering specialized knowledge

The circular economy is thus more dependent on people being involved in the economy in different ways but not only because it involves more labor-intense processes but also because of the circular business models like sharing, renting and leasing. Such business models require more interactions between people since the number of touch points increase over the lifetime of a product (Stahel, 2016; Ellen MacArthur Foundation, 2015c). All the while as sharing economies becomes more widespread, they will affect the incumbent industries in different ways (Stuchtey, Enkvist and Zumwinkel, 2016). Two examples are the automotive and hotel industries. The sales of new cars and hotel nights will ultimately begin to decrease as more consumers will choose to access and share cars rather than owning one. Likewise, more people will opt to stay in lodgings provided by private citizens rather than in hostels or hotels. Increased utilization would also mean that offices would be used differently. Such a development would likely affect today's real-estate owners. Already now, coworking spaces are surfacing increasingly more often as alternate office spaces (Kojo and Nenonen, 2017). Some services, such as UK-based Airbnb-style website Vrumi, arranges for users to book a desk space in a stranger's apartment or home for a few hours during the day with average rental rates ranging between USD24 and USD366 per day (Roos, 2016). This does not necessarily mean that jobs will be lost, rather that the labor market will change and new jobs and business models will evolve (Stuchtey, Enkvist and Zumwinkel, 2016).

According to the Ellen MacArthur Foundation (2015c), circular economy implies an increased economic growth, net material cost savings, job creation and increased innovation. Their study shows that if one were to use digitalization as an enabler, the circular economy could increase the resource productivity up to three percent annually in Europe. This, in turn, would generate a primary resource benefit of €0.6 trillion (≈ USD671.9 billion) annually by 2030, with an additional €1.2 trillion (≈ USD1.35 trillion) in other benefits, such as non-resource and externality benefits (Ellen MacArthur Foundation, 2015c). This means that the annual total benefits in the coming years would be €1.8 trillion (≈ USD2 trillion) as opposed to what it is today.

The Ellen MacArthur Foundation (2015c, p. 14) conducted a review of 65 academic studies on circular economy and its impact on labor, concluding that "existing studies point to the positive employment effects occurring in the case that a circular economy is implemented". The report states that the result is both due to increased spending because of the overall expected lower prices and due to recycling and remanufacturing activities requiring human labor. However, the study also found that some companies would most likely not benefit from increased labor opportunities if they are too slow to adapt and act.

Wijkman and Skånberg (2015) investigated the social benefits of moving toward a circular economy using the Dutch, Finnish, French, Spanish and Swedish economies as test cases. They found that the transition to a circular economy for the investigated countries would imply reductions in carbon emissions as well as new job formations. More specifically, the authors looked at three key steps to enabling a circular scenario. They were: (1) increasing energy efficiency; (2) increasing the percentage of renewable energy (by cutting fossil-fuel use in half and substituting it with renewable energy sources, as for example wind, solar and biofuels); and (3) increasing material efficiency (extending wealth, minimizing waste and maximizing the reuse and recycling of materials). The result showed that if all of the key steps previously listed were introduced in the investigated economies, carbon emissions in Spain would potentially be cut structurally by two-thirds or more (by almost 70%). The new job opportunities would increase to over 75,000 in Finland, 100,000 in Sweden, 200,000 in the Netherlands, 400,000 in Spain and 500,000 in France. As a result, the unemployment rates could be reduced by: a third (and possibly more) in Sweden and the Netherlands; 15–20% in Spain; a third in Finland and almost by a third in France. The trade-balance improvements were predicted to approximately 1.5% of GDP in all of the studied countries.

A report by the European Commission (2018a) investigated the impacts of circular economy policies on the labor market and found that it is possible to become more resource-efficient and increase employment at the same time since the circular economy will reduce negative environmental impacts and result in higher employment levels. The report studied trends within circular economy across sectors in Europe and found that taking steps toward a circular economy would result in an increase in GDP by approximately 0.5% by 2030, compared to the baseline case. It would also result in a 700,000 net increase in jobs compared to the baseline due to a higher labor demand from recycling plants and repair services.

Taking care of the product, i.e., repairing, maintaining, upgrading and remanufacturing, are all activities that require human labor. On the other hand, activities for mining and manufacturing products do not necessarily require human work but is nowadays often automatized (Wijkman and Skånberg, 2015). In this way, a tax shift is necessary in order to allow for a socially and ecologically sustainable society to evolve. As it were, taxation in the industrialized countries of today is, by and large, much too dominated by taxes on labor, when there is a need for taxes to rather be based on the extraction, use and waste of natural resources (Wijkman and Skånberg, 2015; Pomerleau, 2014; Mendoza, Razin and Tesar, 1993). Lowering taxes on labor and increasing taxes on the consumption of virgin materials would reduce the use of these resources and speed up the shift toward a circular economy (Wijkman and Skånberg, 2015).

Policy-making around resource productivity has been very rare while labor productivity has been the priority historically (Wijkman and Skånberg, 2015; Bin and Vassallo, 2016; OECD, 2011b). Wijkman and Skånberg (2015) argue that the focus should be on reducing energy and material throughput in society and that circular economy needs to be considered both as an environmental issue and as an integral part of jobs and competitiveness strategies. They further contend that the level of resource use is rarely considered in a society and they mean that the climate change mitigation strategies should be more holistic by having resource efficiency as the key instrument.

Repairing and recovering has not been a standard in the linear economy where oftentimes used products are seen as waste. In order for a circular economy to emerge it is thus necessary to create a market and a demand for the used materials and products. This means that new industries, prepared for using second hand products and spill materials to create something valuable and usable, will emerge. Taking a product and/or material and creating a new product that has equal or higher value than the original product is called up-cycling (Lund University et al., 2018). Up-cycling is a way of creatively finding new applications for traditional materials or products and in that way keep the value, and in some cases even extending the value of the product or material as far as possible. An example of this is the Finnish company Pure Waste (2019) that takes textile production spill and uses it to manufacture clothes. Up-cycling could thus be performed both through automatized processes but also through human labor. Nevertheless, the up-cycling industry could create new businesses and job opportunities. Hermann Erdmann, CEO of Redisa mentions that:

> On a planet of finite resources, the circular economy is not optional, it is inevitable. Its implementation will provide world economies with unprecedented opportunities, through the creation of reverse logistics networks, new processes, and new industries using the recovered resources. Resource efficiency will allow us to rethink the concept of urban mining. Countries will be able to create industries in fields that were previously not viable. Relatively simple changes to existing legislation can enable this shift in mindset on short timescales. Restructuring economies to become circular will moreover bring with it enormous environmental benefits.
>
> (Ellen MacArthur Foundation, 2015c, p. 8)

Several researchers argue that the new and more sustainable economy will need to be based on more local farming and manufacturing in order to supply the demand with fluctuating resource prices and scarce supply (Lund University et al., 2018; Dunbar, 2017; Faludi, Cline-Thomas and Agrawala, 2017; OECD, 1998; Ijomah et al., 2007). Using 3D-printing in order to manufacture spare parts and whole products will enable for small-scale local manufacturing, which, in turn, creates local job opportunities (Faludi, Cline-Thomas and Agrawala, 2017).

According to research presented by Lund University et al. (2018), the value created when transforming to a circular economy can be compiled into three areas:

- Securing global resource availability
- Preserving the ability of natural systems to deliver goods and services to society
- Spurring development of new technologies, new norms and new institutions that can support and stimulate society.

In other words, circular economy fosters socio-economic development and stimulates employment, economic growth, and flow-on[3] social benefits concerning boosting of the recovery, recycling and upgrading of materials and creates new business models, while also securing more man-hours and more job opportunities.

4. Circular economy and digitalization

Transforming society and industry is not easy, and even though circular economy proposes a way to that, it implies difficulties, obstacles and uncertainties. However, one tool has been found to be an integral aspect to the acceleration of the circular economy; digitalization (Antikainen, Uusitalo and Kivikytö-Reponen, 2018).

We are currently in the midst of the fourth industrial revolution, or the fifth technological revolution, depending on how you categorize, called the age of Information Technology and Telecommunications (Perez, 2018). This means that digital technologies are cheaper to use and more wide spread than before. Today digital technology is being used by a majority of the western organizations and institutions (OECD, 2018b). Oftentimes, digitalizing companies means less manual work and more efficient working processes with the potential to transform processes and businesses and enable new ways of working. In many cases, digital technology has been used with the purpose of increasing efficiency in our current economy. Streamlining linear companies would most likely mean speeding up consumption, thereby increasing waste and increasing the extraction of new resources. This, in turn, perpetuates a vicious circle, as it only serves to expedite the development toward creating a society that is ultimately unsustainable.

According to various scholars, such as Rifkin (2011, 2014) this new industrial revolution is reshuffling our most fundamental system. This new revolution would present us with endless opportunities, had it only been that we had learned how to adjust and reap the benefits. The point of a new industrial revolution has been further reiterated by Klaus Schwab, the founder of the influential NGO *World Economic Forum* (WEF), in his 2018 guide, by drawing on contributions by more than 200 of the world's leading technology, economic and sociological experts (Schwab and Davis, 2018). Also, Perez (2018) elaborates on the current industrial revolution in her research about techno-economic paradigm shifts in society. She finds that each historical revolution shows the same pattern of development. Based on that pattern, we are currently in the right place to form the outcome of the Information Technology and Telecommunications (ICT)-revolution. She means that the time is right to use technology to reshape the future and enable a more sustainable and circular society that benefits business, people and the environment (Perez, 2018). However, Perez' research also shows that if we reach the maturity stage of the revolution without having used the technology in the right way, this opportunity will be lost.

In a commentary, the European Policy Center has stated that:

> In short, the transition towards a smarter use of resources will only make headway if Europeans make the most of digitalisation. This implies understanding the linkages and exploiting the synergies between the digital and the circular economy agendas. The EU must promote the smart use of data and

> digital solutions with a view to encouraging the transition towards a circular economy. If Europe does not move in this direction, opportunities to close material loops and improve processes will be missed.
>
> (Pardo, 2018, p. 2)

Presenting their digital roadmap for a circular economy, the European Policy Center adds that digitalization enables information transfer; sustainable business models; circular products, services and processes; as well as the ability for companies to scale up and profit from the circular economy (European Policy Center, 2019).

In addition, Stuchtey, Enkvist and Zumwinkel (2016, p. 31) contend that digital technology has enormous potential to help transform society into a more sustainable one, stating that, "We have the most powerful tool in history right at our fingertips". However, they also emphasize that the technology disruption will not foster a sustainable society on its own referring to "the technology disruption is a beast in need of taming" (Stuchtey, Enkvist and Zumwinkel, 2016, p. 199). Thus, realizing the value that digitalization can bring with in terms of positive environmental impact and using it to foster just that is of great essence.

Digital technology and sustainability overall, are two big and complex areas that are, in fact, interlinked. What we know about the environment today, and how we predict what it will look like in the future, is to a great extent thanks to measurements, smart sensors, open data and analytics. Digitalization has thus increased our knowledge and understanding, allowing for more informed decisions based on accurate data. Also, while digitalization of the industrial sector increases resource efficiency, digitalization may also help close the loop of material cycles and contribute to keeping goods/materials in use for a longer period of time (Wilts and Berg, 2018). For instance, intelligent solutions enable the reduction of energy consumption, while optimizing the logistics chains in addition to providing for a more efficient use of capacity (Marinescu, 2015). Moreover, through digitalization it is possible to acquire access to material-specific data and resource consumption (Austin, 2016; Woetzel et al., 2017). This, in turn, enables the optimization of product life cycles for circular-economy solutions. It is important to remember that circular economy is by and large based on distribution and access (Markopolu et al., 2019; Lund University et al., 2018). This means that digital technology is vital in order to keep track of the information flow, transactions, logistics and communication between all actors in the value chains.

To illustrate in more detail how digital technology could be used to catalyze a circular economy, the following section is divided into how digitalization in this way can facilitate the design phase, the use phase and the end-of-life phase of the product/service.

4.1. Design

In order for businesses to create more circular products fit for longer lifetimes and multiple life cycles, data is necessary. Secure data and increased transparency could potentially be provided with blockchain (Ellen MacArthur Foundation,

McKinsey & Co and Google, 2019). Further, several databases and digital material banks are being set up to provide the data necessary to enable finding the right information about a product or material. One such example is the EU Science Hub called Raw Materials Information System[4] set up by the European Commission (2019). In a similar way, the industries have developed various self-assessment standards for assessing environmental and social sustainability throughout the supply chain. These indexes are based on large quantities of data, which enable direct feedback in the form of a score/index of how sustainable a material or product is depending on selected parameters. One such example is the Higg Index from the Sustainable Apparel Coalition, which can measure and score a company's or product's sustainability performance in the textile industry (Sustainable Apparel Coalition, 2019). Data can also be gathered from the reverse logistics and the take-back of products which allow for capturing more information and feedback about the products. This feedback is valuable when designing the products. This is presented in more detail later on in the end-of-life section.

4.2. Use/reuse

In order to achieve a circular economy, it is important for the user to accept alternative means of consumption. Specifically, this entails renting and sharing products rather than buying and owning them. In order to foster and facilitate a change in behavior, digital solutions such as digital platforms or marketplaces could be of assistance. Digital technology has made sharing and virtualization much more attractive over the last decade and has resulted in a better utilization of products (Stuchtey, Enkvist and Zumwinkel, 2016). Already today, several sharing economy platforms such as Airbnb, Hygglo, Sunfleet, DriveNow as well as consignment platforms, such as VSP Consignment, and secondhand platforms such as Etsy, eBay, Craigslist, Thredup, The Real Real, Sellpy and Blocket, etc. constitute examples of how to effectively increase the utilization level of products through digital solutions. A prerequisite for sharing economies, consignment and secondhand platforms to function is also having matching algorithms and dynamic pricing (Ellen MacArthur Foundation, McKinsey & Co and Google, 2019).

In order to help the consumer use the product in a more sustainable way, and prolong the lifetime of the product, it is important to both provide the right information on how to do so, as well as to facilitate the process. Digital solutions could increase the communication between business and consumer. When the physical or non-physical product is connected to the cloud, it creates an opportunity for the business to enhance customer value through better communication and information sharing (Ellen MacArthur Foundation, 2016). Further, a digital tag/sensor could gather data of how the product is used and give feedback to the user on how to handle the product more sustainably. Researchers have been discussing the possibility of including digital tags in products that could function as a "material passport", allowing for tracking and tracing, while being able to identify the condition of the product at all times (Ellen MacArthur Foundation, 2016; Guldager Jensen and Sommer, 2016). To date, several companies have been investigating the possibilities of incorporating these chips in different types of products (Ellen MacArthur Foundation, 2019b; Turntoo, 2019; Heinrich and Lang, 2019; C2C-Centre, 2019). Increased digital communication also fosters the ability to offer services to the user instead of merely selling products. Ultimately, this means that the interactions between business and consumer will increase.

Moreover, repair processes are today oftentimes optimized through the use of connected products and predictive maintenance (Ellen MacArthur Foundation, McKinsey & Co and Google, 2019; Ellen MacArthur Foundation, 2016). Connected products allow for proactively sending out automatic notifications if a product needs repair or if a product is beginning to break (Ellen MacArthur Foundation, McKinsey & Co and Google, 2019). Predictive maintenance can be enhanced by the use of artificial intelligence (AI) (Lund University et al., 2018; Ellen MacArthur Foundation, McKinsey & Co and Google, 2019). Also, new technologies such as 3D printing, allows for printing spare parts to facilitate repairs (Ellen MacArthur Foundation, McKinsey & Co and Google, 2019). In addition, repairing products by yourself have been made easier through digital communication tools (Lund University et al., 2018). Through platforms like YouTube, Fixperts and Instructables, knowledge and skills could be shared globally.

Dematerialization has been achieved and gained popularity through the use of online streaming services i.e., for streaming videos and music (Lund University et al., 2018). This has almost eliminated the need for producing CDs and DVDs. Famous examples are Spotify, YouTube, Viaplay and Netflix.

4.3. End-of-life

Digitalization has shown to be crucial for all activities concerning the take-back of products and valorizing them after initial use. The take-back of products means that the coordination of information flow and materials/products are vital to the circular economy (Masi et al., 2017; Ellen MacArthur Foundation, 2016). That is to say, the information regarding the quality and quantity of products, and the raw materials the products contain, must be efficiently gathered, stored and used. To this end, it is essential that this can be done in a reliable and transparent manner (Collot d'Escury, 2015). If not, a closed-loop system, where the value of the product and material is recovered, would not be possible. Thus, when looping the product in different or the same life cycle multiple times, circular-economy systems with interconnected cycles generally contain large quantities of data (Earley and Goldsworthy, 2018; Bressanelli et al., 2018; Pagoropoulos, 2017). In this way, digitalization provides for new means of collecting and using the data in real time (OECD/Eurostat, 2018).

Internet of Things (IoT) can keep track of valuable products and materials at a much lower cost than in the past, thus increasing recovery opportunities (Ellen MacArthur Foundation, 2015c). The digital material passport previously mentioned could be used to read information also in the end-of-life phase. The information and feedback gathered when taking back products can be utilized whenever there is a need to reach various decisions about the different phases of a particular product's life cycle, logistical arrangements, reuse of waste materials and/or the operators required in the value network (Ellen MacArthur Foundation, 2016). The data and feedback gathered from the recovered products also allow for analyzing and optimizing the design and composition of the product (Ellen MacArthur Foundation, 2016; Ellen MacArthur Foundation, McKinsey & Co and Google, 2019). As the use of big data and IoT is expected to increase, Cukier contends that "The internet of things could become the 'soul' that animates objects in the circular economy" (Ellen MacArthur Foundation, 2016, p. 21).

It is however important to add that using large amounts of data in order to design circular products and components could prove very advanced for the human mind to handle alone, as it entails processing very large quantities of information from the products as well as from the consumers. AI could enable decision-making designers on how to design for disassembly-capabilities, reuse, easier repair and longevity based on the complex data and feedback gathered in the take-back phase (Ellen MacArthur Foundation, McKinsey & Co and Google, 2019). As an example, an agile, continuous feedback process where designers test and refine AI-generated design suggestions based on collected data would potentially lead to more optimized design outcomes faster, than if not using AI (Ellen MacArthur Foundation, McKinsey & Co and Google, 2019). In addition, AI could also analyze new materials fast; its composition, structure, quality and other properties. Ultimately, AI could enable for circular products to remain fit for longer lifetimes through repair, reuse and multiple life cycles. According to Ellen MacArthur Foundation, McKinsey & Co and Google, (2019), 30% of plastic packaging is in need of a complete redesign; an innovation process to which AI could provide a powerful tool. A pressing issue since, as previously mentioned, plastic packaging is of particular concern today due to the fact that 95% of plastic packaging material value is lost and 32% is not recollected (World Economic Forum, 2016).

In fact, AI has already demonstrated some of its main benefits in how it can create value in realizing circular material flows and in enabling enhanced valorization of materials and products (Ellen MacArthur Foundation, McKinsey & Co and Google, 2019; Selvan Ramadoss, Alam and Seeram, 2018; Pagoropoulos, 2017). AI can be used when sorting post-consumer mixed material streams using visual-recognition techniques. For instance, the company ZenRobotics (2019) deals in robotic waste separation, where robots scan the waste stream, after which it analyzes the data in real time and then finally determines (autonomously) how to sort the waste. These robots are controlled by an AI (based on imagery input) and they can reportedly reach an accuracy level of 98% in sorting myriad material streams, from plastic packaging to construction waste (Ellen MacArthur Foundation, McKinsey & Co and Google, 2019).

Looping products and materials in the same or different life cycles ultimately results in more complex supply chains than if the products would have only one life cycle as is generally the case today. As stated previously, the complexity increases as the supply chain needs to cover larger information flows. However, adding to this complexity is the increased amount of transactions necessary when looping products and materials. Ensuring secure transactions is thus also prioritized in a circular economy. To this extent, blockchain technology may offer a practical solution for ensuring both reliable data and secure transactions (Abeyratne and Monfared, 2016; Herzberg, 2015). As a concrete example, the company Bext360 (2019) uses blockchain to provide reliable data and secure transactions in the supply chains of for instance coffee, cotton and palm oil. Bext360 describes it like physical assets in the supply chain are represented as tokens which are stored in the blockchain to facilitate payments, yield smart contracts and track assets through the whole supply chain.

Digital technology thus helps us manage products more sustainably, it helps us reuse, repair and share as well as take back the products and close the loop of the business. However, it is also necessary to investigate how digitalization and circular economy can create value together on a wider level in order for a societal shift to circular economy to occur. It is estimated that roughly two people move into cities every second, or approximately 180,000 people every day (Ellen MacArthur Foundation, 2016; Hollis, 2013). This means there is a pressing need to find new ways of enabling more people to live on less physical space with a lower overall impact on the environment. Smart circular cities with smart streetlights, optimized route planning, autonomous vehicles and urban farming are some topics currently investigated by many as solutions to this problem (Ellen MacArthur Foundation, 2016; City of Amsterdam, 2016; Ellen MacArthur Foundation and Arup, 2019).

Given that the threat to our future existence on this planet is so clear and present, it is crucial that we ensure that we ascertain the right information to make the right decisions on how to steer the situation in a different direction. This means we need to understand the consequences of our actions and predict the long-term impact they carry. These decisions need to be made both on a small scale in human's everyday lives, as well as in everyday business. Still, and maybe even more importantly so, we need to understand the larger-scale ramifications of our actions. Our whole ecosystems and biodiversity have been severely impacted in a negative way due to our actions in our industries. It is thus important to both measure the actions we have taken in order to determine if they were fruitful, and also to predict what actions will have the most impact and are of most benefit to our entire ecosystem for future use. Digitalization and the digital transformation provide the tools for doing exactly that.

In March 2019, more than 30 Swedish digitalization consultancies and *Fossil Free Sweden* [Swe: Fossilfritt Sverige] submitted a roadmap to the Swedish government as a first step to highlight the potential for digitalization to act as a catalyst for sustainability (Fossil Free Sweden, 2019). Among other things, this proposal called for the government to launch an inquiry to review collected data on global sustainability in order to investigate predictions and assessments on the different sustainability initiatives available. One company that tries to facilitate sustainability work to this extent is ClimateView, who has developed a digital tool for visualizing factors such as emissions, potentials and policies within different areas. By using open data, ClimateView (2019) operates with the intention of securing the development toward CO_2 reduction. Another example is developed by the Gothenburg Region (GR) [Swe: Göteborgs Regionen] and IVL Swedish Environmental Research Institute [Swe: IVL Svenska Miljöinstitutet] who have created a digital tool for calculating the expected CO_2 emissions for a specific residential area (Gothenburg Region, 2019). There are also digital tools for managing land areas, its current condition and how they are changing (Sanborn, 2019).

Today, we can also simulate large-scale ecosystems and use machine learning and data analytics in order to understand and predict the impacts of our actions, something that was not possible a few years ago (Xprize, 2017). In this way, digitalization can help us think in systems and understand the complex contexts that we as humans function in. As Webster describes it, "The circular economy uses understanding the system to give a better overall result" (Waldegrave, 2017, para.5). Webster means that we need to understand a larger context over longer time periods to see patterns and gain insights and in that way understand the causalities of our actions when conducting business in the contemporary world, as well as in the future world. Also Steffen et al., (2018) argue that we need to see the earth as one system and carefully treat it as our life-support system. The authors argue that, "The Stabilized Earth trajectory requires deliberate management of humanity's relationship with the rest of the Earth System if the world is to avoid crossing a planetary threshold" (Steffen et al., 2018, para.51). Seeing patterns through the large quantities of data that these systems entails is what AI and machine learning can enable us to do (Sas, 2019).

Other important factors of digital technology that can facilitate favorable sustainability work are the methods used when working with digital solutions. When designing digital solutions, service design and demand-driven design is often used and when creating digital solutions, the process is agile, with continuous feedback loops, iterations and incremental changes to the existing product. This is to ascertain that the solution always is feasible, desirable and viable, while answering to the demand of the user. In this way, this design process opens up for new and different ways of answering to the same demand – as a circular economy aims to do (Maslin and Shayler, 2016; Ideo and Ellen MacArthur Foundation, 2019). Design thinking and service design also carries the potential to design entire product life cycles and not only the product in itself (Maslin and Shayler, 2016; Ideo and Ellen MacArthur Foundation, 2019). In this way, service design could be seen as crucial in order for a circular economy to evolve since, as previously mentioned, it is necessary to design entire life cycles, flows and processes in a circular economy (SB Insight, 2019).

In the same way, the aforementioned roadmap by Fossil Free Sweden, urged companies and governments to form strategies and structures that evolves from the user-demand rather than from strategies following a conventional way of delivering solutions (Fossilfree Sweden, 2019). The report stated that we need a shift toward demand-driven structures and strategies in order to find solutions to the negative environmental impact caused by our industries. Furthermore, it emphasized the need for establishing zones for quickly testing digital solutions for a fossil-free society i.e., testing new innovations, business models, technology, cooperation and regulation. In this way, sustainable innovation would evolve from agile working methods.

However, oftentimes today, "circular" initiatives stem from the sustainability department of the business, which tends to favor long-term projects and long-term goals. These "circular" initiatives neither work in an agile manner nor with design-thinking processes. The innovative nature of circular economy is often neglected or overlooked when in fact circular economy ought to signify the very essence of an innovation inasmuch that it fundamentally transforms current processes and rethinks the way businesses have traditionally been run. Thus, design thinking and agile working methods could facilitate, enable and spur circular solutions. Many of the new popular solutions on the market today (especially smaller innovative start-ups) have merged digitalization, sustainability (circular economy or sharing economy) and agile ways of working and have in this way managed to solve the demand of the users through sustainable innovation. Thus, the incumbent firms could benefit from using service design and design thinking in their sustainability work, to keep up with the competitors and disruptors and find new ways of operating within the planetary bounds.

The Quadruple Helix Model is an initiative that uses agile thinking to tackle complex ideas and focus on sharing innovation and data among many stakeholders, calling it Open Innovation 2.0 (Edwards, 2018). It is discussed as an important part of the research that investigates climate action, health and well-being (Edwards, 2018; Selada, 2017). The European Commission identifies The Quadruple Helix Model as an important part of their Digital Single Market Policy, stating:

> Open Innovation 2.0 (OI2) is a new paradigm based on a Quadruple Helix Model where government, industry, academia and civil participants work together to co-create the future and drive structural changes far beyond the scope of what any one organization or person could do alone. This model encompasses also user-oriented innovation models to take full advantage of ideas' cross-fertilisation leading to experimentation and prototyping in real world setting.
>
> (European Commission, 2018b, para.2)

In short, circular economy should not only be part of the "sustainability work", but should also be seen as an innovation that encompasses all parts of business and society. However, it needs optimal working methods in order to realize its potential. Suitable methods for fulfilling the potential of innovative solutions are commonly used in the context of designing digital solutions. This, in turn, would support the argument of merging the two megatrends of digitalization and circular economy.

5. Moving forward with the circular economy and digitalization

Even though circular economy has gained some political traction internationally in its endeavors to resolve the increasing threats to the planet, a 2019 report showed that only 9% of the world was indeed circular in 2017 and 2018 (De Wit et al., 2019). In fact, 2018 indicated an increased negative impact on the climate as opposed to the preceding year. Even so, circular economy is now present on the agenda more than ever before, as it frequently occurs in discussions covering both enterprise and policy. It is evident that the shift is difficult in many ways and even though digitalization provides ways of facilitating the transformation to a circular economy, several other aspects and requirements need to be in place in order to successfully carry out the shift from linear to circular. It is also necessary to account for the potential obstacles that arise when using digital technology, which in turn, makes it relevant to also consider other aids that may facilitate such a transformation.

To begin, one of the foremost greatest challenges toward implementing a circular economy on a wider level is the fact that it lacks standardization and official strategic guidelines, indicators and/or certifications (such as ISO etc.) (Ivanovic, 2018). Thus, the application of the circular model varies immensely depending on the market and assets, which in itself is a vicious circle as it makes any form of generalization, standardization or even harmonization, imprecise at best (Ivanovic, 2018; Kirchherr, Reike and Hekkert, 2017). Research and innovation at all levels (social, technological and commercial) are thus necessary to enable the transformation (Stahel, 2016).

Policies, laws and legislations are rooted in the linear economy and waste policies, which oftentimes means that circular initiatives are systemically obstructed, making them infeasible to implement. In this way, it would be paradoxical for an economy that seeks to eliminate waste to have legislations based on waste management, as in such a system, waste is not regarded as a resource, but rather as a matter that has no further value. By extension, the new job opportunities that a circular economy could bring, brought forth by ways of digitalization, may also be scuttled if not properly supported by political establishment. OECD (2019) has claimed that, in the absence of policy change, global material use will more than double by the year 2060, due to the cumulated effects of economic growth, structural change and technological change. If world leaders decided to shift toward a more circular economy, managing the transition would have to be a top priority (Ellen MacArthur Foundation, 2015b).

According to the OECD Deputy Director of the Environment Directorate, Anthony Cox, the organization recently initiated a new policy experiment (Cox, 2019). The experiment was to impose a material tax, while redistributing the revenue from that material tax into subsidies for recycling and enabling lower labor taxes. In this way, it was a budget-neutral policy experiment. The anticipated results for 2040 indicate a dramatic impact on the use of resources, such as aluminum, copper, iron and steel, non-ferrous metals (i.e., not containing iron) and non-metallic minerals. The impact was driven through efficiency gains, i.e., scaling up or down production. Impact was to a large extent also propelled by trade. The conclusion drawn from this experiment was that international trade can contribute to less material being used. Cox described it as though trade is the "glue" that will make circular economy feasible and that trade policies will need to be amended as to work in favor of a circular economy rather than (as the case is now) counteracting it. The international trade flows will be affected by a transition to circular economy as it will result in an increase in areas such as trade-in services, secondhand goods, goods left for refurbishment or remanufacturing, waste for recovery and in secondary raw materials. To this end, the importance of digitalization was stressed upon in order to enable and facilitate the transition.

In 2018, the European Commission presented their implementation steps from their previously presented, Circular Economy Action Plan (European Commission and switch2green, 2018). The Action Plan contains actions promoting a closed-loop production, consumption and waste management as well as creating a secondhand market for recycled/reused materials. The report describes legislative proposals on waste with targets for recycling, measures for reducing landfill, how to promote reuse and stimulate industrial symbiosis, as well as the economic incentives for producers to put greener products on the market. The Action Plan should be fully implemented by 2019 (European Commission and switch2green, 2018).

An example of a policy that promotes circular economy and is already implemented in the EU is the Waste Electrical and Electronic Equipment Directive (WEEE), which stipulates that all producers of phones need to accommodate for a take-back system (Lund University et al., 2018). One example of a phone producer working more circular and sustainable is Fairphone. Fairphone is a social enterprise company seeking to develop smartphones designed and produced with minimal environmental impact by having modular, upgradeable phones and take-back programs (Fairphone, 2019). Fairphone's Resource Efficiency Manager, Miguel Ballester, stated that the new policy developments are evolving in the right direction for them but not fast enough (Lund University et al., 2018). Ballester further suggested that an efficient way to change behavior is through taxation. At the same time, he expressed some skepticism as to whether or not the politicians would actually dare to propose taking such a course of action. Additional laws and regulations that have been discussed in terms of them having a positive impact on circular processes are discussed by Lund University et al. (2018) to be:

- Circular product design guidelines
- Lower VAT on repair service
- Acquiring available spare parts
- Recycled content mandates
- Banning planned obsolescence
- Higher tax on material and less tax on labor

In fact, France has already in 2015 pressed a law that bans planned obsolescence in order to promote longer lifetimes of products and move away from the take-make-dispose model (SGS Offices & Labs, 2015). In 2016, France became the first country in the world to ban supermarkets from disposing or destroying unsold food, forcing these businesses to donate the food to charity instead (Chrisafis, 2016). Following its approval by the French Parliament in July 2019, the same legislation aims to also include non-food items such as clothes, electronics and plastic products, making it illegal for supermarkets to throw away unsold items starting in 2021 or 2023 (depending on the sector) (Samuel, 2019).

Important to add though is that, when discussing environmental policies, it is necessary to address the potential impact those policies might have on the labor market (OECD, 2016). In an empirical analysis from 2014, OECD concluded that increasing stringency in environmental policies will not be harmful to productivity levels (Albrizio et al., 2014). In order to make this conclusion, OECD collected data on selected environmental policies over countries and time allowing for creating a proxy of environmental policy stringency[5] (EPS) and to analyze its effect on economic performance. Another report that analyzed the relationships between environmental policy stringency, productivity and innovation also concluded that "the stringency of environmental policies can be increased without harming economy-wide productivity" (De Santis and Jona-Lasinio, 2015, p. 20). The findings also indicated that a tighter environmental regulation will channel innovation and productivity.

As previously discussed in Section 3, circular business models might acquire additional skilled employees, leading to higher cost when processes such as reuse, repair, refurbishment and remanufacturing are necessary. A consequence of the high-level taxation on labor today is that firms may dismiss a circular transition as financially unviable. Added expenditures associated with product take-back could also make firms reluctant, given the fact that virgin materials are often cheaper than non-virgin materials (Lund University et al., 2018; Koren and Bisesi, 2003; Wijkman and Skånberg, 2015).

Additional challenges include securing an organized collaboration between different partners, and pooling different areas of competence, the latter of which carries its own challenge, as there are still many organizations that lack the sufficient expertise in regards to the basic concepts of the circular economy (Cumming, 2018; Ivanovic, 2018). To enable the circular economy, a network of information flows and knowledge sharing as well as cooperation between a wide range of actors on many levels are necessary (Stahel, 2016). A circular economy requires more dependencies across borders, as an example, it is necessary to share resources amongst businesses so that one business's waste can become another business's resource.

In order to use digitalization to catalyze circular economies, the knowledge about how to use it is crucial. Ellen MacArthur Foundation, McKinsey and Company and Google (2019) discuss this matter by contending that the use of AI is meaningless if people cannot reach an overall consensus of what defines relevant output and input variables. Another important factor when using digitalization in this way is to set the supporting structures for doing so. Government support that could favor this purpose were investigated in a report by Fossil Free Sweden (2019) and include examples such as: appointing a digital transformation commission, clarifying responsibilities for digitalization and sustainability in every department and/or increasing the knowledge nationally around the aforementioned issues.

Stuchtey, Enkvist and Zumwinkel (2016) argue that for technology disruption to be developed and used as a foundation for circular economies, direction and support is crucial. They explain that innovation is to a large extent the result of combining existing technologies and business models in new ways to meet a customer need. They contend that disruption and innovation is sprung from the process of adding together technology building blocks, allowing that technology to ripen and evolve to a point at which it is able to meet an existing, specific customer need, thus bringing innovation to fruition. In that way, the authors believe that the technology development can, to a large extent, be controlled or at least steered. Specifically, they state "setting direction for the technology disruption is one of the most crucial tasks for environmental and business policy makers in the next decade" (Stuchtey, Enkvist and Zumwinkel, 2016, p. 212). They continue saying, "If we want a circular material bank or high productivity delivery systems, we need to identify the required technologies to get us there, and the building blocks of those technologies in turn, and make sure there is momentum in developing those technologies" (Stuchtey, Enkvist and Zumwinkel, 2016, p. 207).

Another challenge associated with digitalization is the potential pollution that comes from using digital technology. Training an AI model as well as validating transactions in the blockchain results in heavy carbon emissions (Hao, 2019; Sexton, 2019). Other challenges and costs associated to digitalization relate to data sharing, data ownership, data integration, collaboration, competence and business models (Antikainen, Uusitalo and Kivikytö-Reponen, 2018; Finger et al., 2016). As an example, the availability and ownership of data/assets is of crucial importance to circular economy since circular economy advocates performance over ownership (Ellen MacArthur Foundation, 2016). The question around ownership of data could thus potentially be problematic in a future circular scenario, enabled by digitalization (Stahel, 2016). Other challenges include the sharing of data between competitors, in addition to the protection of privacy, intellectual property rights and trust. It is also necessary to integrate the large quantities of data owned by various operators, as the management of data flows also constitutes a big challenge due to the magnitude of the responsibilities involved (Kelly et al., 2017).

Making a transition to a circular economy thus entails high transition costs, including research and development and asset investments, stranded investments, subsidy payments to promote market penetration of new products, and public expenditure for digital infrastructure (Ellen MacArthur Foundation, 2015b). It could thus be argued that the economic as well as political cost following a transition to a circular economy will be too high for some companies.

Though it should be stated, that even though circular economy is the way forward, this model in and of itself does not unequivocally entail that it presents itself as a more environmentally friendly alternative to any given instance. That is to say, there are possible rebound effects, such as there being less demand for secondhand market products due to them being perceived of having lower quality, reuse and/or recycling processes causing a high negative impact on the environment and reverse logistics requiring substantially more transportations and by extension, also more in the way of pollution (Lund University et al., 2018). In addition, sharing economies might not always imply a reduced consumption and a higher utility rate, as in some cases they can in fact increase consumption. Thus, sharing economies does not imply circular systems/products in itself. Hence, if the material and products are not taken care of after they have been used to exhaustion, a circular system is not upheld. The same predicament arises when using digitalization as a tool. As mentioned, mining blockchain could have a substantial negative environmental impact due to the high energy use (Truby, 2018; De Vries, 2018).

Therefore, it is important to measure and compare the outcome of different alternatives in a circular economy. Traditionally, life-cycle analysis and cradle-to-cradle certifications have been used and could continue being used for this purpose. However, there is a lack of KPIs measuring success in a circular economy, which is why it can be difficult to measure and calculate the circularity of a product and its overall impact, both negative and positive (Franconi et al., 2016; Tuppen, 2016). Circular product indicators and measurements that could potentially give such indications/measurements/scores are currently being investigated, however, thus far only prototypes have surfaced (Franconi et al., 2016; Cayzer, Griffiths and Beghetto, 2017). Consequently, there is at present no universally accepted means of measuring the success of a business in undertaking the transformation from a linear to a circular practice (Ellen MacArthur Foundation, 2015a; Munholland, 2018).

6. Technological unemployment or more human jobs?

It is important to note that as digital technology advances and is increasingly used in business, there is a possibility of technology taking over certain human jobs. This is by many referred to as technological unemployment (Kim, Kim and Lee, 2017; Peters, 2017). However, while that might be true, many also argue that while in some cases technology replaces human labor, it will create new job opportunities elsewhere. On this topic, Brynjolfsson and McAfee (2014) argue that:

> Rapid and accelerating digitization is likely to bring economic rather than environmental disruption, stemming from the fact that as computers get more powerful, companies have less need for some kinds of workers. Technological progress is going to leave behind some people, perhaps even a lot of people, as it races ahead. As we'll demonstrate, there's never been a better time to be a worker with special skills or the right education, because these people can use technology to create and capture value. However, there's never been a worse time to be a worker with only 'ordinary' skills and abilities to offer, because computers, robots, and other digital technologies are acquiring these skills and abilities at an extraordinary rate.
>
> (Brynjolfsson and Mcafee, 2014, pp. 10–11)

Considering the findings of this chapter the previous comment could be analyzed to be both true and false. As investigated in this chapter, also Brynjolfsson and

McAfee agree that digital technology will spur a shift in the workforce where skilled labor will still be sought for, and the jobs required less-skilled workers could be automated. As exemplified in previous sections, activities for mining and manufacturing products do not necessarily require human work but is nowadays often automatized whereas repairing, maintaining, upgrading and remanufacturing are all activities that require human labor (Wijkman and Skånberg, 2015). However, when digitalization is used to catalyze circular economies it will not only bring an economic disruption in terms of new markets and job opportunities but also an environmental disruption. In this way, the first part of the statement by Brynjolfsson and McAfee would not hold true in regards to the findings of this chapter. Digitalization could indeed propel and enable an environmental disruption if it is used to lay the foundation for a circular economy, together with other tools and aids.

Stuchtey, Enkvist and Zumwinkel (2016, p. 202) argue that throughout history, the economic and technological revolutions have all brought with them a shift of workforce, rather than an "end of work". The authors however highlight that the scale and pace of change that defines the digital revolution might affect the labor market differently than what can be seen in historical contexts. If the change happens too fast, it might be detrimental to those who are affected by it, making the change more arduous, or at least (paradoxically so) slower to implement.

Perez (2018) explains that technological revolutions have traditionally been shown to cause destruction of jobs and skills – and in some cases whole regions. However, she argues that each revolution has brought about needs of new lifestyles that require new services and products, which in turn opens up for new employment. In the case of a transformation to a circular economy enabled by digitalization, the new lifestyles would mean new needs for not only new services and products, but also new business models, new consumption patterns, new behavior and new interactions. In conclusion, utilizing tech to drive the transformation of society into a circular economy would naturally automatize some processes, but it will foremost give way to new employments requiring human labor, elsewhere.

7. Beyond GDP – measuring growth, prosperity and well-being

When stating that circular economy can lead to an increased growth, job opportunities and development, it is important to also highlight the definition of growth since it is indeed subject to much debate. Ultimately this new circular economy proposes a more environmentally friendly society without impeding humans' social development and citizen's well-being since it also proffers new job opportunities and new ways of capturing value from resources. The circular economy thus proposes a new way of thinking in regards to industrial/human growth and human well-being, as it decouples economic growth from resource use (Lund University et al., 2018; Ellen MacArthur Foundation, 2015b).

Historically, Gross Domestic Product (GDP) has been used as a measurement of a country's welfare and as an index of the well-being and development of a society

(Tukker et al., 2014; Jackson, 2009). OECD (2018a, para.1) writes that GDP "at market prices is the expenditure on final goods and services minus imports: final consumption expenditures, gross capital formation, and exports less imports". The mean standard human growth and human social development has thus been closely linked to consumption. Even though, as previously discussed in this chapter, GDP is by many predicted to increase with a transition to a circular economy, the fact that growth is measured on the basis of consumption is disputed. As stated, the growth in a circular economy will be based on human capital instead of extraction of new resources (Groothuis, 2015). Among others, Stahel (2016) discusses the contradiction of using GDP as a measurement of growth in a circular economy. He means that GDP measures a financial flow over a period of time whereas circular economy preserves physical stocks. Stahel (2016) argues that wealth and well-being should be measured in stock instead of flow and in capital instead of sales since materials should be seen as assets to be preserved instead of continually consumed. He urges policymakers to use "resource-miser" indicators such as value-per-weight and labor-input-per-weight ratios rather than GDP.

It has further been debated whether GDP is an accurate measurement to use for measuring welfare and well-being in the context of high-income countries on the basis that consuming more products does not necessarily mean increased well-being when reaching a certain level of consumption (Tukker et al., 2014; Jackson, 2009). Tukker et al., (2014) argue that Human Social Development Index (HDI) would be a more accurate index to use. While instead measuring HDI, researchers find that the relation between well-being, human social development and resource usage is not linear since human social development stagnates after a certain level of resource usage (Tukker et al., 2014). When measuring the global resource footprint of nations, Tukker et al., (2014) found that Japan's legislations to reduce waste in the country resulted in the lowest material footprint observed for highly developed countries. In this way Japan set an example of the possibility of decoupling standards of living from environmental impacts and resource usage. Jackson (2009) discussed this matter already back in 2009, when he argued that well-being encompasses much more than material concerns and consumption:

> [Well-being] resides in the quality of our lives and in the health and happiness of our families. It is present in the strength of our relationships and our trust in the community. It is evidenced by our satisfaction at work and our sense of shared meaning and purpose. It hangs on our potential to participate fully in the life of society. Prosperity consists in our ability to flourish as human beings within the ecological limits of our planet. The challenge for our society is to create the conditions under which this is possible. It is the most urgent task of our times.
>
> (Jackson, 2009, p. 16)

New Zealand has promised to introduce a tool and framework in 2019, for how to measure its economic success through the well-being of its inhabitants, thus being the first country in the world to do so (Walters, 2018). The prime minister stated that, "We want New Zealand to be the first place in the world where our budget is not presented simply under the umbrella of pure economic measures, and often inadequate ones at that, but one that demonstrates the overall wellbeing of our country and its people" (Walters, 2018, para.4). It should be noted however, that other countries have previously sought to implement alternate measurements to the GDP. For instance, Bhutan has implemented a Gross National Happiness (GNH) index, which is more of a philosophy that guides the national government in its aspiration to measure the collective happiness and well-being of its population (Kelly, 2012; Dorji, 2012). Regional/local variations of GNH exists in multiple places across the world (Chatterjee, 2011; Rattiwan, 2016).

Questioning the traditional measurements of growth, Stuchtey, Enkvist and Zumvinkel (2016) present Solow's theory around growth, explaining that Solow quantitatively described growth as the output of two input factors: (1) labor and (2) capital. Solow explained that labor accounts for 70% of growth and capital the remaining 30%, in most developed countries. Challenging this notion, the authors remarked that for the observed GDP growth rate over time, for many countries and during a longer time series, the empirical data on input and productivity development only accounted for half of that GDP. The remaining part of the growth, Solow referred to as a result of "technological progress and a synergy effect from combining better labor practices with more productive capital investments" (Stuchtey, Enkvist and Zumwinkel, 2016, p. 58). The authors conclude that, "To this date, there is little quantitative microeconomic understanding of this other half of growth and how it can be managed. More than half of our growth engine remains a black box" (Stuchtey, Enkvist and Zumwinkel, 2016, p. 58). Instead, Stuchtey, Enkvist and Zumvinkel (2016) propose that energy and resources play a crucial role in the equation of calculating growth and should be seen as inputs. They posit that when measuring economic growth, it is necessary to take into account how much energy and other resources are put into the economy and how productively they are used. In summary, the authors mean to say that resource productivity does indeed matter when measuring economic growth.

Perez (2018) states that most people believe growth is about mass production. Perez instead argues that growth could involve anything but products, such as intangibles and services. She calls it "smart green growth". She further contends that the smart green growth is built up by a new lifestyle based on an aspirational good life and new jobs to cater for it. Thus, she implies that both these things should determine the level of growth and job creation. Perez (2018, 00:14:26) states that a smart green growth is, "[a] constant increase in the proportion of intangibles in both GDP and lifestyles".

Taking the previous discussion in consideration, it could be concluded that we have reached a point where it is necessary to measure growth and well-being in a different way. It is yet unknown what measurement(s) will replace or complement GDP as a standard. However, it is evident that some sort of index for how productively resources are used is necessary, as well as to measure human development and well-being decoupled from resource use. Some of the primary findings in extant research on this topic is that the present economy needs to change its guiding principle and allow it to be driven by something other than the classic market-based economic growth. Thus, there is a need for a new economy that can function within the planetary bounds without impeding economic growth. Mary Robinson, the former president of Ireland and United Nations high commissioner for human rights, captures this predicament by asking: "What does prosperity mean in a world of 9 billion people living under the threat of climate change and resource scarcity? One thing is absolutely clear. It cannot mean business as usual. It cannot mean more of the same" (Jackson, 2009, p. xvi). Also Steffen (2017) highlights this issue, saying:

> Particularly in the Anglo-Saxon world we have narrowed down so much of life and tried to stick it in a market-based economic system, that we are headed for failure if we don't realise that. We have to put limits on what is in an economic system and what we manage elsewhere, outside, based on fundamental principles and values. So, we have to understand that we have to manage ourselves in such a way that we take pressure off the Earth system.
>
> (Steffen, 2017, para.9)

8. Conclusion

In conclusion, circular economy is fundamentally reshuffling our current economic system. It will change how materials and resources are handled and consumed, what the labor market looks like and how growth and well-being are defined and measured. It will not stop at the business level but will affect all aspects of society, people, cities and the planet. It is a socioeconomic as well as a techno-economic paradigm shift that requires system changes on all levels.

The transformation to a circular economy will ultimately not only benefit the environment and our future lives on this planet, as it will also have positive impact on labor. With the shift follows increased jobs and new industries. A circular economy is more dependent on people overall, both due to the fact that skilled jobs in local businesses are needed and because users are invited to take part in the economic system of sharing, renting, lending and communicating with and influencing products and goods. However, there are several obstacles on the way toward a circular economy and the transformation specifically requires digital technology as an enabler; to handle complex information flows, enable new business models, provide and increase transparency, secure transactions and increase communication.

Digitalization (along with the digital transformation) has, in and of itself, already begun to fundamentally change all dimensions of society at large. Thus, it is predicted to have a disruptive effect on the labor market. Circular economy and digitalization are fully compatible with one another. That is to say, with the aid of other tools and methods, the union between circular economy and digitalization has the potential to facilitate the economic revolution that would bring about a financial and environmentally sustainable society. It is true that one may argue that there will be an increased threat of technological unemployment once the digitalization process reaches its full effect. However, this chapter has concluded that if digitalization is used the right way, e.g., to help us derive, handle, analyze and understand data and complex information flows, it has the potential to both catalyze sustainable societies and circular economies while at the same time create new industries, new jobs and new innovations and opportunities. Specifically, fears of unemployment are unwarranted provided that digitalization is used to propel an economic system that relies on human skills and human labor (what signifies a true circular system).

That being said, digitalization should not be seen as the end-all and be-all solution to circular economy, as digitalization could never facilitate a circular economy in total seclusion. It should, however, be seen as one means among many in reaching that state. Several other enablers are needed for the transformation to emerge, such as changed mindsets, changed policies and legislations, changed measurements and indicators, changed supply chains, along with more research and innovation.

There are, indubitably, also potential risks of using digitalization to enable the shift to a circular economy, such as privacy and ownership issues. However, the fact that it is a key-enabler, if not even a prerequisite, for us to change to a more sustainable world should be enough of an argument for using it. If digital technology is instead used to bolster the linear economies, it will be much too late to transform our current economic system into one that can accommodate and function within the planetary bounds. The reason is that the way our businesses are run today cannot be optimized in terms of maximizing profit and sustainability, nor can it provide for the necessary sustenance of tomorrow. The potential risks then rather lie in the cost of not doing what we can to change the status quo.

As evidenced in this chapter, the transition cost of turning the current economic system as we know it into a circular system, will no doubt be considerable. However, each and every day we experience the constantly increasing opportunity costs of *not* transitioning, with global warming, with increasingly polluted oceans and air, with more animals becoming extinct, more icebergs melting and waste piles that keep growing. The linear businesses will ultimately have to pay more for conducting business in the traditional way as it is anticipated that the demand of goods and services will continue to increase all the while resources are becoming scarcer. When the circular transformation not only promises to deliver change on the aforementioned parameters, but also to revamp and provide prosperity to the labor market while fostering economic growth, the transition cost becomes substantially more affordable.

At the end of the day, there is a pressing need for a new economic system that has the ability to provide for human's modern way of life while at the same time respecting and accounting for the limitations of the earth's resources. With the advancement of new technology, the circular economy presents itself as a new economic model that can deliver on all of these accounts and provide society with a sustainable economic growth. It is thus necessary to put circular economy on the agenda in all aspects of business and society and treat it as the innovation and disruption it actually is. Circular economy should not only be a strategy for the sustainability department, but for the whole business, and the whole society. It is especially important to discuss circular economy and how it can be achieved in conjunction with the digital strategies of both businesses and societies. It is crucial to provide the right supporting structures for the two revolutions (digital and circular) to be synergized. However, disruptive ideas are already now making use of technology. This paves the way for commitment to change, while also showing that we truly have the right tools and means to achieve a more durable society that embraces a new direction of growth. Ultimately, circular economy is not an activity for the sustainability department, it is a mindset, an innovation and a business strategy for competitiveness and survival. A circular economy can aid in accomplishing a society that decouples resource use and environmental impact from economic activity and human well-being and the digital transformation can help provide the means from turning this theory into practice.

Notes

1 The authors contributed equally to this work.
2 Reverse logistics signifies a process of recollecting goods after they have been used for the purpose of capturing as much value as possible. After they have been recollected, the products could for example get reintroduced to the market, remanufactured, refurbished or recycled.
3 Flow-on entails an increase in wage or another benefit that a group of people receive because a similar group receives it.
4 A web-based knowledge platform with information on non-fuel, non-agricultural raw materials from primary and secondary sources.
5 Stringency indicates a higher implicit or explicit price placed on the relevant environmental damage produced by firms or consumers.

References

Abeyratne, S.A. and Monfared, R.P., 2016. Blockchain ready manufacturing supply chain using distributed ledger. *International Journal of Research in Engineering and Technology*, 5(9), pp. 1–10.

Albrizio, S., Botta, E., Koźluk, T. and Zipperer, V., 2014. *Do Environmental Policies Matter for Productivity Growth? Insight from New Cross-Country Measures of Environmental Policies*. [online] Available at: <www.oecd-ilibrary.org/economics/do-environmental-policies-matter-for-productivity-growth_5jxrjncjrcxp-en> [Accessed 21 Sep. 2019].

Antikainen, M., Uusitalo, T. and Kivikytö-Reponen, P., 2018. Digitalisation as an enabler of circular economy. *Procedia CIRP*, 73, pp. 45–49.

Apple, 2019. *Environmental Responsibility Report: 2019 Progress Report, Covering Fiscal Year 2018*. [online] Available at: <www.apple.com/uk/environment/resources> [Accessed 21 Sep. 2019].

Aurich, J.C., Fuchs, C. and Wagenknecht, C., 2006. Life cycle oriented design of technical Product-Service Systems. *Journal of Cleaner Production*, 14(17), pp. 1480–1494.

Austin, T., 2016. Towards a digital infrastructure for engineering materials data. *Materials Discovery*, 3, pp. 1–12.

Batra, A., 2017. Resource security and regional cooperation in South Asia. In: N. Nanda and S. Ganeshan, eds., *India's Resource Security: Trade, Geopolitics, and Efficiency Dimensions*. New Delhi, India: The Energy and Resources Institute (TERI), pp. 59–70.

Bettencourt, L., 2018. *The Circular Economy Could Save Life on Earth – Starting with Our Cities*. [online] World Economic Forum. Available at: <www.weforum.org/agenda/2018/03/circular-economy-in-cities> [Accessed 21 Sep. 2019].

Bext360, 2019. *Bext360 – Every. Single. Step.* [online] Available at: <www.bext360.com> [Accessed 21 Sep. 2019].

Bin, P. and Vassallo, M., 2016. The growth path of agricultural labor productivity in China: A latent growth curve model at the prefectural level. *Economies*, 4(4), p. 13, pp. 1–20.

Bonciu, F., 2014. The European economy: From a linear to a Circular Economy. *Romanian Journal of European Affairs*, 14(4), pp. 78–91.

Botsman, R. and Rogers, R., 2010. *What's Mine is Yours: The Rise of Collaborative Consumption.* New York, NY: HarperBusiness.

Bressanelli, G., Adrodegari, F., Perona, M. and Saccani, N., 2018. Exploring how usage-focused business models enable circular economy through digital technologies. *Sustainability*, 10(3), p. 639, pp. 1–21.

Brynjolfsson, E. and Mcafee, A., 2014. *The Second Machine Age: Work, Progress, and Prosperity in a Time of Brilliant Technologies*. New York, NY: W.W. Norton & Co.

C2C-Centre, 2019. *Maersk Cradle to Cradle® Passport.* [online] Available at: <www.c2c-centre.com/library-item/maersk-cradle-cradle®-passport> [Accessed 21 Sep. 2019].

Cayzer, S., Griffiths, P. and Beghetto, V., 2017. Design of indicators for measuring product performance in the circular economy. *International Journal of Sustainable Engineering*, 10(4–5), pp. 289–298.

Chatterjee, R., 2011. *Measuring Happiness in Victoria, British Columbia.* [online] PRI Public Radio International. Available at: <www.pri.org/stories/2011-10-03/measuring-happiness-victoria-british-columbia> [Accessed 21 Sep. 2019].

Chrisafis, A., 2016. *French Law Forbids Food Waste by Supermarkets.* [online] The Guardian. Available at: <www.theguardian.com/world/2016/feb/04/french-law-forbids-food-waste-by-supermarkets> [Accessed 21 Sep. 2019].

City of Amsterdam, 2016. *Circular Amsterdam a Vision and Action Agenda for the City and Metropolitan Area*. Amsterdam, Netherlands.

ClimateView, 2019. *A Global Climate Platform.* [online] Available at: <www.climateview.global> [Accessed 21 Sep. 2019].

Collot d'Escury, A., 2015. *Why Transparent Communications Is a Key Driver of the Shift to the Circular Economy.* [online] Huffington Post. Available at: <www.huffingtonpost.com/roland-jonkhoff/why-transparent-communica_b_8406154.html> [Accessed 21 Sep. 2019].

Cox, A., 2019. *Overview of World Trade and the Circular Economy*. [online] OECD. Available at: <www.sitra.fi/en/articles/international-trade-circular-economy-goods-services> [Accessed 21 Sep. 2019].

Cumming, P., 2018. Framing circularity at an organisational level. In: M. Charter, ed., *Designing for the Circular Economy*. London, UK: Routledge, pp. 35–43.

De Santis, R. and Jona-Lasinio, C., 2015. Environmental Policies, Innovation and Productivity in the EU. *SSRN Electronic Journal.* [online] Available at: <www.ssrn.com/abstract=2696266> [Accessed 21 Sep. 2019].

De Vries, A., 2018. Bitcoin's Growing Energy Problem. *Joule*, 2(5), pp. 801–805.

De Wit, M., Verstraeten-Jochemsen, J., Hoogzaad, J. and Kubbinga, B., 2019. *Circularity Gap Report.* [online] Available at: <https://docs.wixstatic.com/ugd/ad6e59_ba1e4d16c64f44fa94fbd8708eae8e34.pdf> [Accessed 21 Sep. 2019].

Dorji, T., 2012. *The Story of a King, a Poor Country and a Rich Idea – Earth Journalism Network.* [online] Earth Journalism Network. Available at: <http://web.archive.org/web/20190308081400/https://earthjournalism.net//stories/6468> [Accessed 20 Sep. 2019].

Dunbar, D., 2017. *Renewing the Balance*. Denver, CO: Outskirts Press.

Earley, R. and Goldsworthy, K., 2018. Circular textile designs: Old myths and new models. In: M. Charter, ed., *Designing for the Circular Economy*. London, UK: Routledge, pp. 175–185.

Edwards, J., 2018. *The Quadruple Helix Model of Open Innovation*. [online] Innocentive. Available at: <https://blog.innocentive.com/quadruple-helix-model-of-open-innovation> [Accessed 21 Sep. 2019].

Ellen MacArthur Foundation, 2015a. *Circularity Indicators: An Approach to Measuring Circularity*. [online] Available at: <www.ellenmacarthurfoundation.org/assets/downloads/insight/Circularity-Indicators_Project-Overview_May2015.pdf> [Accessed 21 Sep. 2019].

Ellen MacArthur Foundation, 2015b. *Delivering The Circular Economy – A Toolkit For Policy Makers*. [online] Ellen MacArthur Foundation. Available at: <www.ellenmacarthurfoundation.org/resources/apply/toolkit-for-policymakers> [Accessed 21 Sep. 2019].

Ellen MacArthur Foundation, 2015c. *Growth Within: A Circular Economy Vision for a Competitive Europe*. [online] Ellen MacArthur Foundation. Available at: <www.ellenmacarthurfoundation.org/publications/growth-within-a-circular-economy-vision-for-a-competitive-europe> [Accessed 21 Sep. 2019].

Ellen MacArthur Foundation, 2015d. *Towards a Circular Economy: Business Rationale for an Accelerated Transition*. [online] Ellen MacArthur Foundation. Available at: <www.ellenmacarthurfoundation.org/assets/downloads/TCE_Ellen-MacArthur-Foundation_9-Dec-2015.pdf> [Accessed 21 Sep. 2019].

Ellen MacArthur Foundation, 2016. *Intelligent Assets: Unlocking the Circular Economy Potential*. [online] Available at: <www.ellenmacarthurfoundation.org/publications/intelligent-assets> [Accessed 21 Sep. 2019].

Ellen MacArthur Foundation, 2019a. *Cities and Circular Economy For Food*. [online] Available at: <www.ellenmacarthurfoundation.org/assets/downloads/Cities-and-Circular-Economy-for-Food_280119.pdf> [Accessed 21 Sep. 2019].

Ellen MacArthur Foundation, 2019b. *Using Product Passports to Improve the Recovery and Reuse of Shipping Steel*. [online] Available at: <www.ellenmacarthurfoundation.org/case-studies/using-product-passports-to-improve-the-recovery-and-reuse-of-shipping-steel> [Accessed 21 Sep. 2019].

Ellen MacArthur Foundation and Arup, 2019. *Circular Economies in Cities: Project Guide*. [online] Available at: <www.ellenmacarthurfoundation.org/assets/downloads/CE-in-Cities-Project-Guide_Mar19.pdf> [Accessed 21 Sep. 2019].

Ellen MacArthur Foundation, McKinsey & Co and Google, 2019. *Artificial Intelligence and the Circular Economy*. [online] Available at: <www.ellenmacarthurfoundation.org/assets/downloads/Artificial-intelligence-and-the-circular-economy.pdf> [Accessed 21 Sep. 2019].

European Commission and switch2green, 2018. *Circular Economy Action Plan*. [online] Available at: <www.switchtogreen.eu//wordpress/wp-content/uploads/wp-post-to-pdf-enhanced-cache/1/circular-economy-strategy.pdf> [Accessed 21 Sep. 2019].

European Commission, 2014. *Communication from the Commission to The European Parliament, The Council, The European Economic and Social Committee and The Committee of The Regions: Towards a circular economy: A zero waste programme for Europe*. [online] COM/2014/0398 Final. Available at: <http://eur-lex.europa.eu/resource.html?uri=cellar:50edd1fd-01ec-11e4-831f-01aa75ed71a1.0001.01/DOC_1&format=PDF> [Accessed 21 Sep. 2019].

European Commission, 2018a. *Impacts of Circular Economy Policies on the Labour Market.* [online] Available at: <https://circulareconomy.europa.eu/platform/sites/default/files/ec_2018_-_impacts_of_circular_economy_policies_on_the_labour_market.pdf> [Accessed 21 Sep. 2019].

European Commission, 2018b. *Open Innovation 2.0 | Digital Single Market.* [online] Available at: <https://ec.europa.eu/digital-single-market/en/open-innovation-20> [Accessed 21 Sep. 2019].

European Commission, 2019. *Raw Materials Information System.* [online] Available at: <http://rmis.jrc.ec.europa.eu> [Accessed 21 Sep. 2019].

European Policy Center, 2019. *Sustainable Prosperity for Europe.* [online] European Policy Center. Available at: <www.epc.eu/prog_forum.php?forum_id=77&prog_id=2> [Accessed 21 Sep. 2019].

Fairphone, 2019. *Fairphone.* [online] Available at: <www.fairphone.com/en/> [Accessed 21 Sep. 2019].

Faludi, J., Cline-Thomas, N. and Agrawala, S., 2017. 3D printing and its environmental implications. In: OECD, ed., *The Next Production Revolution: Implications for Governments and Business*. Paris, France: OECD Publishing, pp. 171–214.

Finger, M., Bert, N., Razaghi, M. and Kupfer, D., 2016. The challenges of digitalization and the use of data. *Network Industries Quarterly*, 18(3), pp. 1–22.

Food and Agriculture Organization of the United Nations, 2019. *Save Food: Global Initiative on Food Loss and Waste Reduction.* [online] Resources. Available at: <www.fao.org/save-food/resources/keyfindings/en> [Accessed 21 Sep. 2019].

Fossil Free Sweden, 2019a. *Färdplan för fossilfri konkurrenskraft – Digitaliseringskonsultbranschen [Roadmaps for fossilfree competitiveness].* [online] Fossilfree Sweden. Available at: <http://fossilfritt-sverige.se/wp-content/uploads/2019/03/digitaliserings-konsultbranschen.pdf> [Accessed 21 Sep. 2019].

Fossil Free Sweden, 2019b. *Fossil Free Sweden initiative.* [online] Available at: <http://fossilfritt-sverige.se/in-english> [Accessed 21 Sep. 2019].

Franconi, E., Bridgeland, B., Graichen, P., Yao, M., Steigenberger, M., Stuchtey, M., Rossé, M., Pawlyn, M., Lovins, H.L., Nasr, N., Webster, K., Helbing, D., Lämmer, S., Zils, M., Hawkins, P., Hopkinson, P., Batty, M. and Tuppen, C., 2016. *A New Dynamic 2 Effective Systems in a Circular Economy*. Cowes, UK: Ellen MacArthur Foundation Publishing.

Fu, W., Turner, J.C., Zhao, J. and Du, G., 2015. Ecological footprint (EF): An expanded role in calculating resource productivity (RP) using China and the G20 member countries as examples. *Ecological Indicators*, 48, pp. 464–471.

Gallaud, D. and Laperche, B., 2016. *Circular Economy, Industrial Ecology and Short Supply Chain*. Hoboken, NJ: John Wiley & Sons.

Gothenburg Region, 2019. *Hållbarhetsverktyget – Göteborgsregionen (GR) [Sustainability Tool – The Gothenburg Region (GR)].* [online] Available at: <https://goteborgsregionen.se/toppmenyn/dettajobbargrmed/miljoochsamhallsbyggnad/samhallsbyggnad/hallbarhetsverktyget.4.711b2c90161f214330e2de9f.html> [Accessed 21 Sep. 2019].

Groothuis, F., 2015. *The Social Power of the Circular Economy.* [online] Circulate. Available at: <https://circulatenews.org/2015/11/the-social-power-of-the-circular-economy> [Accessed 21 Sep. 2019].

Guldager Jensen, K. and Sommer, J., 2016. *Building a Circular Future*. 2nd ed. Copenhagen, Denmark: GXN Innovation.

H&M Group, 2018. *H&M Group Sustainability Report.* [online] Available at: <https://hmgroup.com/sustainability/sustainability-reporting.html> [Accessed 21 Sep. 2019].

Hannon, E., Magnin, C. and Rosenfield, J., 2016. *Why the Circular Economy is All About Retaining Value*. [online] McKinsey & Company. Available at: <www.mckinsey.com/business-functions/sustainability-and-resource-productivity/our-insights/why-the-circular-economy-is-all-about-retaining-value> [Accessed 21 Sep. 2019].

Hao, K., 2019. *Training a single AI model can emit as much carbon as five cars in their lifetimes*. [online] MIT Technology Review. Available at: <www.technologyreview.com/s/613630/training-a-single-ai-model-can-emit-as-much-carbon-as-five-cars-in-their-lifetimes/amp> [Accessed 21 Sep. 2019].

Heinrich, M. and Lang, W., 2019. *Materials Passports – Best Practice*. Munich, Germany: Technical University of Munich.

Herzberg, B., 2015. *Blockchain: The Solution for Transparency in Product Supply Chains*. [online] Provenance. Available at: <www.provenance.org/whitepaper> [Accessed 21 Sep. 2019].

Hollis, L., 2013. *Cities are Good for You: The Genius of the Metropolis*. London, UK: Bloomsbury Publishing.

Horvath, B., Mallinguh, E., Fogarassy, C., Horvath, B., Mallinguh, E. and Fogarassy, C., 2018. Designing business solutions for plastic waste management to enhance circular transitions in Kenya. *Sustainability*, 10(5), p. 1664, pp. 1–20.

Ideo and Ellen MacArthur Foundation, 2019. *Methods*. [online] Available at: <www.circulardesignguide.com/methods> [Accessed 21 Sep. 2019].

Ijomah, W.L., McMahon, C.A., Hammond, G.P. and Newman, S.T., 2007. Development of design for remanufacturing guidelines to support sustainable manufacturing. *Robotics and Computer-Integrated Manufacturing*, 23(6), pp. 712–719.

Ivanovic, O.M., 2018. Ecological responsibility and sustainable development as preconditions for development of the concept of circular economy. In: V. Sima, ed., *Organizational Culture and Behavioral Shifts in the Green Economy*. Hershey, PA: IGI Global, pp. 1–20.

Jackson, T., 2009. *Prosperity Without Growth: Economics for a Finite Planet*. Hoboken, NJ: Earthscan.

Kalmykova, Y., Sadagopan, M. and Rosado, L., 2018. Circular economy – From review of theories and practices to development of implementation tools. *Resources, Conservation and Recycling*, 135, pp. 190–201.

Kaplan, S., 2012. *The Business Model Innovation Factory: How to Stay Relevant When The World is Changing*. Hoboken, NJ: John Wiley & Sons.

Kavadias, S., Ladas, K. and Loch, C., 2016. *The Transformative Business Model*. [online] Harvard Business Review. Available at: <https://hbr.org/2016/10/the-transformative-business-model> [Accessed 21 Sep. 2019].

Kelly, A., 2012. *Gross National Happiness in Bhutan: The Big Idea From a Tiny State That Could Change the World*. [online] The Guardian. Available at: <www.theguardian.com/world/2012/dec/01/bhutan-wealth-happiness-counts> [Accessed 21 Sep. 2019].

Kelly, T., Liaplina, A., Tan, S.W. and Winkler, H., 2017. *Reaping Digital Dividends: Leveraging the Internet for Development in Europe and Central Asia*. Washington, DC: World Bank Group.

Kessler, S., 2015. *The "Sharing Economy" Is Dead, and We Killed It*. [online] Fast Company. Available at: <www.fastcompany.com/3050775/the-sharing-economy-is-dead-and-we-killed-it> [Accessed 21 Sep. 2019].

Kim, Y.J., Kim, K. and Lee, S., 2017. The rise of technological unemployment and its implications on the future macroeconomic landscape. *Futures*, 87, pp. 1–9.

Kirchherr, J., Reike, D. and Hekkert, M., 2017. Conceptualizing the circular economy: An analysis of 114 definitions. *Resources, Conservation and Recycling*, 127, pp. 221–232.

Kojo, I. and Nenonen, S., 2017. Evolution of co-working places: Drivers and possibilities. *Intelligent Buildings International*, 9(3), pp. 164–175.

Koren, H. and Bisesi, M.S., 2003. *Handbook of Environmental Health, Volume II: Pollutant Interactions in Air, Water, and Soil*. 4th ed. Boca Raton, Fl: Lewis Publishers.

Lahti, T., Wincent, J., Parida, V., Lahti, T., Wincent, J. and Parida, V., 2018. A definition and theoretical review of the circular economy, value creation, and sustainable business models: Where are we now and where should research move in the future? *Sustainability*, 10(8), p. 2799, pp. 1–19.

Levenson, E. and Miller, B., 2018. *2018 is on Pace to be the 4th-Hottest Year on Record*. [online] CNN. Available at: <https://edition.cnn.com/2018/07/28/us/2018-global-heat-record-4th-wxc/index.html> [Accessed 21 Sep. 2019].

Lewandowski, M., 2018. Public sector and circular business models: From public support towards implementation through design. In: L. Moratis, F. Melissen and S.O. Idowu, eds., *Sustainable Business Models: Principles, Promise, and Practice*. Cham, Switzerland: Springer, pp. 85–104.

Linder, M., 2017. Ripe for disruption: Reimagining the role of green chemistry in a circular economy. *Green Chemistry Letters and Reviews*, 10(4), pp. 428–435.

Lund University, International Institute for Industrial Environmental Economics, VITO, Ghent, N.T.U. of and Delft University of Technology, 2018. *Circular Economy – Sustainable Materials Management*. [online] Coursera. Available at: <www.coursera.org/learn/circular-economy> [Accessed 21 Sep. 2019].

Maitre-Ekern, E., 2018. Exploring the spaceship earth. In: E. Maitre-Ekern, C. Dalhammar and H.C. Bugge, eds., *Preventing Environmental Damage from Products*. Cambridge, UK: Cambridge University Press, pp. 23–56.

Marinescu, C.M., 2015. Smart energy-intelligent solutions for residential buildings. *Procedia Technology*, 19, pp. 708–714.

Markopolu, A., Garmulewicz, A., Van Sprang, H., Van de Glind, P. and Rueda, S., 2019. *System Reset*. [online] DIF. Available at: <www.thinkdif.co/sessions/system-reset> [Accessed 21 Sep. 2019].

Masi, D., Day, S. and Godsell, J., 2017. Supply chain configurations in the circular economy: A systematic literature review. *Sustainability*, 9(9), p. 1602, pp. 1–22.

Maslin, R. and Shayler, M., 2016. *Service Design for a Circular Economy*. [online] The Great Recovery. Available at: <www.greatrecovery.org.uk/resources/service-design-for-a-circular-economy> [Accessed 21 Sep. 2019].

Material Economics, 2018. *The Circular Economy A Powerful Force For Climate Mitigation*. [online] Available at: <https://materialeconomics.com/publications/the-circular-economy> [Accessed 21 Sep. 2019].

Mendoza, E.G., Razin, A. and Tesar, L.L., 1993. *A Comparative Analysis of the Structure of Tax Systems in Industrial Countries*. IMF Working Paper, No. 39/14. Washington, DC.

Morgan, J. and Mitchell, P., 2015. *Employment and the Circular Economy: Job Creation in a More Resource Efficient Britain*. [online] Green Alliance. Available at: <www.wrap.org.uk/sites/files/wrap/Employment and the circular economy summary.pdf> [Accessed 21 Sep. 2019].

Munholland, G., 2018. *Is the Circular Economy a Myth? – The Cargo Carousel System*. [online] 2 way supply chain. Available at: <https://2waysupplychain.com/2018/06/01/is-the-circular-economy-a-myth> [Accessed 21 Sep. 2019].

Murray, A., Skene, K. and Haynes, K., 2017. The Circular Economy: An interdisciplinary exploration of the concept and application in a global context. *Journal of Business Ethics*, 140(3), pp. 369–380.

NASA, 2018. *May 2018 was Fourth Warmest May on Record.* [online] NASA's Goddard Institute for Space Studies. Available at: <https://climate.nasa.gov/news/2750/may-2018-was-fourth-warmest-may-on-record> [Accessed 21 Sep. 2019].

OECD, 1998. *Agricultural Policy Reform and the Rural Economy in OECD Countries.* Paris, France: OECD Publishing.

OECD, 2011a. *Perspectives on Global Development 2012 Social Cohesion in a Shifting World: Social Cohesion in a Shifting World.* Paris, France: OECD Publishing.

OECD, 2011b. *Resource Productivity in the G8 and the OECD – A Report in the Framework of the Kobe 3R Action Plan.* [online] Available at: <www.oecd.org/env/waste/resourceproductivityintheg8andtheoecd.htm> [Accessed 21 Sep. 2019].

OECD, 2016. *How Stringent are Environmental Policies? Policy Perspectives.* [online] OECD. Available at: <www.oecd.org/economy/greeneco/How-stringent-are-environmental-policies.pdf> [Accessed 21 Sep. 2019].

OECD, 2018a. *Gross Domestic Product (GDP).* [online] Available at: <https://data.oecd.org/gdp/gross-domestic-product-gdp.htm> [Accessed 21 Sep. 2019].

OECD, 2018b. *Transformative Technologies and Jobs of the Future.* [online] Background report for the Canadian G7 Innovation Ministers' Meeting: Montreal, Canada 27–28 March 2018. Available at: <www.oecd.org/innovation/transformative-technologies-and-jobs-of-the-future.pdf> [Accessed 21 Sep. 2019].

OECD, 2019. *Global Material Resources Outlook to 2060: Economic Drivers and Environmental Consequences.* Paris, France: OECD Publishing.

OECD/Eurostat, 2018. *Oslo Manual 2018: Guidelines for Collecting, Reporting and Using Data on Innovation: The Measurement of Scientific, Technological and Innovation Activities.* 4th ed. Paris, France/Luxembourg, Luxembourg: OECD Publishing & Eurostat.

Pagoropoulos, A., 2017. The Emergent Role of Digital Technologies in the Circular Economy: A Review. *Procedia CIRP*, 64, pp. 19–24.

Pardo, R., 2018. *How the Circular Economy Can Benefit from the Digital Revolution.* [online] European Policy Center. Available at: <www.epc.eu/pub_details.php?pub_id=8469&cat_id=4> [Accessed 21 Sep. 2019].

Perez, C., 2018. *Carlota Perez – Techno-economic Paradigm Shifts, 2018 Summit – YouTube.* [online] 2018 Summit. Available at: <www.youtube.com/watch?v=dhNd3tVR1hI> [Accessed 21 Sep. 2019].

Peters, M.A., 2017. Technological unemployment: Educating for the fourth industrial revolution. *Educational Philosophy and Theory*, 49(1), pp. 1–6.

Pomerleau, K., 2014. *A Comparison of the Tax Burden on Labor in the OECD.* [online] Tax Foundation. Available at: <https://taxfoundation.org/comparison-tax-burden-labor-oecd> [Accessed 21 Sep. 2019].

Pure Waste, 2019. *Home Page.* [online] Available at: <www.purewaste.org> [Accessed 21 Sep. 2019].

Rattiwan, S., 2016. *Gross National Happiness Center Established in Thailand.* [online] NNT National News Bureau of Thailand. Available at: <https://web.archive.org/web/20170202044010/http://thainews.prd.go.th/website_en/news/news_detail/WNPOL5910120010021> [Accessed 21 Sep. 2019].

Rifkin, J., 2011. *The Third Industrial Revolution: How Lateral Power is Transforming Energy, the Economy, and the World.* New York, NY: St. Martin's Press.

Rifkin, J., 2014. *The Zero Marginal Cost Society: The Internet of Things, the Collaborative Commons, and the Eclipse of Capitalism*. New York, NY: Palgrave Macmillan.

Roos, D., 2016. *Like Airbnb But for Work: Rent Your Home as Office Space*. [online] Howstuffworks. Available at: <https://money.howstuffworks.com/second-career-ideas/airbnb-work-rent-your-home-office-space.htm> [Accessed 21 Sep. 2019].

Samuel, H., 2019. *France to Ban Unsold Clothes and Electronics from Being Destroyed in 'World First'*. [online] The Telegraph. Available at: <www.telegraph.co.uk/news/2019/06/04/france-ban-unsold-clothes-electronics-destroyed-world-first> [Accessed 21 Sep. 2019].

Sanborn, D., 2019. *Land Use Mapping*. [online] Available at: <www.sanborn.com/land-use-land-cover-mapping> [Accessed 21 Sep. 2019].

Sas, 2019. *Artificial Intelligence What It Is and Why It Matters*. [online] Sas. Available at: <www.sas.com/en_gb/insights/analytics/what-is-artificial-intelligence.html> [Accessed 21 Sep. 2019].

SB Insight, 2019. *The Nordic Market For Circular Economy*. [online] Available at: <www.sb-insight.com/sb-reports> [Accessed 21 Sep. 2019].

Schroeder, P., Anggraeni, K. and Weber, U., 2019. The relevance of circular economy practices to the sustainable development goals. *Journal of Industrial Ecology*, 23(1), pp. 77–95.

Schwab, K. and Davis, N., 2018. *Shaping the Future of the Fourth Industrial Revolution: A Guide to Building a Better World*. London, UK: Penguin.

Selada, C., 2017. Smart Cities and the Quadruple Helix Innovation Systems Conceptual Framework: The Case of Portugal. In: D.O. Monteiro, S. Paulina and E.G. Carayannis, eds., *The Quadruple Innovation Helix Nexus*. New York, NY: Palgrave Macmillan, pp. 211–244.

Selvan Ramadoss, T., Alam, H. and Seeram, R., 2018. Artificial Intelligence and Internet of Things enabled Circular economy. *The International Journal of Engineering and Science*, 7(9), pp. 55–63.

Sengupta, S., 2018. *2018 Is Shaping Up to Be the Fourth-Hottest Year. Yet We're Still Not Prepared for Global Warming*. [online] *The New York Times*. Available at: <www.nytimes.com/2018/08/09/climate/summer-heat-global-warming.html> [Accessed 21 Sep. 2019].

Sexton, C., 2019. *Bitcoin has the Carbon Footprint of a Small Country*. [online] Earth.com. Available at: <www.earth.com/news/bitcoin-carbon-footprint-small-country> [Accessed 21 Sep. 2019].

SGS Offices & Labs, 2015. *Built to last? A law in France to combat planned obsolescence for appliances*. [online] Available at: <www.sgs.com/en/news/2015/07/built-to-last-a-law-in-france-to-combat-planned-obsolescence-for-appliances> [Accessed 21 Sep. 2019].

Stahel, W.R., 2016. The circular economy. *Nature*, 531(7595), pp. 435–438.

Steffen, W., 2017. *Will Steffen on Living with Climate Change*. [online] The University of Auckland. Available at: <www.uabsknowledge.ac.nz/en/research-and-comment/video/short-takes-series/will-steffen-on-living-with-climate-change.html> [Accessed 21 Sep. 2019].

Steffen, W., Rockström, J., Richardson, K., Lenton, T.M., Folke, C., Liverman, D., Summerhayes, C.P., Barnosky, A.D., Cornell, S.E., Crucifix, M., Donges, J.F., Fetzer, I., Lade, S.J., Scheffer, M., Winkelmann, R. and Schellnhuber, H.J., 2018. Trajectories of the Earth System in the Anthropocene. *Proceedings of the National Academy of Sciences of the United States of America*, 115(33), pp. 8252–8259.

Stuchtey, M., Enkvist, P.-A. and Zumwinkel, K., 2016. *A Good Disruption: Redefining Growth in the Twenty-First Century*. London, UK: Bloomsbury Publishing.

Sustainable Apparel Coalition, 2019. *The Higg Index*. [online] Available at: <https://apparelcoalition.org/the-higg-index> [Accessed 21 Sep. 2019].

Truby, J., 2018. Decarbonizing Bitcoin: Law and policy choices for reducing the energy consumption of Blockchain technologies and digital currencies. *Energy Research & Social Science*, 44, pp. 399–401.

Tukker, A., Bulavskaya, T., Giljum, S., De Koning, A., Lutter, S., Simas, M., Stadler, K. and Wood, R., 2014. *The Global Resource Footprint of Nations*. Melk, Austria: The Netherlands Organisation for Applied Scientific Research.

Tuppen, C., 2016. Circularity indicators. In: K. Webster, ed., *A New Dynamic 2: Effective Systems in a Circular Economy*. Isle of Wight, UK: Ellen MacArthur Foundation Publishing, pp. 195–210.

Turntoo, 2019. *Material Passport*. [online] Available at: <http://turntoo.com/en/material-passport> [Accessed 21 Sep. 2019].

UN environment and International Resource Panel, 2018. *Global Resources Outlook 2019: Natural Resources for the Future We Want*. [online] UN environment, International Resource Panel. Available at: <www.resourcepanel.org/reports/global-resources-outlook> [Accessed 21 Sep. 2019].

Waldegrave, L., 2017. *What Is Systems Thinking?* [online] Circulate. Available at: <https://circulatenews.org/2017/09/what-is-systems-thinking> [Accessed 21 Sep. 2019].

Walters, L., 2018. *NZ Government to lead world in measuring success with wellbeing measures | Stuff.co.nz*. [online] Stuff. Available at: <www.stuff.co.nz/national/politics/101066981/nz-government-to-lead-world-in-measuring-success-with-wellbeing-measures> [Accessed 21 Sep. 2019].

Wastling, T., Charnley, F. and Moreno, M., 2018. Design for circular behaviour: Considering users in a circular economy. *Sustainability*, 10(6), p. 1743, pp. 1–22.

Wijkman, A. and Skånberg, K., 2015. *The Circular Economy and Benefits for Society*. [online] The Club of Rome. Available at: <www.clubofrome.org/wp-content/uploads/2016/03/The-Circular-Economy-and-Benefits-for-Society.pdf> [Accessed 21 Sep. 2019].

Wilts, H. and Berg, H., 2018. Digital cycles and digital scrap: How digitization improves resource efficiency in the circular economy – and where it tends to cost resources. *Journal of Waste Recycling*, 3(1), p. 5, pp. 1–4.

Woetzel, J., Sellschop, R., Chui, M., Ramaswamy, S., Nyquist, S., Robinson, H., Roelofsen, O., Rogers, M. and Ross, R., 2017. *How Technology is Reshaping Supply and Demand for Natural Resources*. [online] McKinsey Global Institute. Available at: <www.mckinsey.com/business-functions/sustainability-and-resource-productivity/our-insights/how-technology-is-reshaping-supply-and-demand-for-natural-resources> [Accessed 21 Sep. 2019].

World Economic Forum, 2014. *Towards the Circular Economy: Accelerating the Scale-up Across Global Supply Chains*. [online] Available at: <http://www3.weforum.org/docs/WEF_ENV_TowardsCircularEconomy_Report_2014.pdf> [Accessed 21 Sep. 2019].

World Economic Forum, 2016. *The New Plastics Economy: Rethinking the Future of Plastics*. [online] Available at: <http://www3.weforum.org/docs/WEF_The_New_Plastics_Economy.pdf> [Accessed 21 Sep. 2019].

Xprize, 2017. *AI for Good – Sustainability – YouTube*. [online] YouTube. Available at: <www.youtube.com/watch?v=mJ6rjJiIHyo> [Accessed 21 Sep. 2019].

ZenRobotics, 2019. *How Does Robotic Recycling With ZenRobotics Recycler work?* [online] Available at: <http://web.archive.org/web/20190320162610/https://zenrobotics.com/solutions/how-it-works> [Accessed 21 Sep. 2019].

Part IV
Conclusion

17 Conclusion

The digital transformation of labor – where do we go from here?

Anthony Larsson

1. Introduction

The digitalization and digital transformation of labor is in many respects illustrated as a "disruptive" development inasmuch that it fundamentally alters the way in which people earn their upkeep, and indeed who gets to keep doing the job they were trained to do in the way they learned how. The convergence of digital technologies is reshaping the future of the workplace, workforce and work processes (Frost & Sullivan, 2019).

Digital transformation in business generally tends to be a complex and unwieldy process to implement fully. As technology evolves exceptionally quickly, organizations are pressured into keeping an even pace or risk falling by the wayside. Thus, organizations today are by and large faced with two main challenges when it comes to digital transformation. First, they must put digital transformation/digitalization on their roadmap. Second, they must ensure that they possess the agility to deploy new technologies before they are rendered irrelevant. Thus, there is no denying that digital technologies fundamentally transform organizations. In order to succeed and stay relevant on the market, organizations must have a coherent and viable strategy, employers must have a plan to reskill workers and citizens must keep themselves informed, motivated and prepared for swift changes. While other technological "revolutions" throughout history, such as the Industrial Revolution and the Agricultural Revolution, etc., have played out over a longer period of time, the digital transformation operates on a radically shorter time trajectory, which prompts businesses to take quick and decisive action. The digital transformation also affects governments inasmuch that they need to prepare their citizens for a digital future, while also dealing with potential inequalities, wage deflation or possibly even social unrest stemming from the digital transformation of society and its labor market.

A 2019 report released by the European Commission argued that the EU was on the "right path towards modernising our labour and social policies" (European Commission, 2019, para.2), while issuing recommendations to actors on the labor market "to reduce structural skill gaps, especially for women in science, technology, engineering and mathematics (STEM), workers at risk of automation and the low-skilled" (European Commission, 2019, para.3). By the same token, figures from the same European Commission also shows that two-fifths (or approximately 40%) of the EU workforce have little or no digital skills (and conversely, approximately 60% are digitally literate) (Panopoulos, 2019). The United States displays more favorable figures, with approximately 84% of all adults being digitally literate (Mamedova and Pawlowski, 2018).

While digital literacy can no doubt be improved further, it is undeniable that digitalization carries several ramifications on the labor market. A recurrent topic in the ongoing debate is whether or not digitalization and digital transformation creates or destroys labor. Some authors and researchers contend adamantly that the digital development is happening to the detriment of human labor, with people being made redundant for most processes and tasks in the future (Ford, 2015; Kaplan, 2015). Others take a more positive outlook and argue that it will in fact create more jobs than it replaces (Jones, 2018).

The truth is that not much is known about the future in this field as predictions thus far have been fragmented and often contradictory, more often than not leaving many fundamental questions unanswered, such as how automatization of work affects the tax-paying ability, and by extension, the very fundament of the welfare state? Why are people eager to pursue technology that could potentially deprive them of their livelihood? Can all workers be reskilled? If not, what becomes of those who cannot? These topics and many others have been explored in this book

Throughout this book, a team of researchers and specialists have investigated various areas of the labor sector and how it is impacted by the digital transformation/digitalization, and what the overall ramifications may be for society at large. In particular, this chapter will seek to expound on the conclusions drawn by the authors throughout their respective chapters, while endeavoring to build up an overall conclusion based on the findings and predictions made in each study.

2. Chapter findings

This following section will seek to elaborate upon the conclusions drawn by the different chapters in regards to the different themes of the digital transformation of labor as stipulated by this book's structure.

The first theme, "Practical utilization of new technologies", looked at how the development of new technology can be applied in practice to enable people to work in ways they have not previously been able to. That is to say, what kind of new opportunities and practices that can be expected for the future, but also the risks and impediments this development might incur on some people, actors or labor traditions.

The next theme, "The role of the digital welfare state", looked at how the transformation of labor markets can affects the welfare state, in particular regard to the funding of the state and the tax revenue system. While this is a broad and complex subject area in and of itself, it is not the focal point of this book. It is nonetheless an important subject to bring up and discuss in context to the ever-changing labor market that the welfare state relies upon to a great extent via various forms of taxation (income tax, payroll tax and so forth) in order to keep the welfare state funded.

The final theme, "Digital disruption of status quo", looked at how digitalization and the digital transformation (or any particular part of it) can affect certain group or actors. That is to say, will the digital development enable any particular actor to advance their positions on the labor market, or conversely, will it act to challenge any particular actors? Moreover, this theme seeks to understand how the digital developments may disrupt the current status quo and prompt actors to change their mode of operation in the future, if at all.

2.1. Practical utilization of new technologies

Technologies are bound to be utilized in new and different ways. Sometimes new technologies are devised for a specific purpose, and sometimes technologies originally intended from one purpose evolve through time to meet the need of a different purpose. Already at the outset of this book, Bard, Söderqvist and Larsson (Chapter 2, this volume) outlined the most important developments through history of mankind that has resulted in various types of labor and crafts. The chapter also looked at how and why mankind has always been drawn to technology, while asserting that the need of communication has always been at the heart of the human condition and in shaping societies. The authors concluded that a new social, cultural and economic paradigm is forming and that this paradigm will take on a fluid form rather than a concrete equivalent. This indicates that the future of labor will develop alongside the evolution of society and that a whole new set of norms will be developed in the future that is not restricted to mere social norms.

On the more tangible, physical side of the spectrum, Van der Zande, Siri, Teigland and Teigland (Chapter 3, this volume) provided an overview of digitalization and automation along with the three underlying technologies of AI, machine learning and robotics. The aim was to explore the potential of these technologies to replace human capabilities in the workplace while discussing some factors beyond the realms of technology that carry an influence on the pace and scope of job automation. The authors argue that while most future jobs will be affected by automation to some extent in terms of individual activities, most jobs will not be completely automatized. Specifically, full automation of non-routine tasks involving autonomous mobility, creativity, problem-solving, and complex communication, appears unlikely even in the future. However, the authors posit that the nature of jobs will change as mundane tasks become substituted and people are necessitated to work closer together with machines. Industry prime candidates for labor substitution are food and accommodation services, transportation and warehousing, retail trade, wholesale trade and manufacturing. While the authors admit that short-term reskilling will likely be required to enable the reemployment of displaced labor, they do not anticipate that the automatization will incur any long-term unemployment. To cope with the pace of automatization, the authors recommend an increased focus on education and training for individuals, as well as for organizations, regions and nations.

While certain types of labor may be at risk of being automatized in the future, the digital transformation may also offer some new labor opportunities. Wood, Graham and Anwar (Chapter 4, this volume) discuss how the rise of the "gig economy" has enabled internet users to find new work previously unavailable to them, particularly so via "online labor platforms", which effectively serves as a "global remote gig economy" and provides workers with access to worldwide labor opportunities. In their concluding words, the authors caution against letting the online gig work function as an unregulated labor market. To this end, the authors argue that there must be stipulations made in regards to matters such as the minimum wage (which in some countries may present itself as an obstacle and devaluing the work done by the workers).

To this end, digitalization is also changing the type of labor done in the government and academic sectors as well. Seemingly, the digital development in Europe is that academia and the public sector are gravitating toward a greater sense of transparency and information accessibility inasmuch that policies concerning Open Access and Open Data are being increasingly implemented on a more formal level (Kisjes, 2015; Toelch and Ostwald, 2018). To this extent, the national states and the governments will continue to have an important role in ensuring that the digital transformation is facilitated throughout the academic sector (Maire, Chapter 5, this volume). However, it should be noted that the digital transformation of sciences and the public sector occurs on many levels, with the government playing one essential key role (Asgarkhani, 2005; Tolbert and Mossberger, 2006). In addition, the digital transformation and the development toward Open Access-based knowledge also occurs through bottom-up evolution and top-down policy implementation (Maire, Chapter 5, this volume).

Of course, the issue of transparency and information accessibility is of equal importance also in other workplaces outside of academia and the public sector, as this generates trust. As posited by Bernhardtz (Chapter 6, this volume), trust in cognitive computer systems are key factors in the successful digitalization of labor. Moreover, the author, much like Van der Zande, Siri, Teigland and Teigland (Chapter 3, this volume), contends that digitalization is not likely to prompt automated replacement of laborers en masse any time soon. One of the subsets of labor is of course the organizations creating them. They are in many cases run by corporate boards, and thus it is relevant to understand how the digital transformation affects these boards as there could be far-reaching implications for the labor market on a broader level. Torre, Teigland and Engstam (Chapter 7, this volume) argue that the rapid advancements in AI will lead to corporate boards being challenged in their decision-making process in ways they have not experienced in the past. Specifically, they contend that AI, and the "big data" will become one of the most prominent topics for corporate boards to deal with within the next decade and propose a leadership matrix to assist boards in how to better develop their competence within AI implementation.

2.2. The role of the digital welfare state

Considering what effects the digital transformation has on labor in the context of the welfare state, Blix (Chapter 8, this volume) and Greve (Chapter 9, this volume) posit some welfare states appear better prepared than others for the digital transformation and that in recent years, labor markets have become more polarized whereas in some countries (like Sweden) the effects have hitherto been limited. Among the attributed reasons is the safety net for its citizens as well as the well-endowed structural funds for retraining of displaced workers (Blix, Chapter 8, this volume). However, even in the cases of well-developed and prepared welfare states, the country may still be put at risk in light of the impending digital transformation, particularly in places outside of large urban/metropolitan areas. That is to say that all labor sectors can and will be affected to some degree or another, and most likely they will be affected simultaneously.

Another aspect raised by Larsson and Sabolová (Chapter 10, this volume), is that the "gig economy" may have far-reaching implications for the welfare state through the emergence of "gig patients". These are workers in the "gig economy" who cannot afford to take time off to seek medical attention until their condition reaches a point that makes it absolutely necessary. This makes the condition more resource-demanding to treat, and also, these workers will rarely have the means to pay for their treatment in full, which causes a strain on the welfare sector that will worsen as the "gig patients" grow in numbers.

Typically, advanced welfare societies will have high taxes placed also on middle-income earners (OECD, 2019; Lindbeck, 1986). This, in combination with a shortage of skilled workers in key-segments of the labor market could cause the tax bases to gradually erode. Given the high level of taxes supporting the welfare spending, this would create even stronger incentives for firms to automate work or to buy services on global gig markets via platform-based labor markets. Blix (Chapter 8, this volume) argues, from the perspective of the welfare state, that this may result in further labor market polarization and more strained financing of comprehensive social welfare. Upholding the social contract in the welfare state is already becoming a challenge. Hence, the outcome of the welfare state is contingent on the policy responses of governments, social partners, trade unions and employer organizations (Blix, Chapter 8, this volume; Larsson and Sabolová, Chapter 10, this volume). That is to say, trade unions that are able to adapt to the digital transformation, while providing new forms of support and safety to its members could still remain relevant to its members. As such, it could serve as a counterweight to some of the increases in income uncertainty. Governments could also consider broadening the tax bases to support welfare ambitions, especially for the self-employed. Digitalization carries challenges in all welfare states, but interestingly for different reasons. As illustrated by Greve (Chapter 9, this volume), the challenges dealt to Southern and Eastern Europe stem from the transformation of the industrial production that has been slower in these countries, resulting from a lower wage level than in other more mature welfare states. The challenges to the Nordic and Continental welfare states, however, emanate from expenditure and the capability of financing the welfare states. In conclusion, the authors argue that education and labor market policy are important for the future employability of workers (Greve, Chapter 9, this volume) and that in order for the welfare state to survive in the way that we know it, judicious reforms are needed (Blix, Chapter 8, this volume).

2.3. Digital disruption of status quo

The topic of "privacy" is a concern that features prominently amidst the discussion of digitalization. That is to say, trust and confidence in a specific new technology is fundamental in order for it to develop further, and the same goes for digitalization by and large, as was further elaborated by Bernhardtz (Chapter 6, this volume). If technology is not considered secure and/or trustworthy, it will not gain the support of its prospective users. As information access is becoming increasingly more widespread and easily accessible, citizens have become increasingly more concerned with how their personal data is processed. This has prompted new legal frameworks. In the EU, the General Data Protection Regulation (GDPR) was launched on 25 May 2018. Larsson and Lilja (Chapter 11, this volume) have investigated how the implementation of GDPR may affect the future types of businesses, or more specifically asking the question: *In what way may GDPR influence the labor market of tomorrow, and what businesses are at risk*? The authors conclude that there is still much work to be done by companies seeking to secure full GDPR compliance, even though there is steady progress being made in the area. Going forward, the authors argue that while GDPR challenges all organizations dealing in large volumes of personal data, it will primarily hit the smaller studios for online game developing, while "big tech" companies are likely to not be too inconvenienced by GDPR. On the other hand, GDPR will also provide opportunities for other types of businesses, such as for consultancy in digital strategy and for professionals in analytics and software architecture.

While prior chapters of this volume has asserted that the digital transformation and automation does not run the risk of entirely replacing all types of labor in the near future, it may change to some extent in regards to individual activities (Van der Zande et al., Chapter 3, this volume; Bernhardtz, Chapter 6, this volume). On this note, an argument raised in regards to automation of labor is that it induces inequality amongst certain individual groups of society. The assertion is that while automatization may replace one person's job, it enables someone else to make more money on their assets. Castronova (Chapter 12, this volume) argues that "these unequal slices are part of a growing pie". As previously mentioned by Van der Zande et al. (Chapter 3 this volume), reskilling of labor will likely be required in order to enable the reemployment of displaced labor. However, Castronova (Chapter 12, this volume) highlights the problem of what is to become of the multitude of low-skill laborers who will find their jobs automatized and who perhaps have no obvious alternative routes to meaningful work at their disposal.

Particularly so, young men with low-skilled jobs may feel so resigned that they opt to stay at home playing video games rather than look for work and/or reskill themselves due to the perceived lack of incentives of putting in the extra effort. On this account, digitalization may provide additional opportunities to this group of people, as the author predicts that there in the future will be a way for them to earn revenue by being paid in some capacity by game companies. Essentially, the author describes a development toward a gamification process in which young, (predominantly) male unemployed low-skilled workers are able to play the companies' video games in order to earn some small point-based incentives that can be liquidated in the form of purchasing power. As this point-based system develops in the coming decades, it will become increasingly more viable and may in many cases present an alternative form to money as legal tender. At the same time, wealthier citizens who also play the same video games may utilize the "pay-to-win" features so often present in online games, meaning they pay for items with real-world money in order to gain access to features in the game that are otherwise closed to the player (or sometimes even necessitated to buy in order to progress further through the game). Thereby these "big spenders" are effectively funding the low-skilled workers now being "paid" to play. The author contends that this may help curb the inequalities placed upon the unemployed low-skilled workers, as within a generation, playing games for money would come to be seen as a legitimate occupational choice, while also securing a way for income gains from technological progress to be distributed to the low-skill population.

Graham and Anwar (Chapter 13, this volume) extrapolated upon the "planetary labor market" in digital work. Specifically, this signifies that digital technologies have been deployed in order to bring about a labor market that can operate at a planetary scale. The authors argued that the concept of geography will not be made irrelevant even though technological advancements exist to facilitate a "planetary labor market". Rather, the planetary labor market utilizes digital technology and geography by helping clients to operate across geographical borders at minimal cost. That is to say, although the planetary labor market allows workers to sell their labor power globally, these workers are still tethered to the physical place where they live and work on a day-to-day basis.

Another recurrent question has been how digitalization affects women's future career opportunities in regards to the "digital gender divide". Larsson and Viitaoja (Chapter 14, this volume) highlight the relative lack of women in the Western countries undertaking science, technology, engineering and mathematics (STEM) subjects, which results in there ultimately being fewer women working with information and communication technology (ICT) related jobs. The authors argue that while digitalization may provide for more flexible working conditions that benefit women, the digital development also incurs a risk of bringing about more atypical and cynical work arrangements, to the detriment of women. Moreover, there is an added risk that the emergent automatization will begin to harvest some of the professions traditionally dominated by women. The chapter concludes that women's disposition in tech appears to be more about attitudinal issues and less about legal/regulatory issues. While the situation is gradually improving, more coordination among different efforts is needed in order to successfully safeguard the future position of women in the digital era.

The digital transformation as indicated previously by Larsson and Lilja (Chapter 11, this volume), consultancy in digital strategy is a business venture that can be expected to capitalize on the digitalization and digital transformation processes. However, this naturally raises the question how consultancies themselves will cope with the digital developments. Larsson, Andersson, Markowski, Nilsson and Mayor (Chapter 15, this volume) have investigated this topic closer and found that the development of analytics tools will be quintessential and that digitalization carries the greatest advantages during the analytical phases of consulting. Nevertheless, an impending risk of digitalization is that organizations may be tempted to try optimizing their performance by having in-house data scientists take on more of the consultants' traditional tasks. This may result in suboptimal results, or in some cases, the consequences may even be dire. For that reason, the authors recommend a tighter future collaboration between consultants and data scientists so that their efforts may be synergized.

One may deduce that the digital transformation carries profound impact on specific job functions, but another important issue is how it affects the overall sustainability of the labor market and society at large. Larsson and Lindfred (Chapter 16, this volume) investigated how labor conditions are affected by the introduction of circular economy in order to achieve circular businesses and societies and what role digitalization plays in this. The authors conclude that circular economy is fundamentally reshuffling our current economic system and that it may facilitate the creation of new jobs in a variety of different fields, such as in manufacturing, marketing, sales, refurbishing and reverse logistics, etc. A variety of other enterprises would change the way they conduct business as well and there will be a greater overall shift toward a sharing economy. In this way, circular economy should be regarded as an innovation and a strategy for businesses as well as for society at large. The authors also contend that digitalization will continue to be a vital instrument in making the circular economy come to pass.

3. Concluding comments

As mentioned in the introductory chapter (Larsson, Chapter 1, this volume), this book endeavored to explore the changes and impact that digital technology could have on the future of labor. The overall results of the studies have concluded that the digital transformation of the labor market is an ongoing process that will have a profound overall effect on labor, but to varying extents when it comes to particular aspects of it. Mankind's pursuit of technological advancements is a constant that will never change. As such, technology will always continue to evolve and the labor landscape is never completely static. Looking at the three different overarching themes this book has explored, one can summarize it in the following ways.

3.1. Practical utilization of new technologies

In regards to the theme of "Practical utilization of new technologies", new digital technologies such as AI, machine learning and robotics will indeed help automatize and streamline various processes. This may cause certain types of labor to become replaced, especially low-skilled, menial labor, which means that some of the workers in this category will likely need to undergo some reskill training (Van der Zande, et al., Chapter 3, this volume; Bernhardtz, Chapter 6, this volume). On higher, executive levels, digital technology such as AI may be used in ways that alters how corporate boards operate, although Torre et al., (Chapter 7, this volume) argue that corporate boards they investigated tended to be cognizant of the importance of AI implementation as a key competitive advantage and did not view AI as threatening to replace job positions in the corporate boardroom.

On a broader, more mundane level, is the digital economy's enablement of the "gig economy". Through the development of "online labor platforms", workers will have access to worldwide labor opportunities (Wood et al., Chapter 4, this volume). However, on the flipside, cost of labor will oftentimes be a decisive factor when contracting workers (all else being equal), meaning that especially low-skilled workers would perpetually have to work at minimum wage level, in the event that their country uses such a system (Todolí-Signes, 2017; Stewart and Stanford, 2017). In other cases, worker wages risk becoming a "race to the bottom", provided the country does not have a high density of labor-union membership with labor unions that have the power to make wage stipulations for its members (Singh and Zammit, 2004; Chan, 2003; Stewart and Stanford, 2017). Even so, it is not certain if these labor unions can indeed uphold their standing in a "gig economy", as this has proven challenging so far (O'Connor, 2019; Meyer, 2016; Minter, 2017). Moreover, a high density of labor union membership may not be possible in certain countries due to political reasons or otherwise. The idea of forcing a raised minimum wage internationally has also been discussed but it remains a highly contentious topic with much of the debate centered on the sheer feasibility of implementing such in practice (Badham, 2017; Worstall, 2017). Needless to say, this will be a much debated subject over the next decade. However, as Maire (Chapter 5, this volume) pointed out, the digital development will also prompt the spread of information to become more detailed and freely accessible in the academic community. This will in turn no doubt foster further research on the effects on the gig economy in regards to its opportunities and its limitations as well as help bolster people's awareness around this complex issue.

3.2. The role of the digital welfare state

The following theme, "The role of the digital welfare state", indicates that while the functions of the welfare states will be impacted differently by the digital transformation (depending on how advanced/mature the welfare is in each respective country), all labor sectors of the welfare state will be affected to some extent

(Blix, Chapter 8, this volume; Greve, Chapter 9, this volume). The high taxation levels will serve as an incentive for companies to automatize as much as possible of their work, or purchase services on gig markets with for the purpose of paying as low wages as possible (and by extension, less tax). This risks creating a highly polarized labor market, for even if not every single aspect of all types of work can be automatized, and even if workers can reskill themselves to get other proficiencies in a digitalized society, as suggested by Van der Zande, et al. (Chapter 3, this volume) and Bernhardtz (Chapter 6, this volume), it does not change the fact that the "gig economy" is making its way into the labor market. The "gig economy" may lead to a "race to the bottom" in wage setting, at least in some economies. This, in turn, will have detrimental effects on the state's ability to generate any sizable taxation income on labor, which on no small measure will worsen as the occurrence of "gig patients" grow in numbers (Larsson and Sabolová, Chapter 10, this volume). Thus, the challenges of upholding the "social contract" will become increasingly more difficult, as the ability to provide the same standard of welfare as today will likely decrease substantially.

3.3. *Digital disruption of status quo*

The final theme, "Digital disruption of status quo" has shown how different types of sectors are affected. On a larger level, the "gig economy" and the "planetary labor market" has less to do with fostering more job opportunities across geographical borders, but more to do with using digital technology to enable businesses to operate across countries at minimal cost (Graham and Anwar, Chapter 13, this volume). To this end, Castronova (Chapter 12, this volume) argues that a new type of "point-based" economy may gain a foothold in the digital era that caters to disillusioned low-skilled workers who are unable to reskill themselves, and/or unable to undertake "gig work". This new type of economy would allow this group of people to gain some kind of "trade-in" benefits through "gamification" of their hobbies. This would ensure that this group of people is not kept entirely passive, while also enabling companies to earn revenue through the "labor" done by this group of people. As such, the economic system may evolve too to some extent alongside the labor market, fostering the emergence of other forms of "alternative" and "virtual" currencies not too unlike cryptocurrency.

Of course, there will always be privacy concerns, and policy regulations that seek to address these will indubitably have far-reaching consequences on the labor market as well. For instance, the implementation of new legal frameworks, such as GDPR, means that there will be a greater need in the future for specific types of professions, particularly so for consultancy in digital strategy and for professionals in analytics and software architecture (Larsson and Lilja, Chapter 11, this volume). To that end, it is prudent to ask what becomes of consultancy in the future and how is it expected to change following digital advancements. The importance of analytics tools will continue to grow, and consultants will need to collaborate more closely with data scientists (Larsson, Andersson,

Markowski, Nilsson and Mayor, Chapter 15, this volume). It is to that possible end that we may see more data scientists entering the "gig economy" taking on freelance work at various occasions to collaborate with consultancies at different stages during different projects. While the "gig economy" may indeed serve to drive worker remuneration down, it is not given that this is the case for specialized trades such as data scientists. That is to say, the skills possessed by various types of specialists are not necessarily mutually interchangeable with another specialist. They may be, but the skill possessed is often individual and different competencies may be better suited for different tasks. However, a possible consequence of this is that specialists, such as data scientists seeking to perform "gig jobs" may have to market themselves harder to become more widely recognized for their individual and unique specialization in their field in order stay ahead of the competition. Alternatively, the consultancies may expand their businesses to also employ data scientists to their team of specialists to a greater extent than what is done today.

There are, however, a few other issues apart from the development toward a "gig economy". Castronova (Chapter 12, this volume) raises the issue of inequality induced by digitalization to certain individual groups of society. Although Castronova focuses on the young male low-skilled workers, other groups may be adversely affected as well. For example, the "digital gender divide" may be further widened as automatization of professions traditionally dominated by women start escalating and reskilling is made difficult due to the fact that there is an underrepresentation of women studying STEM subjects, resulting in fewer women working with ICT-related jobs (Larsson and Viitaoja, Chapter 14, this volume). To this end, working toward attitudinal change on a societal level is paramount in order to ensure that women have a future position and, if need-be, the same preconditions as men to reskill themselves.

Research has indicated that a digital transformation provides the preconditions for societies to adopt a circular economy (Wilts and Berg, 2017; Hobson and Lynch, 2016). As elaborated by Larsson and Lindfred (Chapter 16, this volume), this does not only serve to achieve sustainability in the economy and the resources of society, but one may also expect to see the creation of new types of work. Specifically, specialists in component remanufacturing and product refurbishment, product remarketers and various types of specialists in reverse logistics may find a boost in a circular economy. Still, more importantly, is the way in which a circular economy could change existing business practices in the future. For instance, in recent years, an increasing number of companies have transitioned from traditional workplaces to open-plan office spaces. This has largely been attributed to the evolvement of mobile technology and "knowledge work" (i.e., workers whose main specialty is knowledge, e.g., lawyers, scientists, programmers, etc.) becoming a more prominent feature in the modern labor market. This means that in recent years people have become less bound to a specific workplace than in the past, allowing for alternate, more cost-reducing and space-saving solutions such as "hot-desking", open workstations, group workstations and so forth (Peterson, 2014; Davis, Leach and Clegg, 2011). While these solutions may be cost-saving, they may also have adverse effects on productivity as well as staff morale (especially so for "hot-desking") due to factors such as stress, conflicts and noise-levels, etc. (Christou, 2018; Myers, 2016; Stillman, 2018).

In a circular economy, yet another iteration of the workplace may appear. Larsson and Lindfred (Chapter 16, this volume) argue that in the future, the coworking spaces may turn to a more "Airbnb-style" of workplace, where companies can book an apartment or home for a few hours for whenever an "office" is needed. Due to mobile technology and "knowledge work" it may be possible for people to sit at various locations and still manage to work together with colleagues, at times remotely and at times together at a coworking space. There are no proper studies yet available that show how this setting would affect the staff's productivity and/or morale, and more research into this space is encouraged.

3.4. Synthesis – where do we go from here?

The point of departure for this book was to explore some of the overarching themes in which digitalization and digital transformation can be expected to impact the labor conditions to some degree or another. Through a series of analytical accounts provided by several experts in their respective field, this book has provided an anthology of potential future scenarios and developments for the future of labor, with each study weighing in on the possible opportunities and challenges respectively that the future development stand to offer. Given the results from all the chapters covered in this volume, the ten most important overall takeaway predictions for the future of the labor market can be summarized in Table 17.1, with the advantage/opportunities listed on the left-hand side and the disadvantages/challenges listed on the right-hand side:

The first sentence of this book's introductory chapter (Larsson, Chapter 1, this volume) drew upon the ancient Chinese proverb that "a journey of a thousand miles begins with a single step" (Keyes, 2006, p. 107). At the end of these thousand miles and upon starting a new journey of a thousand miles, the lingering question is: "where do we go from here?" The future of the labor market as we know it is contingent on many different things. It is by and large an iterative process in which we may only know the true outcome by repeated evaluations and follow-ups of each implemented change and/or innovation. Therefore, securing a system of necessary checks-and-balances will be of paramount importance to ensure a successful digital transformation. Notwithstanding, the digital transformation of labor affects everyone to some extent and thus the issue must be broached by politicians and policy-makers. Naturally, additional research is needed to assess, for instance, the opportunity costs of reskilling workers, the effects on productivity and morale in adopting "Airbnb-style" workplaces (as mentioned previously), ramifications of parallel economies (such as using a point-based system), and any number of ethical and/or political ramifications associated to this development. Hence, the time has come for a proper debate on the digital transformation of labor and what direction it should take in the future.

Table 17.1 The ten most prominent advantages and challenges brought forth by the digital transformation of labor covered in this book.

	Advantages (+)	*Challenges (–)*
1	Automatization will not replace all types of labor.	Certain types of labor and processes will be more affected than others.
2	Automatization and AI will continue to develop and expand into additional types of professions, creating new job opportunities.	Automatization and AI will continue to develop and expand into additional types of professions, eliminating job opportunities.
3	Workers affected by automatization may undergo reskill training to stay active and/or relevant on the labor market.	Not all workers are able or willing to reskill.
4	Workers unable or unwilling to reskill may find sustenance through new types of alternative economies.	Digitalization induces inequality to certain individual groups of society.
5	The "gig economy" will become increasingly more widespread and will have fundamental impact on the labor market, creating new and freer opportunities.	The "gig economy" will become increasingly more widespread and will have fundamental impact on the labor market, with risks of more atypical and cynical work arrangements.
6	Digitalization of labor will have consequences on the welfare states. Some states are better prepared for this.	Digitalization of labor will have consequences on the welfare states. Some states are less prepared for this.
7	The "gig economy" provides new opportunities and freedom for people to work whenever and wherever they like, choosing the assignments of their own liking.	The "gig economy" can lead to "gig patients" and may have detrimental long-term effects on the welfare system.
8	More flexible working conditions for women.	Risk that automatization will harvest women-dominated professions.
9	The workplace for office work may shift to become more itinerant, with employees working at different places rather than at one specific address, especially for those workers whose professions qualify as "knowledge work". This can help keep costs down.	Unclear how itinerant workplace for office work affects the staff's productivity and/or morale.
10	The circular business model could help companies save resources while still securing sustainability and profitability.	The linear business model is consolidated into the mindsets of most businesses, making it difficult to change.

References

Asgarkhani, M., 2005. Digital government and its effectiveness in public management reform. *Public Management Review*, 7(3), pp. 465–487.

Badham, V., 2017. *Worker Exploitation: Without An International Minimum Wage, We're All Worth 10 Bucks*. [online] The Guardian. Available at: <www.theguardian.com/commentisfree/2017/may/01/worker-exploitation-without-an-international-minimum-wage-were-all-worth-ten-bucks> [Accessed 26 Sep. 2019].

Chan, A., 2003. Racing to the bottom: International trade without a social clause. *Third World Quarterly*, 24(6), pp. 1011–1028.

Christou, L., 2018. *The Effectiveness of Hot-Desking: Is This Really the Future of the Workplace?* [online] Verdict. Available at: <www.verdict.co.uk/effectiveness-of-hot-desking> [Accessed 26 Sep. 2019].

Davis, M.C., Leach, D.J. and Clegg, C.W., 2011. The physical environment of the office: Contemporary and emerging issues. In: G.P. Hodgkinson and J.K. Ford, eds., *International Review of Industrial and Organizational Psychology, Volume 26*. Chichester, UK: John Wiley & Sons, Ltd., pp. 193–237.

European Commission, 2019. *High-level Group Publishes Report on Impact of Digital Transformation on EU Labour Markets*. [online] Employment, Social Affairs & Inclusion. Available at: <https://ec.europa.eu/social/main.jsp?langId=en&catId=1226&furtherNews=yes&newsId=9344> [Accessed 26 Sep. 2019].

Ford, M., 2015. *Rise of the Robots: Technology and the Threat of a Jobless Future*. New York, NY: Basic Books.

Frost and Sullivan, 2019. *Digital Transformation of the Future of Work, 2019*. [online] Research and Markets. Available at: <www.researchandmarkets.com/research/3vt8hg/digital?w=4> [Accessed 26 Sep. 2019].

Hobson, K. and Lynch, N., 2016. Diversifying and de-growing the circular economy: Radical social transformation in a resource-scarce world. *Futures*, 82, pp. 15–25.

Jones, D., 2018. *Digitisation Will Create More Jobs Than It Replaces*. [online] CEO Magazine. Available at: <www.theceomagazine.com/business/innovation-technology/digitisation-will-create-more-jobs-than-it-replaces> [Accessed 26 Sep. 2019].

Kaplan, J., 2015. *Humans Need Not Apply: A Guide to Wealth and Work in the Age of Artificial Intelligence*. New Haven, CT: Yale University Press.

Keyes, R., 2006. *The Quote Verifier: Who Said What, Where, and When*. New York, NY: St. Martin's Press.

Kisjes, I., 2015. *The Evolution of Open Science – How Digitization is Transforming Research*. [online] Elsevier. Available at: <www.elsevier.com/connect/the-evolution-of-open-science-how-digitization-is-transforming-research> [Accessed 26 Sep. 2019].

Lindbeck, A., 1986. Limits to the welfare state. *Challenge*, 28(6), pp. 31–36.

Mamedova, S. and Pawlowski, E., 2018. *A Description of U.S. Adults Who Are Not Digitally Literate*. [online] Statistics in Brief. Available at: <https://nces.ed.gov/pubsearch/pubsinfo.asp?pubid=2018161> [Accessed 26 Sep. 2019].

Meyer, J., 2016. *Unions Hate the Gig Economy, but Workers Love It*. [online] Foundation for Economic Education. Available at: <https://fee.org/articles/unions-hate-the-gig-economy-but-workers-love-it> [Accessed 26 Sep. 2019].

Minter, K., 2017. Negotiating labour standards in the gig economy: Airtasker and Unions New South Wales. *The Economic and Labour Relations Review*, 28(3), pp. 438–454.

Myers, D., 2016. *New Economic Thinking and Real Estate*. Hoboken, NJ: John Wiley & Sons, Ltd.

O'Connor, S., 2019. *Gig Economy Agreements Promise a Brighter Future for Trade Unions*. [online] Financial Times. Available at: <www.ft.com/content/25271b48-38eb-11e9-b856-5404d3811663> [Accessed 26 Sep. 2019].

OECD, 2019. *Under Pressure: The Squeezed Middle Class*. Paris, France: OECD Publishing.

Panopoulos, D., 2019. *The Role of Digital Transformation in Labour Markets and Business Ecosystem*. [online] European Business Review. Available at: <www.europeanbusinessreview.eu/page.asp?pid=2889> [Accessed 26 Sep. 2019].

Peterson, K., 2014. *Why Companies are Doing Away With Assigned Desks*. [online] CBS News. Available at: <www.cbsnews.com/news/why-companies-are-doing-away-with-assigned-desks> [Accessed 26 Sep. 2019].

Singh, A. and Zammit, A., 2004. Labour standards and the 'race to the bottom': Rethinking globalization and workers' rights from developmental and solidaristic perspectives. *Oxford Review of Economic Policy*, 20(1), pp. 85–104.

Stewart, A. and Stanford, J., 2017. Regulating work in the gig economy: What are the options? *The Economic and Labour Relations Review*, 28(3), pp. 420–437.

Stillman, J., 2018. *New Harvard Study: Your Open-Plan Office Is Making Your Team Less Collaborative*. [online] Inc. Available at: <www.inc.com/jessica-stillman/new-harvard-study-you-open-plan-office-is-making-your-team-less-collaborative.html> [Accessed 26 Sep. 2019].

Todolí-Signes, A., 2017. The 'gig economy': Employee, self-employed or the need for a special employment regulation? *Transfer: European Review of Labour and Research*, 23(2), pp. 193–205.

Toelch, U. and Ostwald, D., 2018. Digital open science – Teaching digital tools for reproducible and transparent research. *PLOS Biology*, 16(7), p. e2006022, pp. 1–11.

Tolbert, C.J. and Mossberger, K., 2006. The effects of e-Government on trust and confidence in government. *Public Administration Review*, 66(3), pp. 354–369.

Wilts, H. and Berg, H., 2017. The digital circular economy: Can the digital transformation pave the way for resource-efficient materials cycles? *International Journal of Environmental Science & Natural Resources*, 7(5), p. 555725, pp. 1–4.

Worstall, T., 2017. *An Amazingly Bad Idea – An International Minimum Wage*. [online] Forbes. Available at: <www.forbes.com/sites/timworstall/2017/05/01/an-amazingly-bad-idea-an-international-minimum-wage> [Accessed 26 Sep. 2019].

Afterword

Impact of digitalization on employment and working conditions

John Øvretveit

1. Introduction

How will the digital transformation impact workers in different countries? Will it increase existing inequalities? Will it cause greater political unrest? These are just some of the questions that this book addresses. Some readers will be asking, "will my job disappear, or be made easier or more demanding?" And, "what advice should I give my adolescent daughter or son about the education, skills and occupation to seek so as to be sure of a job or income?" Not all chapters agree on the answers, and the speed of change is so rapid that predicting the future would be risky, even for the next five years. But there are some indications, and this short afterword both risks some predictions and gives reflections on the chapters.

2. Effects on employment

There are different views about whether digitalization will result in more jobs, and if so how quickly, or conversely, if it will lead to significant unemployment. One conclusion that several chapters are in agreement about is that digitalization has been associated with fewer traditional full-time employment jobs and with more alternate job arrangements, often with the worker as a self-employed contractor that takes on various "gigs", a trend that is likely to continue and spread. However, association is not causation and there is disagreement about how much the move from the traditional employment structure to this new "gig" structure is caused by digitalization and how much is caused by certain dynamics of capitalism. It is likely to be both, with digitalization making it possible and easier to move full-time to part-time employment and create a new, contracted "reserve army of labor" with (for the employee) less favorable employee-type benefits. As with most predictions and generalizations in this field, much depends on which occupation and which type of market, state economy and digital infrastructure we are considering.

While some chapters support the contention that more new jobs will be created in the emerging digital economies of most countries, it remains uncertain how soon and how many, and to what extent, these jobs will make up for the work that is automated-away. What is clear is that most new jobs will require new skills, and retraining will be needed for many workers to be able to perform with digital technologies as a greater or lesser part of the new labor requirements. This transformation of work is likely to cause social disruption as in previous technological change eras. But the speed of change and the nature of the change could be faster and more significant as the changes from farm to factory, from horse to vehicle and from steam to oil-based civilizations. Adding uncertainty to these predictions is the advent of artificial intelligence (AI) in its various forms, and how this will replace some work and occupations and transform many others. That is to say, speech recognition and instruction made possible by AI is beginning to impact many occupations in different sectors around the world.

3. Effects on working conditions

For most occupations, digital technologies will be an increasing part of the work day and work demand. For some health-care workers this has reduced the time they spend with patients and increased the intensity of work. For some, automation and digital devices have increased the amount of time workers can spend with patients. To this end, much is dependent on design and implementation. For most, mobile devices and widespread internet and broadband has increased remote working and the "always on" phenomena touched upon by some chapters. Remote working has its advantages and disadvantages as we have seen from the research reviewed.

4. Political and economic effects

The role played by government funding of research and digital infrastructure has been an underestimated factor in making possible large fortunes and high incomes. Certainly, the early years of the digital era have led to high incomes for those with the high levels of education and skills required. Concomitantly, the income and job opportunities of many occupations have been reduced, leading to increases in inequity of wealth. This, together with the financial crisis of 2009, has resulted in widespread dissatisfactions, expressed politically in various ways. A case has been made that social media have increased these dissatisfactions and allowed political mobilization and organizing. The impact on traditional democratic processes was underestimated. These movements may lead to strong regulations and possibly pro-labor legislation.

5. Conclusion

What then should be our advice to children? My practical advice is:

- Learn the basics of programming using the new games and toys, so as to understand how this works.
- If you have an aptitude for this, learn more of the principles, but recognize automation will replace much coding work, and consider working with digital networks and systems to be part of the 5G-powered world of the Internet of Things (IoT).

- Better still, learn about user-centered design and working in multi-professional teams to humanize the crude machine-person interfaces we have at present. Make these technologies work for the less literate or people with cognitive difficulties to enable them to benefit from the positive aspects of digitalization and reduce the "digital divide".
- Learn how to work with digital assistance and techniques to perform your profession or occupation – even personal service work such as nursing or psychotherapy, where people both want and can afford a human will be supported by considerable technology.
- Consider plumbing or electrical contracting, or handyman work for maintenance: It will take a long time for robots and digitalization to replace these and some other occupations and trades.

If none of this appeals, find something for income, because the "guaranteed income for all" may take some time. Good luck!

John Øvretveit
Professor
Karolinska Institutet
September 2019

Index

Note: **Boldface** page references indicate tables. *Italic* references indicate figures.

For Product Safety Concerns and Information please contact our EU
representative GPSR@taylorandfrancis.com
Taylor & Francis Verlag GmbH, Kaufingerstraße 24, 80331 München, Germany

www.ingramcontent.com/pod-product-compliance
Lightning Source LLC
LaVergne TN
LVHW021038011026
849160LV00008BA/319

* 9 7 8 1 0 3 2 0 8 4 9 3 0 *